Enzyme Handbook 14

W0107325

Springer-Verlag Berlin Heidelberg GmbH

Attention all "Enzyme Handbook" users:

A file with the complete volume indexes Vols. 1 through 14 in delimited ASCII format is available for downloading at no charge from the Springer EARN mailbox. Delimited ASCII format can be imported into most databanks.

The file has been compressed using the popular shareware program "PKZIP" (Trademark of PKware Inc., PKZIP is available from most BBS and shareware distributors).

This file is distributed without any expressed or implied warranty.

To receive this file send an e-mail message to:
SVSERV@DHDSPRI6.BITNET

The message must be:
GET /CHEMISTRY/ENZ_HB.ZIP

SVSERV is an automatic data distribution system. It responds to your message. The following commands are available:

HELP	returns a detailed instruction set for the use of SVSERV,
DIR *(name)*	returns a list of files available in the directory "name",
INDEX *(name)*	same as "DIR"
CD *\<name>*	changes to directory "name",
SEND *\<filename>*	invokes a message with the file "filename",
GET *\<filename>*	same as "SEND".

D. Schomburg · D. Stephan (Eds.)
GBF– Gesellschaft für Biotechnologische Forschung

Enzyme Handbook 14

Class 2.7 – 2.8 Transferases
EC 2.7.1.105 – EC 2.8.3.14

Springer

Professor Dr. Dietmar Schomburg
Universität zu Köln, Institut für Biochemie
Zülpicher Str. 47, 50674 Köln, FRG

Dr. Dörte Stephan

GBF – Gesellschaft für Biotechnologische Forschung mbH
Mascheroder Weg 1, 38124 Braunschweig, FRG

This collection of datasheets was generated from the database „BRENDA"

ISBN 978-3-642-47773-7

Library of Congress Cataloging-in-Publication Data

Enzyme handbook/ D. Schomburg, M. Salzmann (eds.). v. (1–2,4–10); 23 cm. Vols. 6–7 edited by D. Schomburg, M. Salzmann, D. Stephan. Vols. 9–10 edited by D. Schomburg, D. Stephan.
Loose-leaf.
Includes bibliographical references and indexes.
Contents: 1. Class 4: Lyases – 2. Class 5: Isomerases. Class 6: Ligases – 4–5. Class 3: Hydrolases – 6. Class 1.2–1.4, Oxidoreductases – 7. Class 1.5–1.12, Oxidoreductases – 8. Class 1.13–1.97, Oxidoreductases – 9. Class 1.1, Oxidoreductases, EC 1.1.1.150 – EC 1.1.99.26 – v. 10. Class 1.1, Oxidoreductases, EC 1.1.1.150 – EC 1.1.99.26.
ISBN 978-3-642-47773-7   ISBN 978-3-642-59025-2 (eBook)
DOI 10.1007/978-3-642-59025-2
1. Enzymes-Handbooks, manuals, etc. I. Schomburg, D. (Dietmar)   II. Salzmann, M. (Margit)   III. Stephan, D. (Dörte)
QP601-E5158   1990
660'.634–dc20

© Springer-Verlag Berlin Heidelberg 1997
Originally published by Springer-Verlag Berlin Heidelberg New York in 1997

The use of registered names, trademarks, etc. in this publication does not imply, even in the absence of a specific statement, that such names are exempt from the relevant protective laws and regulations and therefore free for general use.

The publisher cannot assume any legal responsibility for given data, especially as far as directions for the use and the handling of chemicals and biological materials are concerned. This information can be obtained from the instructions on safe laboratory practice and from the manufacturers of chemicals and laboratory equipment.

Media conversion, printing and bookbinding: Brühlsche Universitätsdruckerei, Giessen
Production of the plasticfiles: Lux-Plastik oHG, Murnau
SPIN: 10630679   51/3020 - 5 4 3 2 1 0 - Printed on acid-free paper

Preface

Today, as the large international genome sequence projects are gaining a great amount of public attention and huge sequence data bases are created it becomes more and more obvious that we are very limited in our ability to access functional data for the gene products - the proteins, in particular for enzymes. Those data are inherently very difficult to collect, interpret and standardize as they are highly distributed among journals from different fields and are often subject to experimental conditions. Nevertheless a systematic collection is essential for our interpretation of the genome information and more so for possible applications of that knowledge in the fields of medicine, agriculture, etc.. Recent progress on enzyme immobilization, enzyme production, enzyme inhibition, coenzyme regeneration and enzyme engineering has opened up fascinating new fields for the potential application of enzymes in a large range of different areas.

It is the functional profile of an enzyme that enables a biologist of physician to analyze a metabolic pathway and its disturbance; it is the substrate specificity of an enzyme which tells an analytical biochemist how to design an assay; it is the stability, specificity and efficiency of an enzyme which determines its usefulness in the biotechnical transformation of a molecule. And the sum of all these data will have to be considered when the designer of artificial biocatalysts has to choose the optimum prototype to start with.

The development of an enzyme data information system was started 10 years ago at the German National Research Centre for Biotechnology in Braunschweig (GBF). The present book „Enzyme Handbook" represents the printed version of this data bank. A computer searchable version will be soon available.

The enzymes in this Handbook are arranged according to the Enzyme Commission list of enzymes. Some 3500 „different" enzymes are covered. Frequently enzymes with very different properties are included under the same EC number. Although we intend to give a representative overview on the characteristics and variability of each enzyme the Handbook is not a compendium. The reader will have to go to the primary literature for more detailed information. Naturally it is not possible to cover all the numerous literature references for each enzyme (for some enzymes up to 40000) if the data representation is to be concise as is intended.

It should be mentioned here that the data are extracted from literature and critically evaluated by qualified scientists. On the other hand the original authors' nomenclature for enzyme forms and subunits is retained as is their nomenclature for organisms and strains even if the organism is reclassified in the meantime. The cross references to the protein sequence data bank and to the Brookhaven protein 3D structure data bank are taken directly from their data files without further verification by the authors. In order to keep the tables concise redundant information is avoided as far as possible (e.g. if K_m values are measured in the presence of an obvious cosubstrate, only the name of the cosubstrate is given in parentheses as a commentary without reference to its specific role).

The authors are grateful to the following biologists and chemists for invaluable help in the compilation of data: Cornelia Munaretto, Dr. Astrid Beermann, Dr. Ida Schomburg and Dr. Astrid Haberz. In addition we would like to thank Mrs. C. Munaretto and Dr. I. Schomburg for the correction of the final manuscript.

Köln and Braunschweig
Spring 1997

Dietmar Schomburg, Dörte Stephan

List of Abbreviations

A	adenosine		ER	endoplasmic reticulum
Ac	acetyl		Et	ethyl
ACP	acyl-carrier-protein		EXAFS	extended X-ray absorption
ADP	adenosine 5'-diphosphate			fine structure
Ala	alanine		FAD	flavin-adenine dinucleotide
All	allose		FMN	flavin mononucleotide (ribo-
Alt	altrose			flavin 5'-monophosphate)
AMP	adenosine 5'-monophosphate		FPLC	fast protein liquid chroma-
Ara	arabinose			tography
Arg	arginine		Fru	fructose
Asn	asparagine		Fuc	fucose
Asp	aspartic acid		G	guanosine
ATP	adenosine 5'-triphosphate		GABA	4-aminobutanoic acid
Bicine	N,N'-bis(2-hydroxyethyl)		Gal	galactose
	glycine		GDP	guanosine 5'-diphosphate
C	cytidine		Glc	glucose
cal	calorie		GlcN	glucosamine
CDP	cytidine 5'-diphosphate		GlcNAc	N-acetylglucosamine
CDTA	trans-1,2-diaminocyclo-hexa-		Gln	glutamine
	ne-N,N,N,N-tetra-aceticacid		Glu	glutamic acid
CHAPS	3-[(3-cholamidopropyl)-		Gly	glycine
	dimethylammonio]-1-		Glygly	glycylglycine
	propanesulfonate		GMP	guanosine
CHAPSO	3-[(3-cholamidopropyl)-			5'-monophosphate
	dimethylammonio]-		GSH	glutathione
	2-hydroxy-1-propane-		GSSG	oxidized glutathione
	sulfonate		GTP	guanosine 5'-triphosphate
CMP	cytidine 5'-monophosphate		Gul	gulose
CoA	coenzyme A		h	hour
CTP	cytidine 5'-triphosphate		H_4	tetrahydro
Cys	cysteine		HEPES	4-(2-hydroxyethyl)-1-piper-
d	deoxy-			azineethane sulfonic acid
D- and L-	prefixes indicating		His	histidine
	configuration		HPLC	high performance liquid
Dap	diaminopimelic acid			chromatography
DFP	diisopropylfluorophosphate		Hyl	hydroxylysine
DNA	deoxyribonucleic acid		Hyp	hydroxyproline
DPN	diphosphopyridinium		IAA	iodoacetamide
	nucleotide (now NAD)		Ig	immunoglobulin
DTNB	5,5'-dithiobis(2-nitrobenzoate)		Ile	isoleucine
DTT	dithiothreitol (i.e. Cleland's		Ido	idose
	reagent)		IDP	inosine 5'-diphosphate
e	electron		IMP	inosine 5'-monophosphate
EC	number of enzyme in Enzyme		ir	irreversible
	Commission's system		ITP	inosine 5'-triphosphate
E. coli	Escherichia coli		K_m	Michaelis constant
EDTA	ethylene diaminetetraacetate		L-	see D-
EGTA	ethylene glycol bis (β-amino-		Leu	leucine
	ethylether) tetraacetate		Lys	lysine
EPR	electron paramagnetic		Lyx	lyxose
	resonance		M	mol/l

m-	meta-
Man	mannose
MES	2-(N-morpholino)ethane sulfonate
Met	methionine
min	minute
MOPS	3-(N-morpholino) propane sulfonate
Mur	muramic acid
MW	molecular weight
NAD	nicotinamide-adenine dinucleotide
NADH	reduced NAD
NADP	NAD phosphate
NADPH	reduced NADP
NAD(P)H	indicates either NADH or NADPH
NDP	nucleoside 5'-diphosphate
NEM	N-ethylmaleimide
Neu	neuraminic acid
Nle	norleucine
NMN	nicotinamide mononucleotide
NMP	nucleoside 5'-monophosphate
NTP	nucleoside 5'-triphosphate
o-	ortho-
OMP	orotidine 5-monophosphate
Orn	ornithine
p-	para-
PAPS	3'-phosphoadenylylsulfate
PCMB	p-chloro-mercuribenzoate
PEG	polyethylene glycol
PEP	phosphoenolpyruvate
pH	$-\log_{10}[H^+]$
Ph	phenyl
Phe	phenylalanine
PIXE	proton-induced X-ray emission
PMSF	phenylmethane-sulfonylfluoride
Pro	proline
Q_{10}	factor for the change in reaction rate for a 10° temperature increase
r	reversible
Rha	rhamnose
Rib	ribose
RNA	ribonucleic acid
mRNA	messenger RNA
rRNA	ribosomal RNA
tRNA	transfer RNA
Sar	N-methylglycine (sarcosine)
SDS-PAGE	sodium dodecyl sulfate polyacrylamide gel electrophoresis
Ser	serine
SFK-525A	2-diethylaminoethyl-2,2-diphenylvalerate
sp.	species
T	ribosylthymine
$t_{1/2}$	time for half-completion of reaction
Tal	talose
TDP	ribosylthymine 5'-diphosphate
TEA	triethanolamine
TES	N-tris[hydroxymethyl]-methyl-2-amino-ethanesulfonic acid
THF	tetrahydrofolate
Thr	threonine
TMP	ribosylthymine 5'-monophosphate
Tos-	tosyl-(p-toluenesulfonyl-)
TPN	triphosphopyridinium nucleotide (now NADP)
Tris	tris(hydroxymethyl)-aminomethane
Trp	tryptophan
TTP	ribosylthymine 5'-triphosphate
Tyr	tyrosine
U	uridine
U/mg	$\mu mol/(mg \cdot min)$
UDP	uridine 5'-diphosphate
UMP	uridine 5'-monophosphate
UTP	uridine 5'-triphosphate
UV	ultraviolet
Val	valine
Xaa	symbol for an amino acid of unknown constitution in peptide formula
XAS	X-ray absorption spectroscopy
XTP	xanthosine 5'-triphosphate
Xyl	xylose

Index

(Alphabetical order of Enzyme names)

EC-No.	Name	EC-No.	Name
2.7.8.2	Diacylglycerol cholinephosphotransferase	2.8.3.12	Glutaconate CoA-transferase
2.7.1.107	Diacylglycerol kinase	2.7.7.42	[Glutamate-ammonia-ligase] adenylyltransferase
2.7.7.58	(2,3-Dihydroxybenzoyl)adenylate synthase	2.7.2.13	Glutamate 1-kinase
2.7.7.7	DNA-directed DNA polymerase	2.7.2.11	Glutamate 5-kinase
2.7.7.6	DNA-directed RNA polymerase	2.7.7.39	Glycerol-3-phosphate cytidylyltransferase
2.7.7.31	DNA nucleotidylexotransferase	2.7.1.142	Glycerol-3-phosphate-glucose phosphotransferase
2.7.1.108	Dolichol kinase	2.7.6.5	GTP pyrophosphokinase
2.7.4.20	Dolichyl-diphosphate-polyphosphate phosphotransferase	2.7.3.1	Guanidinoacetate kinase
2.7.4.9	dTMP kinase	2.7.7.45	Guanosine-triphosphate guanylyltransferase
2.8.2.4	Estrone sulfotransferase	2.7.4.8	Guanylate kinase
2.7.7.14	Ethanolamine-phosphate cytidylyltransferase	2.8.2.23	Heparin-glucosamine 3-O-sulfotransferase
2.7.8.1	Ethanolaminephosphotransferase	2.8.2.12	Heparitin sulfotransferase
2.7.4.18	Farnesyl-diphosphate kinase	2.7.7.29	Hexose-1-phosphate guanylyltransferase
2.8.2.25	Flavonol 3-sulfotransferase	2.7.8.7	Holo-[acyl-carrier-protein] synthase
2.7.7.2	FMN adenylyltransferase	2.7.1.109	[Hydroxymethylglutaryl-CoA reductase (NADPH)] kinase
2.7.2.6	Formate kinase	2.8.3.14	5-Hydroxypentanoate CoA-transferase
2.7.7.30	Fucose-1-phosphate guanylyltransferase	2.7.1.119	Hygromycin-B kinase
2.7.7.32	Galactose-1-phosphate thymidylyltransferase	2.7.3.6	Hypotaurocyamine kinase
2.8.2.11	Galactosylceramide sulfotransferase	2.7.1.134	1D-myo-Inositol-tetrakisphosphate 1-kinase
2.7.7.46	Gentamicin 2''-nucleotidyltransferase	2.7.1.140	1D-myo-Inositol-tetrakisphosphate 5-kinase
2.7.1.106	Glucose-1,6-bisphosphate synthase	2.7.1.127	1D-myo-Inositol-trisphosphate 3-kinase
2.7.7.27	Glucose-1-phosphate adenylyltransferase	2.7.1.139	1D-myo-Inositol-trisphosphate 5-kinase
2.7.7.33	Glucose-1-phosphate cytidylyltransferase	2.7.1.133	1D-myo-Inositol-trisphosphate 6-kinase
2.7.7.34	Glucose-1-phosphate guanylyltransferase	2.7.1.116	[Isocitrate dehydrogenase (NADP$^+$)] kinase
2.7.7.24	Glucose-1-phosphate thymidylyltransferase	2.8.2.21	Keratan sulfotransferase
2.7.7.44	Glucuronate-1-phosphate uridylyltransferase	2.7.3.5	Lombricine kinase
		2.7.1.131	Low-density-lipoprotein kinase
		2.7.1.136	Macrolide 2'-kinase

EC-No.	Name

2.8.3.3 Malonate CoA-transferase
2.7.7.13 Mannose-1-phosphate guanylyltransferase
2.7.7.22 Mannose-1-phosphate guanylyltransferase (GDP)
2.7.8.21 Membrane-oligosaccharide glycerophosphotransferase
2.8.1.2 3-Mercaptopyruvate sulfurtransferase
2.7.4.19 5-Methyldeoxycytidine-5'-phosphate kinase
2.7.1.115 [3-Methyl-2-oxobutanoate dehydrogenase (lipoamide)] kinase
2.7.7.57 N-Methylphosphoethanolamine cytidylyltransferase
2.7.7.50 mRNA guanylyltransferase
2.7.1.129 Myosin-heavy-chain kinase
2.7.1.117 Myosin-light-chain kinase
2.7.7.1 Nicotinamide-nucleotide adenylyltransferase
2.7.7.18 Nicotinate-nucleotide adenylyltransferase
2.7.4.6 Nucleoside-diphosphate kinase
2.7.4.4 Nucleoside-phosphate kinase
2.7.4.10 Nucleoside-triphosphate-adenylate kinase
2.7.7.28 Nucleoside-triphosphate-hexose-1-phosphate nucleotidyltransferase
2.7.6.4 Nucleotide pyrophosphokinase
2.7.3.7 Opheline kinase
2.8.3.2 Oxalate CoA-transferase
2.8.3.5 3-Oxoacid CoA-transferase
2.8.3.6 3-Oxoadipate CoA-transferase
2.7.7.3 Pantetheine-phosphate adenylyltransferase
2.7.7.54 Phenylalanine adenylyltransferase
2.7.1.137 1-Phosphatidylinositol 3-kinase

2.7.7.41 Phosphatidate cytidylyltransferase
2.7.8.20 Phosphatidylglycerol-membrane-oligosaccharide glycerophosphotransferase
2.7.8.13 Phospho-N-acetylmuramoyl-pentapeptide-transferase
2.7.1.121 Phosphoenolpyruvate-glycerone phosphotransferase
2.7.3.9 Phosphoenolpyruvate-protein phosphotransferase
2.7.1.105 6-Phosphofructo-2-kinase
2.7.4.17 3-Phosphoglyceroyl-phosphate-polyphosphate phosphotransferase
2.7.2.3 Phosphoglycerate kinase
2.7.2.10 Phosphoglycerate kinase (GTP)
2.7.8.9 Phosphomannan mannosephosphotransferase
2.7.4.7 Phosphomethylpyrimidine kinase
2.7.4.2 Phosphomevalonate kinase
2.7.7.19 Polynucleotide adenylyltransferase
2.7.4.1 Polyphosphate kinase
2.7.7.8 Polyribonucleotide nucleotidyltransferase
2.8.3.1 Propionate CoA-transferase
2.7.3.11 Protein-histidine pros-kinase
2.7.3.12 Protein-histidine tele-kinase
2.7.1.112 Protein-tyrosine kinase
2.8.2.20 Protein-tyrosine sulfotransferase
2.8.2.13 Psychosine sulfotransferase
2.7.9.1 Pyruvate,orthophosphate dikinase
2.7.9.2 Pyruvate,water dikinase
2.8.2.28 Quercetin-3,3'-bissulfate 7-sulfotransferase
2.8.2.26 Quercetin-3-sulfate 3'-sulfotransferase
2.8.2.27 Quercetin-3-sulfate 4'-sulfotransferase

1 NOMENCLATURE

EC number
2.7.7.13

Systematic name
GTP:alpha-D-mannose-1-phosphate guanylyltransferase

Recommended name
Mannose-1-phosphate guanylyltransferase

Synonyms
GTP-mannose-1-phosphate guanylyltransferase
PIM-GMP (phosphomannose isomerase-guanosine 5'-diphospho-D-mannose pyrophosphorylase: bifunctional enzyme which catalyzes both the phosphomannose isomerase (PIM) and guanosine 5'-diphospho-D-mannose pyrophosphorylase (GMP) reaction) [3]
Guanylyltransferase, mannose 1-phosphate
GDP-mannose pyrophosphorylase
GTP-mannose 1-phosphate guanylyltransferase
Guanosine 5'-diphospho-D-mannose pyrophosphorylase
Guanosine diphosphomannose pyrophosphorylase
Guanosine triphosphate-mannose 1-phosphate guanylyltransferase
Mannose 1-phosphate guanylyltransferase (guanosine triphosphate)

CAS Reg. No.
37278-24-3

---

2 REACTION AND SPECIFICITY

Catalyzed reaction
GTP + alpha-D-mannose 1-phosphate →
→ diphosphate + GDPmannose

Reaction type
Nucleotidyl group transfer

Natural substrates
More (enzyme of alginate biosynthetic pathway) [3]

Substrate spectrum
1 GTP + alpha-D-mannose 1-phosphate (r [2, 3], equilibrium constant: 2.5 [1], specific for the mannose moiety [4]) [1–4]
2 dGTP + alpha-D-mannose 1-phosphate (34% of the activity with GTP [4]) [1, 4]

Enzyme Handbook © Springer-Verlag Berlin Heidelberg 1997
Duplication, reproduction and storage in data banks are only
allowed with the prior permission of the publishers

3 ITP + alpha-D-mannose 1-phosphate (r [2], 44% of the activity with GTP [4], ITP is more effective than GTP with mannose 1-phosphate [2]) [1, 2, 4]
4 GDPglucose + diphosphate (r [2], most effective substrate in direction of nucleoside triphosphate formation, in the reverse direction GTP is a better glucose acceptor than ITP) [2]
5 IDPmannose + diphosphate (r, 72% of the activity with GDPglucose) [2]
6 GDPmannose + diphosphate (r, 61% of the activity with GDPglucose) [2]
7 8-Azido-GTP + mannose 1-phosphate [2]
8 8-Azido-GTP + glucose 1-phosphate [2]
9 ITP + glucose 1-phosphate [2]
10 More (not: ATP, CTP, TTP, UTP, ADP-alpha-D-glucose, ADP-alpha-D-mannose, CDP-alpha-D-choline, CDP-alpha-D-glucose, CDP-beta-L-fucose, GDP-alpha-D-glucose, dTPP-alpha-D-glucose, IDP-alpha-D-galactose, UDP-alpha-D-glucose, UDP-N-acetyl-alpha-D-glucosamine, UDP-alpha-D-mannose) [4]

Product spectrum
1 GDPmannose + diphosphate [1, 4]
2 dGDPmannose + diphosphate [1]
3 IDPmannose + diphosphate [1]
4 Glucose 1-phosphate + GTP [2]
5 ?
6 ?
7 8-Azido-GDPmannose + diphosphate [2]
8 8-Azido-GDPglucose + diphosphate [2]
9 ?
10 ?

Inhibitor(s)
Mn^{2+} (activation at 70% of the activity with Mg^{2+}, inhibition at 10 mM and higher) [2]; GDPmannose (inhibits utilization of GDPglucose) [2]; GDPglucose (inhibits utilization of GDPmannose) [2]; Zn^{2+} [2]; Hg^{2+} [2]; Cu^{2+} [2]; GMP [2]; GDP [2]; More (not sensitive to DTT) [3]

Cofactor(s)/prosthetic group(s)/activating agents

Metal compounds/salts
Mg^{2+} (Mg^{2+} or Mn^{2+} required [3], divalent cation required, Mg^{2+} most effective, optimum concentration: 5 mM [2]) [2, 3]; Mn^{2+} (Mg^{2+} or Mn^{2+} required [3], 70% of the activity with Mg^{2+}, inhibition at 10 mM and higher [2]) [2, 3]; Co^{2+} (slight activation at about 10 mM) [2]

Turnover number (min^{-1})

Specific activity (U/mg)
More [1, 3]; 24.1 [4]

K_m-value (mM)
0.0004 (mannose 1-phosphate) [4]; 0.001 (GDP-alpha-D-mannose) [4];
0.0035 (GTP) [4]; 0.0142 (GDPmannose) [3]; 0.02 (GDPmannose) [2];
0.0205 (D-mannose 1-phosphate) [3]; 0.0295 (GTP) [3]; 0.1 (GTP (+ mannose 1-phosphate)) [2]; 0.13 (mannose 1-phosphate (+ GTP)) [2]; 0.33
(diphosphate (+ GDPglucose)) [2]; 0.5 (diphosphate (+ GDPmannose)) [2];
1 (GTP (+ glucose 1-phosphate) [2], diphosphate [4]) [2, 4]; 2.85 (glucose
1-phosphate (+ GTP)) [2]

pH-optimum
5.5–7.5 (GTP synthesis) [2]; 6–7.5 [4]; 6.0–8.5 (GDPmannose synthesis) [2];
8.2 (assay at) [2]

pH-range

Temperature optimum (°C)
37 (assay at) [1–3]

Temperature range (°C)

---

3 ENZYME STRUCTURE

Molecular weight
54000 (Pseudomonas aeruginosa, gel filtration, phosphomannose isomerase-guanosine 5'-diphospho-D-mannose pyrophosphorylase is a bifunctional enzyme catalyzing both the phosphomannose isomerase (PIM) and guanosine 5'-diphospho-D-mannose pyrophosphorylase (GMP) reaction) [3]
412000 (pig, sucrose velocity sedimentation) [4]
450000 (pig, gel filtration) [2]

Subunits
? (x × 37000 + x × 43000, pig, SDS-PAGE) [2]
Monomer (1 × 56000, Pseudomonas aeruginosa, SDS-PAGE) [3]

Glycoprotein/Lipoprotein
–

---

4 ISOLATION/PREPARATION

Source organism
Arthrobacter sp. (NRRL B1973) [1]; Pig [2, 4]; Pseudomonas aeruginosa
(phosphomannose isomerase-guanosine 5'-diphospho-D-mannose pyrophosphorylase is a bifunctional enzyme catalyzing both the phosphomannose isomerase (PIM) and guanosine 5'-diphospho-D-mannose pyrophosphorylase (GMP) reaction) [3]

Source tissue
  Liver [2]; Thyroid tissue [4]

Localization in source

Purification
  Arthrobacter sp. (NRLL B1973) [1]; Pig [2, 4]; Pseudomonas aeruginosa
  (phosphomannose isomerase-guanosine 5'-diphospho-D-mannose pyrophosphorylase is a bifunctional enzyme catalyzing both the phosphomannose isomerase (PIM) and guanosine 5'-diphospho-D-mannose pyrophosphorylase (GMP) reaction) [3]

Crystallization
  –

Cloned
  –

Renatured
  –

5 STABILITY

pH
  6–7.7 (most stable at) [2]

Temperature (°C)
  48 (30 min, 50% loss of activity) [1]

Oxidation

Organic solvent

General stability information

Storage
  Frozen, stable for several months [2]; 4°C, stable for a few days, purified enzyme [4]

6 CROSSREFERENCES TO STRUCTURE DATABANKS

PIR/MIPS code
  PIR2:I57096 (Escherichia coli); PIR2:S67590 (yeast (Saccharomyces cerevisiae)); PIR2:A47415 (37K beta chain pig (fragment)); PIR2:B47415 (43K
  alpha chain pig (fragment))

Brookhaven code

7 LITERATURE REFERENCES

[1] Preiss, J., Wood, E.: J. Biol. Chem.,239,3119–3126 (1964)
[2] Szumilo, T., Drake, R.R., York, J.L., Elbein, A.D.: J. Biol. Chem.,268,17943–17950 (1993)
[3] Shinabarger, D., Berry, A., May, T.B., Rothmel, R., Fialho, A., Chakrabarty, A.M.: J. Biol. Chem.,266,2080–2088 (1991)
[4] Smoot, J.W., Serif, G.S.: Eur. J. Biochem.,148,83–87 (1985)

1 NOMENCLATURE

EC number
2.7.7.14

Systematic name
CTP:ethanolamine-phosphate cytidylyltransferase

Recommended name
Ethanolamine-phosphate cytidylyltransferase

Synonyms
Phosphorylethanolamine transferase
ET [5]
Cytidylyltransferase, ethanolamine phosphate
CTP-phosphoethanolamine cytidylyltransferase
Phosphoethanolamine cytidylyltransferase
Ethanolamine phosphate cytidylyltransferase

CAS Reg. No.
9026-33-9

---

2 REACTION AND SPECIFICITY

Catalysed reaction
CTP + ethanolamine phosphate →
→ diphosphate + CDPethanolamine (ordered reaction mechanism [1, 6])

Reaction type
Nucleotidyl group transfer

Natural substrates
More (catalyzes a central step in phosphatidylethanolamine synthesis) [1, 6]

Substrate spectrum
1 CTP + ethanolamine phosphate (r [1]) [1–7]
2 CTP + 2-aminoethylarsonic acid [2]
3 CTP + 2-aminoethylphosphonate [4]
4 dCTP + ethanolamine phosphate [6]
5 CTP + phosphomonomethylethanolamine [7]
6 CTP + phosphodimethylethanolamine [7]
7 CTP + phosphocholine [7]

Product spectrum

1 Diphosphate + CDPethanolamine [1, 6]
2 Diphosphate + Cytidine-O-PO$_2^-$-O-AsO$_2^-$-CH$_2$-CH$_2$-NH$_3^+$ (spontaneous hydrolysis to CMP and -O-AsO$_2^-$-CH$_2$-CH$_2$-NH$_3^+$) [2]
3 ?
4 Diphosphate + dCDPethanolamine [6] ?
5 ?
6 ?
7 ?

Inhibitor(s)

CTP (at concentrations exceeding that of Mg^{2+}) [1]; Diphosphate [1]; CDPethanolamine [1]; 2-Aminoethylphosphonate (inhibitory power stimulated by Mg^{2+}) [4]; 3-Aminopropylphosphonate (inhibitory power stimulated by Mg^{2+}) [4]; Sphingosine/phosphatidylcholine vesicles (inhibit cytosolic and purified enzyme) [5]; Phosphomonomethylethanolamine (weak, competitive to phosphoethanolamine) [7]; Phosphodimethylethanolamine (weak, competitive to phosphoethanolamine) [7]; Phosphocholine (weak, competitive to phosphoethanolamine) [7]; NEM [5]; Iodoacetamide [5]; PCMB [5]

Cofactor(s)/prosthetic group(s)/activating agents

Reducing agent (like DTT required [1], cysteine or 2-mercaptoethanol are ineffective [1], activity of purified enzyme is dependent on the presence of DTT [6]) [1, 6]; More (addition of lipids does not stimulate the activity of cytosolic enzyme) [6]

Metal compounds/salts

Mg^{2+} (absolute requirement for a divalent cation [1], 5–10 mM most efficient) [1]; Mn^{2+} (activation, optimal concentration: 2–4 mM, 50% of the activity with Mg^{2+}) [1]; Ca^{2+} (activation, 10% of the activity with Mg^{2+}) [1]; Co^{2+} (activation, 10% of the activity with Mg^{2+}) [1]; More (no activation by Ba^{2+}, Zn^{2+}, Cd^{2+}, Ni^{2+}) [1]

Turnover number (min^{-1})

Specific activity (U/mg)

1.13 [1]; 4.38 [6]; 6.505 [5]

K$_m$-value (mM)

0.053 (CTP) [1, 6]; 0.065 (ethanolamine phosphate) [1, 6]; 0.072 (phosphoethanolamine) [7]; 0.11 (phosphomonomethylethanolamine) [7]; 3 (about, 2-aminoethylarsonic acid) [2]; 6.2 (phosphocholine) [7]; 6.8 (phosphodimethylethanolamine) [7]; 30 (2-aminoethylphosphonate, pH 5.5) [4]

pH-optimum

6 (one sharp optimum at pH 7.8 and one with a lower maximal activity around 6) [1, 6]; 6.5 (2 optima: 6.5 and 8.0) [7]; 7.8 (one sharp optimum at pH 7.8 and one with a lower maximal activity around 6) [1, 6]; 8.0 (2 optima: 6.5 and 8.0) [7]

pH-range

Temperature optimum (°C)
37 (assay at) [1, 2, 6]

Temperature range (°C)

3 ENZYME STRUCTURE

Molecular weight
100000–120000 (rat, gel filtration) [1, 6]

Subunits
? (x × 49000–50000, rat, SDS-PAGE [1, 6], x × 49600, rat, SDS-PAGE [5]) [1, 5, 6]

Glycoprotein/Lipoprotein
--

4 ISOLATION/PREPARATION

Source organism
Rat [1–7]

Source tissue
Liver [1, 2, 4–7]; Hepatocytes [5]; Brain [3]

Localization in source
Cytosol (predominantly) [6]; Myelin [3]; More (postmicrosomal supernatant [1, 5], approximately 50% of the total activity is in the supernatant, the remainder being distributed among subcellular localizations [5]) [1, 5]

Purification
Rat [1, 5, 6]

Crystallization
--

Cloned
--

Renatured
--

5 STABILITY

pH
  7 (limited stability below) [6]

Temperature (°C)

Oxidation

Organic solvent

General stability information
  Bovine serum albumin, 2% w/v, 1 mM CTP and 20 mM Mg^{2+}, or 10% v/v glycerol stabilizes [1]; Quite stable towards freezing and thawing [6]; Purified enzyme can be stabilized by the addition of 10% v/v glycerol or 2% bovine serum albumin [6]; Omission of DTT from buffers used in the later steps of purification results in severe loss of activity [1]; Enzyme in crude postmicrosomal supernatant is quite stable towards freezing and thawing, but the highly purified enzyme loses 85–90% of its activity when frozen and thawed twice [1]

Storage
  0–4°C, pH 7.5–9.0, in presence of DTT, stable for weeks [1]; –20°C, stable for at least 4 weeks without significant loss of activity [6]

6 CROSSREFERENCES TO STRUCTURE DATABANKS

PIR/MIPS code

Brookhaven code

7 LITERATURE REFERENCES

[1] Sundler, R.: J. Biol. Chem.,250,8585–8590 (1975)
[2] Visedo-Gonzalez, E., Dixon, H.B.F.: Biochem. J.,260,299–301 (1989)
[3] Kunishita, T., Ledeen, R.W.: J. Neurochem.,42,326–333 (1984)
[4] Plantavid, M., Maget-Dana, R., Douste-Blazy, L.: Biochimie,57,951–957 (1975)
[5] Vermeulen, P.S., Tijburg, L.B.M., Geelen, M.J.H., van Golde, L.M.G.: J. Biol. Chem., 268,7458–7464 (1993)
[6] Tijburg, L.B.M., Vermeulen, P.S., van Golde, L.M.G.: Methods Enzymol.,209,258–263 (1992) (Review)
[7] Vermeulen, P.S., Geelen, M.J.H., van Golde, L.M.G.: Biochim. Biophys. Acta,1211, 343–349 (1994)

1 NOMENCLATURE

EC number
2.7.7.15

Systematic name
CTP:choline-phosphate cytidylyltransferase

Recommended name
Choline-phosphate cytidylyltransferase

Synonyms
Phosphorylcholine transferase
Cytidylyltransferase, choline phosphate
CDP-choline pyrophosphorylase
CDP-choline synthetase
Choline phosphate cytidylyltransferase
CTP-phosphocholine cytidylyltransferase
CTP:cholinephosphate cytidylyltransferase
CTP:phosphorylcholine cytidylyltransferase
Cytidine diphosphocholine pyrophosphorylase
Phosphocholine cytidylyltransferase
Phosphorylcholine cytidylyltransferase
Phosphorylcholine:CTP cytidylyltransferase

CAS Reg. No.
9026-34-0

---

2 REACTION AND SPECIFICITY

Catalyzed reaction
CTP + choline phosphate $\rightarrow$
$\rightarrow$ diphosphate + CDPcholine (random mechanism [5])

Reaction type
Nucleotidyl group transfer

Natural substrates
More (H-form is the active form of enzyme in cytoplasm [14], change in relative distribution of H-form and L-form in cytosol may be important in the regulation of phosphatidylcholine synthesis [14], catalyzes a major rate-limiting
step in the biosynthesis of phosphocholine [13, 20, 21], active lipoprotein
form (H-form) is the membrane-associated form of the enzyme in adult lung
[17], active form of enzyme on the ER, enzyme in cytosol appears to be latent [20], comparison of lipid regulation of yeast and rat enzyme [16]) [13,
14, 16, 17, 20, 21]

Substrate spectrum

1 CTP + choline phosphate (r [9], highly specific for phosphocholine [9])
[1–21]
2 CTP + ethanolamine phosphate (not [13]) [19]
3 CTP + phosphodimethylethanolamine (not [13]) [19]
4 CTP + phosphomonomethylethanolamine [13, 19]

Product spectrum

1 Diphosphate + CDPcholine [1, 5]
2 ?
3 ?
4 ?

Inhibitor(s)

3-sn-Lysophosphatidylcholine [10]; Lysosphingolipids [15]; Sphingolipids
(reversed by activating phospholipids) [1]; 5,5'-Dithiobis(2-nitrobenzoate) [1,
6]; NEM (CTP and phosphocholine protect) [1, 6]; PCMB [1, 6]; Phosphate
[1, 11]; Blue MX-R [4]; Blue H-B [4]; Turquoise H-A [4]; Levafix E-5BNA [4];
Turquoise MX-G [4]; Green H-4G [4]; CDPcholine (competitive to choline
phosphate) [5]; Ca^{2+} [6]; Mn^{2+} [6]; Zn^{2+} [6]; Sphingosine [15]; ATP (not [6,
13], extent of inhibition is dependent on preincubation time, temperature
and Mg^{2+} and Ca^{2+} concentration [11]) [11]; Phosphoethanolamine (competitive to phosphocholine) [19]; Phosphomonomethylethanolamine (competitive to phosphocholine) [19]; Phosphodimethylethanolamine (competitive to phosphocholine) [19]; Chlorpromazine (addition of saturating
amounts of rat liver phospholipid reverses inhibition) [20]; More (not: phosphorylethanolamine [3], tetracaine [21], CDP [6], CMP [6], propanolol [21])
[3, 6, 21]

Cofactor(s)/prosthetic group(s)/activating agents

Lipid (greatest stimulation (8–10fold) with 0.1 mM phosphatidylcholine-oleic
acid vesicles (1:1) [6], 4–5fold stimulation by 0.01 mM phosphatidylglycerol
[6], requires addition of exogenous lipid for maximal activity [1, 2], lysophosphatidylethanolamine is the best lipid activator [2], good activators:
phosphatidylserine, phosphatidylinositol, phosphatidylglycerol, lysophosphatidylcholine, phosphatidylcholine [2], the isolated 45000 MW catalytic
protein has the same lipid requirement and kinetic properties as the purified
enzyme containing both proteins, maximal stimulation by phosphatidylcholine vesicles containing 9 mol% of either oleic acid, phosphatidylinositol or
phosphatidylglycerol [7], stimulation by phospholipid extracted from castor
bean endosperm and phosphatidylcholine-oleate vesicles [13], L-form:
markedly stimulated by liposomes made from rat liver lipids [9], H-form: requires a fraction containing lipids for optimal activity [9], phosphatidylserine
and phosphatidylinositol activate in vitro, the activation by lysophosphatidyl-

ethanolamine may have more physiological significance [10], L-form: dependent on phospholipid for activity [14], H-form: active without lipid [14], phosphatidylinositol and phosphatidylglycerol cause L-form to aggregate into a form similar to H-form [14], phosphatidylcholine/oleic acid (1:1 molar ratio) and oleic acid also aggregate to L-form [14], phosphatidylcholine does not produce aggregation [14], lysophosphatidylcholine causes slight increase of activity [21], comparison of lipid regulation of yeast and rat enzyme [16]) [1, 2, 6, 7, 9, 10, 13, 14, 16, 18, 21]; Fatty acids (activity slightly enhanced by addition of saturated fatty acids, markedly increased by addition of unsaturated fatty acids regardless of chain length and number of double bonds) [5]; Oleate (stimulates) [18]

Metal compounds/salts
Mg^{2+} (required [3, 5, 11], 2–20 mM [5], optimum concentration: around 2 mM [11]) [3, 5, 11]

Turnover number (min^{-1})
2000 (45000 MW catalytic protein, CTP + choline phosphate) [7]

Specific activity (U/mg)
1.518 [13]; 0.612 [12]; 47.5 [1]; 0.1096 [3]; 12.25 [6]; 0.106 [9]; 0.6024 [11]

K_m-value (mM)
0.004 (diphosphate, L-form) [2]; 0.17 (phosphocholine, L-form) [2]; 0.21 (CTP, CDPcholine, L-form) [2]; 0.22 (CTP) [1]; 0.24 (phosphocholine) [1]; 0.3 (choline phosphate) [5]; 0.55 (CTP) [3]; 0.64 (CDPcholine, H-form) [2]; 2.1 (phosphorylcholine) [3]; 4.0 (phosphodimethylethanolamine) [19]; 6.9 (phosphomonomethylethanolamine) [19]; 10 (CTP) [5]; 68.4 (phosphoethanolamine) [19]

pH-optimum
6.0 [2]; 6.3 (assay at) [11]; 7.0 [1, 5, 9]; 7–7.5 [3, 13]; 7.0–9.0 [11]; 7.5 [3, 13]

pH-range
6–8 (6: about 80% of activity maximum, 8: about 60% of activity maximum) [6]

Temperature optimum (°C)
35 [11]; 37 (assay at) [1, 2, 12, 15, 21]

Temperature range (°C)

3 ENZYME STRUCTURE

Molecular weight

97000 (rat, gel filtration, glycerol density gradient centrifugation) [1]

155000 (Ricinus communis, gel filtration) [13]

200000 (rat, L-form, gel filtration, aggregates in the cytosol to form high molecular weight species (H-form) with a median value of 1200000) [2, 9]

284000 (rat, Hep G2 cells, H-form, glycerol density gradient centrifugation) [14]

1300000 (rat, H-form, gel filtration, consists of multiple copies of the L-form, wide range of MWs with a median value of 1300000) [19]

Subunits

Dimer (2×44500, rat, SDS-PAGE [1], 2×45000, rat, Hep G2 cells, L-form, SDS-PAGE [14]) [1, 14]

Tetramer ($2 \times 39000 + 2 \times 48000$, rat, SDS-PAGE [6], 4×40000, Ricinus communis, SDS-PAGE [13]) [6, 13]

? ($x \times 56000$, Pisum sativum, SDS-PAGE [3], $x \times 45000$ (catalytic subunit) + $x \times 38000$ (functional role of this subunit not documented), rat, SDS-PAGE [7], rat, if bound to a detergent micelle or membrane vesicle the purified native enzyme is a dimer composed of two noncovalently linked 42000 MW subunits, in the absence of a membrane or micelle, the dimers self-aggregate in a reversible manner [8]) [3, 7, 8]

Glycoprotein/Lipoprotein

Lipoprotein (H-form [14, 17], H-form appears to be a lipoprotein consisting of an apoprotein (L-form dimer of 45000 MW subunits) complexed with lipids [14], phosphatidylinositol is present in the H-form isolated from Hep G2 cells [14]) [14, 17]

---

4 ISOLATION/PREPARATION

Source organism

Yeast (overexpressed in CHO cells) [16]; Ricinus communis (L- var. Hale) [11, 13]; Rat (overexpressed in CHO cells [16]) [1, 2, 5–10, 12, 14–21]; Pisum sativum [3]; Human [4]

Source tissue

Intestinal mucosa (contains only L-form) [21]; Stem [3]; Liver [1, 2, 6–10, 12, 14, 15, 19, 20]; Lung (L-form and H-form [14]) [14, 17, 18]; Fetal lung-derived fibroblasts [4]; Brain [5]; Endosperm (postgermination [11]) [11, 13]; Cells (Hep G2 cells (L-form and H-form), A549 cells (L-form and H-form), alveolar type II cells (L-form and H-form) [14], COS cells [16]) [14, 16]

Localization in source
Cytosol [1, 2, 5, 6, 9, 12, 14, 16, 20, 21]; Microsomes [7, 17]; Membrane [18]

Purification
Rat (affinity chromatography, 2 forms: L-form and H-form, L-form: major species in fresh cytosol, H-form: consists of multiple copies of L-form [9]) [1, 2, 6, 9, 12, 19]; Pisum sativum [3]; Human (dye-affinity chromatography with Green H-4G-Sepharose CL4B) [4]; Ricinus communis [11, 13]

Crystallization
–

Cloned
[1, 16]

Renatured
–

---

5 STABILITY

pH

Temperature (°C)

Oxidation

Organic solvent

General stability information
Unstable at low protein concentration [13]; Purified enzyme is unstable in presence of DTT. CTP-Mg^{2+} or bovine serum albumin do not significantly stabilize the enzyme [9]; Freezing and thawing once causes more than 90% loss of activity [9]; Instability of the highly purified enzyme is partially overcome by addition of 1 mM DTT and 1% bovine serum albumin or 0.5 mM CTP and 2.5 mM magnesium acetate [12]

Storage
–70°C, 50 mM Tris-HCl, 150 mM NaCl, 1.0 mM EDTA, 2.0 mM DTT, 0.025% NaN_3, pH 7.4, 0.03% Triton X-100, 200 mM phosphate, stable for several months [1]; 4°C, 20% loss of activity per day [1]; –70°C, stable in presence of Triton X-100 and 0.2 M potassium phosphate [6]; 4°C, 30% loss of activity after 1 week [11]; –20°C, stable for more than 1 month in presence of 20% glycerol [11]; 0°C, 80% loss of activity after 6 days [12]; –20°C, enzyme concentrated to a small volume, 50% glycerol, 20% loss of activity after 2 weeks [13]

6 CROSSREFERENCES TO STRUCTURE DATABANKS

PIR/MIPS code
PIR2:A49366 (mouse); PIR2:A36001 (rat); PIR1:XNBYCP (yeast (Saccharomyces cerevisiae))

Brookhaven code

7 LITERATURE REFERENCES

[1] Weinhold, P.A., Feldman, D.A.: Methods Enzymol.,209,248–258 (1992) (Review)
[2] Vance, D.E., Pelech, S.D., Choy, P.C.: Methods Enzymol.,71,576–581 (1981) (Review)
[3] Price-Jones, M.J., Harwood, J.L.: Biochem. Soc. Trans.,13,1243–1245 (1985)
[4] Hunt, A.N., Postle, A.D.: Biochem. Soc. Trans.,14,1279–1281 (1986)
[5] Mages, F., Rey, C., Fonlupt, P., Pacheco, H.: Eur. J. Biochem.,178,367–372 (1988)
[6] Weinhold, P.A., Rounsifer, M.E., Feldman, D.A.: J. Biol. Chem.,261,5104–5110 (1986)
[7] Feldman, D.A., Weinhold, P.A.: J. Biol. Chem.,262,9075–9081 (1987)
[8] Cornell, R.: J. Biol. Chem.,264,9077–9082 (1989)
[9] Choy, P.C., Lim, P.H., Vance, D.E.: J. Biol. Chem.,252,7673–7677 (1977)
[10] Choy, P.C., Vance, D.E.: J. Biol. Chem.,253,5163–5167 (1978)
[11] Wang, X., Moore, T.S.: Arch. Biochem. Biophys.,274,338–347 (1989)
[12] Choy, P.C., Vance, D.E.: Biochem. Biophys. Res. Commun.,72,714–719 (1976)
[13] Wang, X., Moore, T.S.: Plant Physiol.,93,250–255 (1990)
[14] Weinhold, P.A-, Rounsifer, M.E., Charles, L., Feldman, D.A.: Biochim. Biophys. Acta, 1006,299–310 (1989)
[15] Sohal, P.S., Cornell, R.B.: J. Biol. Chem.,265,11746–11750 (1990)
[16] Johnson, J.E., Kalmar, G.B., Sohal, P.S., Walkey, C.J., Yamashita, S., Cornell, R.B.: Biochem. J.,285,815–820 (1992)
[17] Feldman, D.A., Rounsifer, M.E., Charles, L., Weinhold, P.A.: Biochim. Biophys. Acta,1045,49–57 (1990)
[18] Weinhold, P.A., Charles, L.G., Feldman, D.A.: Biochim. Biophys. Acta,1086,57–62 (1991)
[19] Jamil, H., Vance, D.E.: Biochim. Blophys. Acta,1086,335–339 (1991)
[20] Pelech, S.L., Jetha, F., Vance, D.E.: FEBS Lett.,158,89–92 (1983)
[21] Mansbach II, C.M., Arnold, A.: Biochim. Biophys. Acta,875,516–524 (1986)

1 NOMENCLATURE

EC number
    2.7.7.18

Systematic name
    ATP:nicotinate-ribonucleotide adenylyltransferase

Recommended name
    Nicotinate-nucleotide adenylyltransferase

Synonyms
    Deamido-NAD$^+$ pyrophosphorylase
    Adenylyltransferase, nicotinate mononucleotide
    Deamidonicotinamide adenine dinucleotide pyrophosphorylase
    NaMN-ATase
    Nicotinic acid mononucleotide adenylyltransferase
    More (may be identical with EC 2.7.7.1 [3])

CAS Reg. No.
    9026-98-6

---

2 REACTION AND SPECIFICITY

Catalyzed reaction
    ATP + nicotinate ribonucleotide →
    → diphosphate + deamido-NAD$^+$

Reaction type
    Nucleotidyl group transfer

Natural substrates

Substrate spectrum
    1 ATP + nicotinate ribonucleotide (i.e. nicotinate mononucleotide, r [1–3], best substrate (E. coli) [3], reaction at 77% the rate of nicotinamide ribonucleotide (yeast) [3]) [1–3]
    2 ATP + nicotinamide ribonucleotide (i.e. NMN or nicotinamide mononucleotide, r, reverse reaction at 17% the rate of deamido-NAD$^+$-synthesis (E. coli)) [3]
    3 Deoxy-ATP + nicotinate ribonucleotide [3]
    4 Deoxy-ATP + nicotinamide ribonucleotide [3]
    5 ATP + 3-acetyl-pyridine-NAD$^+$ (reaction at 76% the rate of nicotinamide ribonucleotide (yeast), poor substrate (E. coli)) [3]
    6 ATP + 3-pyridinealdehyde-NAD$^+$ (reaction at 28% the rate of nicotinamide ribonucleotide (yeast), poor substrate (E. coli)) [3]

Product spectrum
  1 Diphosphate + deamido-NAD$^+$ (i.e. nicotinic acid adenine dinucleotide)
    [1–3]
  2 Diphosphate + NAD$^+$ [3]
  3 ?
  4 ?
  5 ?
  6 ?

Inhibitor(s)
  NAD$^+$ (deamido-NAD$^+$ as substrate, E. coli) [3]

Cofactor(s)/prosthetic group(s)/activating agents

Metal compounds/salts
  Mg^{2+} (requirement) [1–3]; KCl (activation, 25 mM, NAD$^+$-synthesis, not dea-
  mido-NAD$^+$-synthesis) [3]; NH$_4$Cl (activation, can substitute for KCl) [3]

Turnover number (min^{-1})

Specific activity (U/mg)
  0.00133 (Nicotiana tabacum cv. Xanthi, callus culture) [2]; 0.00269 (Nicotiana tabacum cv. Samsun, callus culture) [2]; 0.00358 (Nicotiana tabacum
  cv. Samsun, root) [2]; 0.0038 (NAD$^+$-synthesis, E. coli) [3]; 0.004 (NAD$^+$ as
  substrate, E. coli) [3]; 0.024 (deamido-NAD$^+$ as substrate, E. coli) [3]; 0.068
  (deamido-NAD$^+$-synthesis, E. coli) [3]; 3.3 (NAD$^+$-synthesis, yeast) [3]; 3.85
  (deamido-NAD$^+$ as substrate, yeast) [3]; 5 (NAD$^+$ as substrate, yeast) [3];
  7.3 (deamido-NAD$^+$-synthesis, yeast) [3]

K$_m$-value (mM)
  0.0045 (deamido-NAD$^+$, E. coli) [3]; 0.029 (deamido-NAD$^+$, yeast) [3]; 0.03
  (nicotinate ribonucleotide) [1]; 0.06 (ATP (+ nicotinate ribonucleotide),
  yeast) [3]; 0.069 (NAD$^+$, yeast) [3]; 0.08 (nicotinate ribonucleotide, E. coli)
  [3]; 0.13 (nicotinate ribonucleotide, yeast) [3]; 0.2 (nicotinamide ribonucleotide, E. coli) [3]; 0.27 (3-acetylpyridine-NAD$^+$, yeast) [3]; 0.37 (NAD$^+$, E. coli)
  [3]; 0.4 (nicotinamide ribonucleotide, yeast) [3]; 0.5 (ATP, E. coli [3]) [1, 3];
  0.65 (diphosphate, yeast) [3]; 0.74 (3-pyridinealdehyde-NAD$^+$, yeast) [3];
  1.1 (diphosphate, E. coli) [3] 1.3 (nicotinate ribonucleotide) [2]; 4 (ATP) [2]

pH-optimum
  7.5 [1]

pH-range

Temperature optimum (°C)
  30 (assay at) [2]; 37 (assay at) [1, 3]

Temperature range (°C)

3 ENZYME STRUCTURE

Molecular weight

Subunits

Glycoprotein/Lipoprotein
    –

---

4 ISOLATION/PREPARATION

Source organism
    Brewer's yeast [3]; E. coli (strains K-12 [1] or B [3]) [1, 3]; Nicotiana taba-
    cum (tobacco, cv. Samsun or cv. Xanthi) [2]

Source tissue
    Cell [1, 3]; Callus culture (from seedlings' roots) [2]; Root (cv. Samsun) [2]

Localization in source

Purification
    E. coli (partial) [1, 3]; Brewer's yeast (partial) [3]

Crystallization
    –

Cloned
    –

Renatured
    –

---

5 STABILITY

pH

Temperature (°C)
    55 (5 min, 54% loss of activity, E. coli) [3]; 60 (5 min, 39% (yeast) or 95% (E.
    coli) loss of activity) [3]; 65 (5 min, 65% loss of activity, yeast) [3]

Oxidation

Organic solvent

General stability information
    Phosphate buffer, 0.08 M, pH 7.5, stabilizes [2]

Storage
    Frozen or in the cold, 24 h [2]; 0–3°C, E. coli enzyme, 31 days stable, yeast
    enzyme: 22% loss of activity within 31 days [3]

6 CROSSREFERENCES TO STRUCTURE DATABANKS

PIR/MIPS code

Brookhaven code

7 LITERATURE REFERENCES

[1] Imsande, J.: J. Biol. Chem.,236,1494–1497 (1961)
[2] Wagner, R., Wagner, K.G.: Planta,165,532–537 (1985)
[3] Dahmen, W., Webb, B., Preiss, J.: Arch. Biochem. Biophys.,120,440–450 (1967)

1 NOMENCLATURE

EC number
2.7.7.19

Systematic name
ATP:polynucleotide adenylyltransferase

Recommended name
Polynucleotide adenylyltransferase

Synonyms
NTP polymerase
RNA adenylating enzyme
Nucleotidyltransferase, polyadenylate
AMP polynucleotidylexotransferase
ATP-polynucleotide adenylyltransferase
ATP:polynucleotidylexotransferase
Poly(A) polymerase
Poly(A) synthetase
Polyadenylate nucleotidyltransferase
Polyadenylate polymerase
Polyadenylate synthetase
Polyadenylic acid polymerase
Polyadenylic polymerase
Terminal riboadenylate transferase [2]
Poly(A) hydrolase [15]
RNA formation factors, PF1
Adenosine triphosphate:ribonucleic acid adenylyltransferase [21]
More (see also EC 2.7.7.6)

CAS Reg. No.
9026-30-6

---

2 REACTION AND SPECIFICITY

Catalyzed reaction
ATP + RNA_n →
→ diphosphate + RNA_{n+1} (mechanism [3])

Reaction type
Nucleotidyl group transfer

Natural substrates

ATP + RNA (overview of biological function [3], synthetic and hydrolytic activities are functions of the same molecule, the level of adenine nucleotides regulates synthesis and degradation of poly(A), the hydrolytic reaction is responsible for poly(A) shortening or turnover, poly(A) itself is a storage form of adenine nucleotides [15], involved in the 3'-end processing of mRNA [22, 25], the enzymatic machinery that catalyzes formation of 3'-ends of polyadenylated mRNAs consists of two distinct factors: a poly(A) polymerase and a cleavage/specificity factor required for the correct cleavage at the poly(A) site of pre-mRNA [24, 25], processing and activation of stored mRNAs after resumption of development [28], 2 enzymes participate in the polyadenylation of chromosomal RNA, by a coupled mechanism. The chromatin bound enzyme adds 120–130 adenosine nucleotides to chromosomal RNA. The nucleoplasmic enzyme completes the polyadenylation by adding 80–90 more AMP units to the polyadenylated end [29]) [3, 15, 22, 24, 25, 28, 29]

Substrate spectrum

1 ATP + RNA (enzyme also catalyzes hydrolysis of poly(A) [15], does not degrade poly(A) associated with poly(A) * poly(U) helical structure [15], enzyme is unable to catalyze pyrophosphorolysis or phosphorolysis reaction [21], enzyme also has cleavage activity [25], catalyzes the synthesis of polyadenylate linked to the 3'-hydroxyl end of the terminal nucleoside of an RNA primer [1], primer required [1–36]: rRNA (16S (E. coli) [2], 23S (E. coli) [2]) [1, 2, 31], mixture of tRNA (not [35]) [1, 31], methionyl-tRNA [1], tRNA lacking terminal adenosine [1], viral RNA MS-2 and QB [2], poly(A) [2, 4], poly(G, U) [35], short poly(U) [35], dinucleoside phosphates having 3'-OH [4], variety of oligoribonucleotides having free 3'-OH [9], various E. coli tRNAs or rRNAs [23], RNA homopolymers [23], oligonucleotides A-A-A-A and A-A-A [31], poly(A) and poly(C) minimal effective [31], poly(A) (short [35]) is the most effective primer [16, 35], mitochondrial RNA at least five times more efficiently used as nuclear RNA [16], polymerase IIa and IIb utilize a variety of natural and synthetic RNAs as well as DNA as primer [18], rather low specificity for primer [3, 5], minimum effective primer length is 4 to 6 nucleotides [8], influence of shape and size on priming efficiency [6], Mg^{2+}-activated calf thymus enzyme uses poly(A), tRNA, small RNA fragments from calf thymus RNA well, but HeLa 18 and 28S rRNA and MS-1 RNA poorly if at all [3], human nuclear enzyme and Vaccinia virus enzyme are able to use both RNA and oligo(A) as primer, human cytoplasmic enzyme is able to use RNA but not oligo(A) [12], chromatin enzyme uses chromosomal RNA as primer, enzyme from nucleoplasm uses poly(A) and hnRNA isolated from chromatin as primer [29], no specificity for the 3'-terminal nucleotides [3, 5, 28] when poly(C) and poly(I), but not poly(U), primes poly(A) synthesis with the Mg^{2+}-activated calf thymus enzyme [3], the Mg^{2+}-activated enzyme from calf thy-

mus or HeLa cells prefers either longer poly(A) or RNAs rather than shorter oligomers of AMP [3], Mn^{2+}-activated enzymes are indifferent to primer
length [3], elongation of the primer is distributive [22, 23], highly specific
for ATP [1–5, 8, 9, 11, 16, 18, 19, 21, 23, 29, 35], ATP is utilized 2000-fold
more than any other nucleoside triphosphate tested [1], other nucleotides
polymerized at less than 1% of the ATP rate [2, 3], adenosine 5'-(beta,
gamma-methylene)triphosphate is efficiently polymerized into poly(A) with
a polymerase from quail oviduct [3], enzyme catalyzes both polyadenylic
acid synthesis in absence of a template and DNA-dependent RNA synthesis [14], not: phage RNA [35], poly(G) [35], poly(C) [35], poly(U) [31],
poly(dT) [31]) [1–36]
2 dATP + RNA (15% of the activity with ATP) [28]
3 CTP + RNA (12% of the activity with ATP, adenylyltransferase A) [31]

Product spectrum
1 Diphosphate + RNA(A)$_n$ (AMP is the predominant product of the hydrolysis, ADP and ATP are also formed [15], polyadenylate sequences of
100–200 AMP residues [6], average length of poly(A) formed is 600 nucleotides [16], polymerase IIa: chain length of the product synthesized is
independent of the primer concentration, polymerase IIb: the length of
the product decreases when RNA concentration increases [18], no apparent length limitation for the poly(A) tail synthesized [23], length of the
poly(A) tail is dependent on incubation time and RNA primer concentration [28]) [1–36]
2 ?
3 ?

Inhibitor(s)
SO_4^{2-} [1]; PO_4^{3-} [1]; Na^+ (NaCl [3, 8], above 50 mM [3], 0.1 M [8]) [3, 8, 19];
K^+ (KCl [3, 5, 11, 18, 28], 80 mM: 50% inhibition [28], above 50 mM [3, 18],
100 mM [5], maximal stimulation at 40 mM, inhibition above 250 mM [11])
[3, 5, 11, 18, 19, 28]; Ca^{2+} [19]; Mg^{2+} [19]; Zn^{2+} [19]; NH_4^+ (ammonium sulfate [5, 13, 16, 28], 50 mM: 50% inhibition [28], 10–40 mM [5], 0.1 M polymerase Ia and Ib completely inhibited, polymerase II: 68% inhibition [13],
maximal activity at 33 mM, inhibition above 150 mM [16]) [5, 13, 16, 19, 28];
Dibasic sodium phosphate [9, 17]; Diphosphate (noncompetitive to ATP
and primer [3]) [3, 9, 11, 16, 18, 21, 31, 35]; Heparin [35]; Polyvinyl sulfate
[35]; Bentonite [35]; Inorganic phosphate (inhibits enzyme from E. coli and
calf thymus nuclei, but not rat liver nuclear enzyme [3]) [3, 21]; Aurintricarboxylic acid [21]; Poly(C) [21]; Poly(U) [31]; Poly(dT) [31, 34]; Calf thymus
DNA [21]; Rose Bengal [11]; ATP (hepatoma enzyme less effective to substrate inhibition than liver enzyme [11], inhibits hydrolytic reaction [15],
above 0.5 mM [28]) [11, 15, 28]; ADP (inhibits hydrolytic reaction) [15]; AMP
(inhibits hydrolytic reaction) [15]; Adenosine 5'-(alpha,beta-methylenetri-
phosphate) [17]; Cordycepin 5'-triphosphate [11, 17]; Cordycepin (not [5])

[31]; GTP [3, 18]; UTP [3, 18]; CTP [3, 18]; 2'-dATP [3, 36]; 3'-dATP [3, 18, 31, 36]; Adenylyl-(3'-5')adenosine [21]; Adenylyl-(3'-5')cytosine [21]; Rifamycin derivatives (some derivatives are effective, others not) [31]; Rifamycin derivative AF/013 (O-n-octyloxime of 3-formylrifamycin SV) [3, 16, 17, 30]; 3'-Acetyl-1'-benzyl-2'-methylpyrrolo[3,2-C]4-desoxy-rifamycin [30]; 3-(4-Ethylpiperazinoiminomethyl)rifamycin SV [30]; Rifamycin B:N,N-dipentylamide [30]; Rifamycin B:N,N-diethylamide [30]; 3-(4-Benzyl-2,6-dimethyl piperazinoiminomethyl)rifamycin SV [30]; Rifamycin SV [30]; Proflavine (only at very high levels [3]) [3, 17]; Rifampicin [14]; NEM (inhibits Mn^{2+}-activated enzyme of rat liver and calf thymus [3]) [3, 11, 18, 23]; Spermine (inhibition if poly(A), nuclear RNA, or tRNA serves as primer, not with short oligonucleotide primers such as $(Ap)_3A$ [3, 33]) [3, 17, 33]; Spermidine [33]; Putrescine [33]; 1,10-Phenanthroline [17]; Sodium vanadate [17]; Polyamines [3, 33]; alpha-Amanitin [11]; Pancreatic ribonuclease [11]; Ionic strength (0.1) [5]; Ribonucleoside triphosphates other than ATP [9]; More (not: alpha-amanitin [5, 13, 16, 18], actinomycin D [5], insensitive to high levels of RNA-polymerase inhibitors [9], 4-(dimethylamino)-4-desoxy rifamycin SV [30], 3-formal rifamycin SV:o-methyloxime [30]) [5, 9, 13, 16, 18, 30]

Cofactor(s)/prosthetic group(s)/activating agents

3',5'-AMP (slight stimulation) [21]; DTT (required) [27]; Poly(U) (stimulates) [34]

Metal compounds/salts

Mg^{2+} (in presence of Mg^{2+} and a specificity factor required for correct cleavage at the poly(A) site of pre-mRNA [21, 25], ATP is utilized 150-fold more with Mn^{2+} than with Mg^{2+} [1], one-fifth of the activity of Mg^{2+} in NTP activation [2], Mg^{2+} or Mn^{2+} required [34], divalent cation requirement may be fulfilled by Mn^{2+}, Mg^{2+} or a combination of the two depending on the source of the enzyme [3], more active in presence of Mg^{2+} than Mn^{2+} (adenylyltransferase A [31]) [5, 31], more active in presence of Mn^{2+} than Mg^{2+} [8, 14–17, 21–23], 10% of the activity with Mn^{2+} [11], HeLa cells contain one enzyme form that is stimulated by Mn^{2+} and also by Mg^{2+}, and a second one that is absolutely dependent on the presence of Mg^{2+} [12], Vaccinia virus enzyme is stimulated by Mn^{2+} and also by Mg^{2+} [12], Mg^{2+} is inactive, maximum activity in presence of both Mn^{2+} and Mg^{2+} [18], NE PAP I (isoenzyme from cytoplasmic fraction) and S100 PAP (isoenzyme from nuclear fraction): higher activity in presence of Mn^{2+} than in presence of Mg^{2+}, NE PAP II: approximately equal levels in presence of Mn^{2+} and Mg^{2+} [20], optimum concentration: 4–6 mM [22], 5 mM [34], 8–10 mM (polymerase I from chromatin, polymerase II from nucleoplasm is inactive in presence of Mg^{2+}) [29], optimum concentration depends on ATP concentration [23], completely inactive in presence of Mg^{2+} [35]) [1–3, 5, 8, 11, 12, 14–18, 20–23, 25, 29, 31, 34]; Mn^{2+} (nonspecific adenylation of RNA in presence of Mn^{2+} [25], ATP is utilized 150-fold more with Mn^{2+} than with Mg^{2+} [1], required for NTP activation

[2], Mn^{2+} or Mg^{2+} required [34], divalent cation requirement may be fulfilled by Mn^{2+}, Mg^{2+} or a combination of the two depending on the source of the enzyme [3], more active in presence of Mg^{2+} than Mn^{2+} (adenylyltransferase A [31]) [5, 31], more active in presence of Mn^{2+} than Mg^{2+} [8, 14–17, 21–23], maximum activity in presence of both Mn^{2+} and Mg^{2+} [18], HeLa cells contain one enzyme form that is stimulated by Mn^{2+} and also by Mg^{2+}, and a second one that is absolutely dependent on the presence of Mg^{2+} [12], Vaccinia virus enzyme is stimulated by Mn^{2+} and also by Mg^{2+} [12], NE PAP I (isoenzyme from nuclear fraction) and S100 PAP (isoenzyme from cytoplasmic fraction): higher activity in presence of Mn^{2+} than in presence of Mg^{2+}, NE PAP II: approximately equal levels in presence of Mn^{2+} and Mg^{2+} [20], required [4, 9, 13, 16, 27, 28, 35], exclusively activated by Mn^{2+} [19], Mg^{2+} can partially replace Mn^{2+} in the reaction with polymerase II [13], absolute requirement [11], optimal concentration: 0.25–0.75 mM [11], 2 mM [27, 34], 2–4 mM [28], 0.50–0.75 mM [15], 0.25–1.0 mM [16], 4 mM (polymerase IIa), 4–8 mM (polymerase IIb) [18], 0.5 mM (at 0.5 mM ATP) [22], 0.8 mM (polymerase I and II) [29], optimum concentration depends on ATP concentration [23]) [1–5, 8, 9, 11–23, 25, 27–29, 31, 34, 35]; KCl (maximal stimulation at 40 mM, inhibition above 250 mM [11], maximal activity at 33 mM, inhibition above 150 mM [16], optimum concentration: 60 mM [23], requirement is dependent on the primer and the divalent cation used [22]) [11, 16, 22, 23]; NH_4^+ (maximal activity at 33 mM, inhibition above 150 mM) [16]; More (poly(A) polymerases purified from different sources, and in some cases even from the same source, respond differently to the presence of Mg^{2+} and Mn^{2+} [3], low ionic strength required for maximal activity [29]) [3, 29]

Turnover number (min^{-1})
  200 (nucleotide polymerized) [1]; 1800 (ATP $(+ rA(pA)_5)$) [9]

Specific activity (U/mg)
  More [5, 11, 13, 16, 21, 22, 27–29, 34]; 28.33 [9]; 0.0939 [19]

K_m-value (mM)
  More (dependence on divalent cation concentration [23]) [3, 21, 23, 26, 27, 31]; 0.002 (RNA primer, E. coli) [3]; 0.0036 (poly(A), Mn^{2+}-activated enzyme) [22]; 0.007 (short poly(A)) [35]; 0.01 (oligo(A), Mn^{2+}-activated enzyme) [22]; 0.028 (ATP) [16]; 0.03 (ATP, polymerase IIa) [18]; 0.04 (ATP) [28]; 0.05 (ATP, E. coli [3], $p(A)_3$ primer, Mn^{2+}-activated calf thymus enzyme [3], oligoadenylate (in presence of Mn^{2+}) [9], ATP, polymerase IIb [18]) [3, 9, 18]; 0.06 (dATP) [28]; 0.07 (ATP, rat liver) [3, 11]; 0.14–0.36 (poly(A), Mg^{2+}-activated enzyme) [22]; 0.15 (ATP, rat hepatoma cells) [11]; 0.2 (oligoadenylate (in presence of Mg^{2+})) [9]; 0.3 (oligo(A), Mg^{2+}-activated enzyme) [22]

pH-optimum

6.4–8.0 (in presence of Mn^{2+}) [15]; 7.0 (Mn^{2+}-activated enzyme) [23]; 7.8–8.2 [16]; 8.0 (vaccinia virus enzyme, human, cytoplasmic Mn^{2+}-dependent enzyme [12], polymerase IIa and IIb [18], Mg^{2+}-activated enzyme [22], polymerase II (nucleoplasm) [29]) [12, 18, 19, 22, 27, 29]; 8–8.5 [28]; 8.2 [2, 5]; 8.3 (human nuclear Mn^{2+}- and Mg^{2+}-activated enzyme [12], Mn^{2+}-activated enzyme [22]) [9, 12, 22]; 8.5 (polymerase I (chromatin)) [29]; 8.6 [8]; 9.5 (adenylyltransferase A) [31]

pH-range

6–9 (pH 6: 47% (polymerase IIa) and 5% (polymerase IIb) of activity maximum, pH 9: 55% of activity maximum) [18]; 7–8.5 (pH 7: about 40% of activity maximum, pH 8–8.5: activity maximum) [28]; 7.0–8.8 (about 50% of activity maximum at pH 7.0 and 8.8) [9]; 7–9 (active in this range [3], about 65% of activity maximum at pH 7.0 and 9.0 [27]) [3, 27]; 7.2–9.2 (pH 7.2: 55% of activity maximum, pH 9.2: 62% of activity maximum) [22]; 7.5–9 (pH 7.5: 50% of activity maximum, pH 9.0: 15% of activity maximum) [21]; 8–10 (pH 8: about 40% of activity maximum, pH 10: about 50% of activity maximum) [8]

Temperature optimum (°C)

35 (assay at) [9]; 37 (assay at) [5]

Temperature range (°C)

---

3 ENZYME STRUCTURE

Molecular weight

43000 (Tetrahymena pyriformis, polymerase Ia and Ib, gel filtration) [13]
45000–60000 (hamster, sedimentation analysis) [3]
47000 (Saccharomyces cerevisiae, gel filtration) [23]
50000–60000 (human, NE PAP I and II, S100 PAP, sucrose gradient sedimentation [20], human, gel filtration [25]) [20, 25]
57000 (and 60000, bovine, 2 major forms of enzyme, gel filtration [22], bovine, glycerol density gradient centrifugation [35]) [22, 35]
58000 (human, nuclear Mg^{2+}- and Mn^{2+}-stimulated enzyme, glycerol density gradient sedimentation [12], human, Mg^{2+}-activated enzyme, sedimentation analysis [3], Pseudomonas putida, adenylyltransferase B, glycerol density gradient sedimentation [31], E. coli, gel filtration [21]) [3, 12, 21, 31]
60000 (bovine, Mn^{2+}-activated enzyme, sedimentation analysis, gel filtration [3], rat, mitochondria, glycerol density gradient centrifugation [16], and 57000, bovine, 2 major forms of enzyme, gel filtration [22]) [3, 16, 22]
62000 (bovine [2, 9], sucrose gradient sedimentation, gel filtration [9]) [2, 9]
63000 (human, Mn^{2+}-activated, sedimentation analysis [3], human, cytoplasmic Mn^{2+}-dependent enzyme, glycerol density gradient sedimentation [12]) [3, 12]

65000 (mouse embryos, sedimentation analysis, gel filtration) [3]
65000–70000 (Triticum aestivum, gel filtration) [26]
70000 (Vaccinia virus, enzyme from infected cytoplasm, glycerol density gradient sedimentation [3, 12], Artemia sp., gel filtration [28]) [3, 12, 28]
76987 (bovine, predicted from nucleotide sequence) [7]
80000 (Vaccinia virus cores [3, 8], mouse L cells [3], sucrose gradient sedimentation [3, 8]) [3, 8]
82400 (bovine enzyme expressed in E. coli, predicted from nucleotide sequence) [10]
95000 (Tetrahymena pyriformis, polymerase II, gel filtration) [13]
120000 (Vigna radiata, gel filtration) [27]
120000–140000 (bovine, Mg^{2+}-activated enzyme, gel filtration) [3]
140000–160000 (bovine, gel filtration) [5]
145000 (hamster, polymerase IIa, gel filtration) [18]
145000–155000 (hamster, gel filtration) [3]
150000 (above, mouse L-cells, gel filtration) [3]
155000 (hamster, polymerase IIb, gel filtration) [18]
185000 (Pseudomonas putida, adenylyltransferase A, glycerol density gradient sedimentation) [31]

Subunits

Monomer (1 × 63000, Saccharomyces cerevisiae, SDS-PAGE [23], 1 × 62000, bovine [2, 9], SDS-PAGE [9], 1 × 64000, Triticum aestivum, SDS-PAGE [26], 1 × 60000, bovine, Mn^{2+}-activated enzyme, denaturing gel electrophoresis [3], 1 × 75000, human, cytoplasmic, Mn^{2+}-dependent enzyme, SDS-PAGE [12], 1 × 50000, human, nuclear Mn^{2+}- and Mg^{2+}-activated enzyme, SDS-PAGE [12], 1 × 60000, rat, SDS-PAGE [16], 1 × 64000, rat, SDS-PAGE [17], 1 × 50000, E. coli, SDS-PAGE [21], 1 × 57000, bovine, SDS-PAGE [35]) [2, 3, 9, 12, 16, 17, 21, 23, 26, 35]
Dimer (1 × 51000 + 1 × 35000, Vaccinia virus, SDS-PAGE [8], 1 × 37000 + 1 × 57000, Vaccinia virus, SDS-PAGE [12], 1 × 85000 + 1 × 60000, hamster, SDS-PAGE [18]) [8, 12, 18]
Tetramer (4 × 30000, Vigna radiata, SDS-PAGE) [27]
? (x × 48000, rat liver nucleoplasm, denaturing gel electrophoresis [3], x × 75000, human, Mn^{2+}-activated, denaturing gel electrophoresis [3], x × 50000, human, Mg^{2+}-activated, denaturing gel electrophoresis [3], x × 50000, E. coli, denaturing gel electrophoresis [3], x × 63000, Vigna unguiculata, SDS-PAGE [4], x × 48000, rat liver, SDS-PAGE [11], x × 60000, rat hepatoma cells, SDS-PAGE [3, 11], x × 70000, Artemia salina, SDS-PAGE [19]) [3, 4, 11, 19]

Glycoprotein/Lipoprotein

Glycoprotein [17, 26]; More (not glycosylated) [22]

4 ISOLATION/PREPARATION

Source organism

Vigna radiata [27]; Maize [1]; Triticum aestivum [26]; Rat [1, 3, 11, 15–17, 29, 30, 32, 33, 36]; Bovine (calf [2, 3, 5–7, 9, 22], 2 forms of enzyme: Mn^{2+}-activated and Mg^{2+}-activated [3], expression in E. coli [10]) [2, 3, 5–7, 9, 10, 22, 35]; E. coli [3, 21]; Mouse [3]; Human (HeLa cells [20, 24, 25], HeLa infected with vaccinia virus [3, 12], 2 forms of enzyme: Mn^{2+}-activated, Mg^{2+}-activated [3], 2 forms: 1. nuclear enzyme, stimulated by Mn^{2+} and Mg^{2+}, 2. cytoplasmic, dependent on Mn^{2+} [12], 2 forms from nuclear fraction: NE PAPs I and II, one form from cytoplasmic fraction: S100 PAP [20]) [3, 12, 20, 24, 25]; Hamster (CHO fibroblasts) [3, 18]; Vaccinia virus [3, 8, 12, 34]; Quail [3]; Vigna unguiculata (2 forms of enzyme with some difference in primer preference) [4]; Tetrahymena pyriformis [13]; Caulobacter crescentus (strain CB15, enzyme catalyzes both polyadenylic acid synthesis in absence of a template and DNA-dependent RNA synthesis) [14]; Artemia salina [19]; Artemia sp. [28]; Saccharomyces cerevisiae [23]; Pseudomonas putida [31]

Source tissue

Liver [1, 3, 11, 15, 17, 29, 30, 32, 33, 35, 36]; Thymus [2, 3, 5, 6, 9, 22]; HeLa cells (infected with vaccinia virus [34]) [3, 12, 20, 24, 25, 34]; Hepatomas (Morris hepatomas 3924A and 7777, relative lack of poly(A) polymerase activity is partly due to decreased level of this enzyme in the tumors, but largely due to the nonavailability of the primer-binding sites on the solubilized enzyme and to occupation of the available binding sites with an ineffective primer [32], Morris hepatoma tumor cells 3924A [11, 16]) [3, 11, 16, 32]; L-cells [3]; CHO fibroblasts [3, 18]; Embryos [3]; Virus cores [3, 8]; Oviduct [3]; Cryptobiotic gastrulae [19]; Cell [23]; Germinating seeds (embryo [26]) [4, 26]; Hypocotyl [27]; Encysted dormant embryos [28]

Localization in source

Soluble [2, 9]; Cytoplasm (2 forms from nuclear fraction: NE PAPs I and II, one form from cytoplasmic fraction: S100 PAP [20], infected cytoplasm of HeLa cells [3, 12]) [3, 9, 12, 19, 20]; Nucleus (polymerase II [13], 2 forms from nuclear fraction: NE PAPs I and II, one form from cytoplasmic fraction: S100 PAP [20], 2 forms: one from chromatin and one from nucleoplasm [29]) [3, 5, 11–13, 17, 20, 25, 29, 33, 35, 36]; Nuclear envelope [17]; Mitochondria [16, 30, 32]; Particulate [31]

Purification

More (homogenous preparation of bovine enzyme [2], purification methods [3], high salt conditions required during purification [21]) [2, 3, 21]; Vigna unguiculata [4]; Bovine (calf [5, 9, 22]) [5, 9, 22, 35]; Saccharomyces cerevisiae [23]; Vaccinia virus (partial, HeLa cells infected with [34]) [8, 12, 34]; Rat (2 forms, one from chromatin and one from nucleoplasm [29]) [11, 16,

17, 29, 32, 33]; Human (2 forms from nuclear fraction: NE PAPs I and II, one form from cytoplasmic fraction: S100 PAP [20]) [12, 20, 25]; Tetrahymena pyriformis (3 forms: Ia, Ib, II) [13]; Triticum aestivum [26]; Caulobacter crescentus [14]; Hamster (polymerase IIa and IIb) [18]; Artemia salina [19]; Artemia sp. [28]; E. coli [21]; Vigna radiata [27]; Pseudomonas putida (adenylyltransferase A and B) [31]

Crystallization

–

Cloned
[7, 10]

Renatured

–

5 STABILITY

pH

Temperature (°C)
40 (5 min, stable) [28]; 45 (5 min, 50% loss of activity) [28]; 50 (5 min, complete inactivation) [28]

Oxidation

Organic solvent

General stability information
Freezing and thawing accelerates inactivation [9]

Storage
–90°C, 50% glycerol, stable for 4 weeks, about 40% loss of activity after 5 months [16]; –70°C, stable for at least 2 months [19]; –80°C, storage for 5 days including 2 cycles of freezing and thawing results in 35% inactivation [22]; –70°C, stable for several weeks [27]; –20°C, stable for at least 3 months [31]; –70°C, stable [35]; –20°C, 50% glycerol, 80% inactivation in the first few weeks, the 20% remaining activity is stable for more than 2 years [9]; –70°C or under liquid nitrogen, extensive inactivation [9]

6 CROSSREFERENCES TO STRUCTURE DATABANKS

PIR/MIPS code
PIR2:S17875 (class I bovine); PIR2:S17925 (class II (version 1) bovine); PIR2:S18642 (class II (version 2) bovine)

Brookhaven code

7 LITERATURE REFERENCES

[1] Mans, K.J., Walter, T.J.: Biochim. Biophys. Acta,247,113–121 (1971)
[2] Bollum, F.J., Chang, L.M.S., Tsiapalis, C.M., Dorson, J.W.: Methods Enzymol.,29E, 70–81 (1974) (Review)
[3] Edmonds, M. in "Enzymes",3rd Ed. (Boyer, P.D., Ed.) 15,217–244 (1982) (Review)
[4] Tarui, Y., Minamikawa, T.: Plant Cell Physiol.,29,835–842 (1988)
[5] Winters, M.A., Edmonds, M.: J. Biol. Chem.,248,4756–4762 (1973)
[6] Winters, M.A., Edmonds, M.: J. Biol. Chem.,248,4763–4768 (1973)
[7] Raabe, T., Bollum, F.J., Manley, J.L.: Nature,353,229–234 (1991) (Review)
[8] Moss, B., Rosenblum, E.N., Gershowitz, A.: J. Biol. Chem.,250,4722–4729 (1975)
[9] Tsiapalis, C.M., Dorson, J.W., Bollum, F.J.: J. Biol. Chem.,250,4486–4496 (1975)
[10] Wahle, E., Martin, G., Schiltz, E., Keller, W.: EMBO J.,10,4251–4257 (1991)
[11] Rose, K.M., Jacob, S.T.: Eur. J. Biochem.,67,11–21 (1976)
[12] Nevins, J.R., Joklik, W.K.: J. Biol. Chem.,252,6939–6947 (1977)
[13] Ueyama, H.: J. Biochem.,86,1301–1311 (1979)
[14] Cheung, K.K., Newton, A.: J. Biol. Chem.,253,2254–2261 (1978)
[15] Abraham, A.K., Jacob, S.T.: Proc. Natl. Acad. Sci. USA,75,2085–2087 (1978)
[16] Rose, K.M., Morris, H.P., Jacob, S.T.: Biochemistry,14,1025–1032 (1975)
[17] Kurl, R.N., Holmes, S.C., Verney, E., Sidransky, H.: Biochemistry,27,8974–8980 (1988)
[18] Pellicer, A., Salas, J., Salas, M.L.: Biochim. Biophys. Acta,519,149–162 (1978)
[19] Roggen, E., Slegers, H.: Eur. J. Biochem.,147,225–232 (1985)
[20] Ryner, L.C., Takagaki, Y., Manley, J.L.: Mol. Cell. Biol.,9,4229–4238 (1989)
[21] Sippel, A.E.: Eur. J. Biochem.,37,31–40 (1973)
[22] Wahle, E.: J. Biol. Chem.,266,3131–3139 (1991)
[23] Lingner, J., Radtke, I., Wahle, E., Keller, W.: J. Biol. Chem.,266,8741–8746 (1991)
[24] Takagaki, Y., Ryner, L.C., Manley, J.L.: Cell,52,731–742 (1988)
[25] Christofori, G., Keller, W.: Mol. Cell. Biol.,9,193–203 (1989)
[26] Kapoor, R., Verma, N., Saluja, D., Lakhani, S., Sachar, R.C.: Plant Sci.,89,167–176 (1993)
[27] Saluja, D., Mathur, M., Sachar, R.C.: Plant Sci.,60,27–38 (1989)
[28] Sastre, L., Sebastian, J.: Biochim. Biophys. Acta,661,54–62 (1981)
[29] Antoniades, D., Antonoglou, O.: Biochim. Biophys. Acta,519,447–460 (1978)
[30] Jacob, S.T., Rose, K.M.: Nucleic Acids Res.,1,1549–1559 (1974)
[31] Blakesley, R.W., Boezi, J.A.: Biochim. Biophys. Acta,414,133–145 (1975)
[32] Jacob, S.T., Rose, K.M., Morris, H.P.: Biochim. Biophys. Acta,361,312–320 (1974)
[33] Rose, K.M., Jacob, S.T.: Arch. Biochem. Biophys.,175,748–753 (1976)
[34] Brakel, C., Kates, J.R.: J. Virol.,14,715–723 (1974)
[35] Ohyama, Y., Fukami, H., Ohta, T.: J. Biochem.,88,337–348 (1980)
[36] Koch, S., Niessing, J.: FEBS Lett.,96,354–356 (1978)

1 NOMENCLATURE

EC number
2.7.7.21

Systematic name
CTP:tRNA cytidylyltransferase

Recommended name
tRNA cytidylyltransferase

Synonyms
tRNA CCA-pyrophosphorylase
tRNA-nucleotidyltransferase [1]
transfer-RNA nucleotidyltransferase [2]
Transfer ribonucleic acid nucleotidyl transferase [2]
CTP(ATP):tRNA nucleotidyltransferase [8]
Adenylyltransferase, transfer ribonucleate
Transfer ribonucleate adenyltransferase
Transfer RNA adenylyltransferase
Nucleotidyltransferase, transfer ribonucleate
ATP (CTP):tRNA nucleotidyltransferase
Ribonucleic cytidylic cytidylic adenylic pyrophosphorylase
Transfer ribonucleate nucleotidyltransferase
Transfer ribonucleic adenylyl (cytidylyl) transferase
Transfer ribonucleic-terminal trinucleotide nucleotidyltransferase
Cytidylyltransferase, transfer ribonucleate
Ribonucleic cytidylyltransferase
Transfer ribonucleate cytidylyltransferase
-C-C-A pyrophosphorylase [3]
ATP(CTP)-tRNA nucleotidyltransferase [3]
tRNA adenylyl(cytidylyl)transferase [16]
ATP(CTP):tRNA nucleotidyltransferase [19]
tRNA adenylyltransferase
ATP:tRNA adenylyltransferase
EC 2.7.7.25 (EC 2.7.7.25 is identical with EC 2.7.7.21)

CAS Reg. No.
52523-59-8; 9026-11-3; 9026-32-8

›

2 REACTION AND SPECIFICITY

Catalyzed reaction

CTP + tRNA$_n$ $\rightarrow$
$\rightarrow$ diphosphate + tRNA$_{n+1}$ (rabbit liver enzyme: rapid equilibrium mechanism [3], stereochemistry [34]);
ATP + tRNA$_n$ $\rightarrow$
$\rightarrow$ diphosphate + tRNA$_{n+1}$

Reaction type

Nucleotidyl group transfer

Natural substrates

More (mitochondria may contain a system for the maturation of tRNA molecules [23], in most organisms tRNA nucleotidyltransferase plays a role both in tRNA biosynthesis and in tRNA repair [3], role in repair of tRNA deprived of terminal CCA [24], regulation of AMP and CMP incorporation into tRNA-A-C-C and tRNA-C [36]) [3, 23, 24, 36]

Substrate spectrum

1 ATP + tRNA [3]
2 tRNA-N + CTP [3]
3 tRNA-C + CTP [3]
4 tRNA-C-C + ATP [3]
5 UTP + tRNA$_n$ (at 2–10% of the rate of CMP incorporation [3], not [18]) [3]
6 tRNA-C-C-A + diphosphate (r) [8]
7 tRNA-C-C + dATP (r) [8]
8 tRNA-C-A + diphosphate (r) [8]
9 tRNA-C + dATP (r) [8]
10 2-Thiocytidine 5'-triphosphate + tRNA$_n$ [10]
11 ATP + tRNA$_n$ (r [3, 8], enzyme catalyzes the incorporation of AMP and CMP residues into tRNA molecules from which all or part of 3'-terminal trinucleotide sequence -C-C-A has been removed [3], tRNA-C-C, tRNA-C and tRNA-N from liver, yeast or E. coli are equally active, all tRNA molecules in a mixed population are active as acceptors, reactions with 5S RNA, rRNA and modified tRNA's such as tRNA-C-A, tRNA-C-U and tRNA-C-C-C occur at much slower rates [5], overview: RNA acceptor specificity [3, 5], untreated yeast tRNA [18], tRNA-C-C [3, 4]) [1–36]
12 CTP + tRNA$_n$ (r [3], snake venom phosphodiesterase treated RNA from Neurospora crassa, E. coli or yeast [3]) [1–36]
13 tRNA-A-C-C + ATP (not CTP or UTP) [21]
14 tRNA-A-C + CMP [21]
15 tRNA-C + UMP [21]

16 More (cytidines in tRNA that are required for activity [26], misincorpora-
    tions: synthesis of sequences other than C-C-A occur when either ATP or
    CTP is ommited from reaction mixture: 1. tRNA-N + ATP→ tRNA-A,
    tRNA-A-A-A, tRNA-A-A-A-A, 2. tRNA-C + ATP→ tRNA-C-A, tRNA-C-A-A, 3.
    tRNA-C-C + CTP→ tRNA-C-C-C, tRNA-C-C-$(C)_n$, 4. tRNA-C-C + CTP, then
    ATP→ tRNA-C-C-A-A, 5. tRNA-C-C-A + CTP→ tRNA-C-C-A-C-C, 6.
    tRNA-C-C-A + CTP, then ATP→ tRNA-C-C-A-C-C-A [5], enzyme also con-
    tains nucleolytic activity which removes terminal CMP residues from
    tRNA-C-C and tRNA-C-C-C, other tRNA molecules (tRNA-C-C-A, tRNA-C-A,
    tRNA-C-U, and tRNA-C) are not substrates [28], removal of AMP from the
    terminus of tRNA proceeds optimally at 1.0 mM diphosphate, incorpora-
    tion of 2'- or 3'-dAMP proceeds optimally at 6.0 mM concentration of
    deoxynucleoside triphosphate [8], replacement of the terminal CCA se-
    quence in yeast tRNA^{Phe} by several unusual sequences [13], extent of
    normal and anomalous nucleotide incorporation [22], low activity of in-
    corporation of CMP into rRNA partially degraded by phosphodiesterase
    [33], not: DNA [18], rRNA [18, 23], GMP [18, 21], dATP [21]) [5, 8, 13,
    18, 21–23, 26, 28, 33]

Product spectrum
     1 tRNA-C-C-A + diphosphate [3]
     2 tRNA-C or tRNA-C-C + diphosphate [3]
     3 tRNA-C-C + diphosphate [3]
     4 tRNA-C-C-A + diphosphate [3]
     5 Diphosphate + tRNA$_{n+1}$ [3]
     6 Diphosphate + tRNA$_{n+1}$ [3, 4]
     7 Diphosphate + tRNA$_{n+1}$ [3]
     8 ATP + tRNA-C-C [8]
     9 tRNA-C-C-dA + diphosphate [8]
    10 ?
    11 ?
    12 ?
    13 ?
    14 ?
    15 ?
    16 ?

Inhibitor(s)
    Proflavine sulfate (complete inhibition at 1 mM, CMP incorporation more
    sensitive than AMP incorporation) [1]; Ethidium bromide (complete inhibition
    at 2 mM, CMP incorporation more sensitive than AMP incorporation) [1];
    Diphosphate (nucleolytic activity [28]) [3, 28, 35]; DTNB [20]; tRNA-N (inhi-
    bition of AMP incorporation into tRNA-C-C) [3]; tRNA-C (inhibition of AMP in-
    corporation into tRNA-C-C) [3]; tRNA-C-C (inhibition of CMP incorporation)
    [3]; 1,10-Phenanthroline (E. coli, effects AMP incorporation, no effect in CMP

incorporation, no inhibition of yeast, Rous sarcoma virus and rabbit liver enzyme [3], inhibition of AMP incorporation [15, 16], no inhibition of CMP incorporation [16], no effect [5]) [3, 15, 16]; Bathophenanthroline [15]; CTP (AMP incorporation [3, 9, 14, 18, 24, 28, 35], competitive [14, 24], CMP incorporation [12], nucleolytic activity [28]) [3, 9, 12, 14, 18, 24, 28, 35]; ATP (CMP incorporation [3, 9, 14, 18, 24, 28, 35], AMP incorporation [12], competitive [14, 24], nucleolytic acitivity [28]) [3, 9, 12, 14, 18, 24, 28, 35]; p-Substituted mercuribenzoate (inhibition of AMP incorporation) [5]; Mersalyl (inhibition of AMP incorporation) [5, 20]; 5,5'-Dithiobis(2-nitrobenzoic acid) (inhibition of AMP incorporation [5], inhibits incorporation of AMP into tRNA-X-C-C but is without effect on the incorporation of CMP or UMP into tRNA-X [32]) [5, 32]; Hg^{2+} ($HgCl_2$ [20]) [5, 20]; PCMB (reversal by 2-mercaptoethanol [12]) [12, 20]; UTP [12]; N-Ethylmaleimide (reversal by 2-mercaptoethanol [12], inhibits incorporation of AMP into tRNA-X-C-C but is without effect on incorporation of CMP or UMP into tRNA-X [32]) [12, 32]; DTT [5, 20]; $(NH_4)_2SO_4$ (AMP incorporation: 50% inhibition at 0.2 ionic strength, CMP incorporation stimulated) [5]; KCl (AMP incorporation: 50% inhibition at 0.2 ionic strength, CMP incorporation stimulated [5], 0.2 M: 50% inhibition of AMP incorporation, 5% inhibition of CMP incorporation [9], nucleolytic activity [28]) [5, 9, 28]; NaCl (AMP incorporation: 50% inhibition at 0.2 ionic strength, CMP incorporation stimulated) [5]; tRNA[Phe] containing iodoacetamide-alkylated 2-thiocytidine [10]; 2,2-Dipyridyl [15]; 2,2,2-Terpyridyl [15]; Spermine (inhibition of AMP incorporation, stimulation of CMP incorporation) [5]; Spermidine (inhibition of AMP incorporation, stimulation of CMP incorporation) [5]; Putrescine (inhibition of AMP incorporation, stimulation of CMP incorporation) [5]; Cadaverine (inhibition of AMP incorporation, stimulation of CMP incorporation) [5]; EDTA (not [5, 15]) [12]; tRNA-X (competitive in AMP attachment to tRNA-X-C-C) [14]; tRNA-X-C-C-A (noncompetitive in CMP attachment to tRNA-X and AMP attachment to tRNA-X-C-C) [14]

Cofactor(s)/prosthetic group(s)/activating agents
2-Mercaptoethanol (activates) [35]; Polyamines (stimulate activity of rabbit liver enzyme [3], decrease requirement for Mg^{2+} (from 10 to 1 mM) [3], absolute requirement for a divalent cation which can be satisfied only by Mg^{2+}, Mn^{2+} or Co^{2+}, in addition a second function for cations has been identified which is carried out most efficiently by polyamines, although additional Mg^{2+} or monovalent cations are also effective [27], neither spermine, nor spermidine (0.1–10 mM) can replace divalent cations for the tRNA nucleotidyltransferase activity [32]) [3, 27, 32]

Metal compounds/salts
Mg^{2+} (required [3, 4, 17, 18, 21, 31, 35], Mg^{2+} or Mn^{2+} required (for nucleolytic activity [28]) [12, 24, 28], highest rate of incorporation of AMP into tRNA-X-C-C or tRNA-X and of CMP into tRNA-X are observed in the presence of Mg^{2+} [32], optimal concentration: 1 mM (Lactobacillus luteus) [3],

5 mM (E. coli A19) [3], 5–10 mM (rabbit liver) [3], 10 mM (E. coli MRE 600) [3], 10–15 mM (mutant enzyme) [17], 5 mM (wild type enzyme) [17], 15–20 mM (higher concentrations inhibit) [18], K_m: 5 mM [18]) [3, 4, 12, 17, 18, 21, 24, 28, 31, 32, 35]; Mn^{2+} (Mg^{2+} or Mn^{2+} required (for nucleolytic activity [28]) [12, 24, 28], can partially replace Mg^{2+} in activation [3, 5, 17, 21], 30% (ATP incorporation), 40–50% (CTP incorporation) of the activity with Mg^{2+} [5], optimal concentration is lower than the level of triphosphate present and higher concentrations strongly inhibit [5], in presence of Mn^{2+} a variety of anomalous reactions catalyzed by tRNA nucleotidyltransferase are stimulated whereas normal reactions are inhibited [3], can partially replace Mg^{2+} in AMP incorporation, inefficient for CMP incorporation, in its presence UMP is incorporated instead of CMP, in presence of optimal Mg^{2+}-concentrations Mn^{2+} decreases the rate of CMP incorporation and to a lower extent of AMP, but increases the rate of UMP incorporation [32], optimal concentration: 4 mM (ATP incorporation) [5], 0.5–1 mM (CTP incorporation) [5]) [3, 5, 12, 17, 21, 24, 28, 32]; Co^{2+} (can partially replace Mg^{2+} in activation [3, 5, 21], 15% (ATP incorporation), 20–30% (CTP incorporation) of the activity with Mg^{2+} [5], optimal concentration: 7.5–10 mM (ATP incorporation) [5], 5 mM (CTP incorporation) [5], low efficiency [32]) [3, 5, 15, 21, 32]; $(NH_4)_2SO_4$ (AMP incorporation: 50% inhibition at 0.2 ionic strength, CMP incorporation stimulated) [5]; KCl (AMP incorporation: 50% inhibition at 0.2 ionic strength, CMP incorporation stimulated) [5]; NaCl (AMP incorporation: 50% inhibition at 0.2 ionic strength, CMP incorporation stimulated) [5]; More (E. coli enzyme is a metalloenzyme [16], absolute requirement for a divalent cation which can be satisfied only by Mg^{2+}, Mn^{2+} or Co^{2+}, in addition a second function for cations has been identified which is carried out most efficiently by polyamines, although additional Mg^{2+} or monovalent cations are also effective [27]) [16, 27]

Turnover number (min^{-1})
  600 (Lupinus luteus, AMP incorporation) [3]; 3600 (rabbit, AMP incorporation) [3]; 7200 (Saccharomyces cerevisiae, AMP incorporation) [3]; 15600 (E. coli A19, AMP incorporation) [3]; 321000 (E. coli B, AMP incorporation) [3, 15]

Specific activity (U/mg)
  280 [15]; 0.252 [35]; 58.3 [4]; 401.67 [18]; 33.33 [20]; 13.2 [19]; More [3, 5, 11, 14, 25]

K_m-value (mM)
  0.000238 (tRNA) [13]; 0.015 (tRNA-C-C) [9]; 0.017 (CTP, E. coli MRE 600) [3]; 0.020 (tRNA-C) [9]; 0.028 (CTP) [1]; 0.029 (CTP, Lactobacillus acidophilus) [3]; 0.03 (CTP, E. coli A19, rabbit [3], CTP, E. coli UT481 (pEC 4) [9]) [3, 9]; 0.033 (ATP [1], CTP, Musca domestica [3], ATP, E. coli UT 481(pEC 4) [9]) [1, 3, 9]; 0.07 (CTP, Lupinus luteus) [3]; 0.2 (ATP, Musca domestica, CTP,

Saccharomyces cerevisiae) [3]; 0.25 (ATP, Lupinus luteus) [3]; 0.31 (ATP, E. coli A19) [3]; 0.4 (tRNA-C-C, Lactobacillus acidophilus) [3]; 0.71 (ATP, Lactobacillus acidophilus) [3]; 1.5–1.7 (tRNA-C-C, E. coli MRE 600) [3]; 4 (tRNA-N, rabbit) [3]; 5.5 (tRNA-C-C, Saccharomyces cerevisiae) [3]; 6 (tRNA-C, rabbit) [3]; 7.8 (tRNA-N, Saccharomyces cerevisiae) [3]; 11 (tRNA-C, Saccharomyces cerevisiae) [3]; 12 (tRNA-C-C, rabbit) [3]; 13 (tRNA-C, E. coli A19) [3]; More [3, 4, 6, 12–14, 17–19, 21, 31, 35]

pH-optimum

9.0–9.4 (E. coli A19 [3], E. coli, AMP incorporation [17]) [3, 17]; 9–10 [4]; 9 (above) [31]; 9.3–10 (rabbit) [3, 5]; 9.4 (AMP incorporation) [9]; 9.5 (Lupinus luteus, Saccharomyces cerevisiae, E. coli MRE 600 [3]) [3, 12, 19]; 10 (CMP incorporation [9, 17], nucleolytic activity [28]) [9, 17, 28]

pH-range

7.1–10 (7.1: 20–30% of activity maximum, 9.3–10: activity maximum, rabbit) [5]; More [12]

Temperature optimum (°C)

30 (assay at) [11]; 35 (assay at) [17]; 37 (assay at) [15, 16, 28, 30]; 45 [12]

Temperature range (°C)

---

3 ENZYME STRUCTURE

Molecular weight

30000 (Musca domestica, gel filtration, sucrose density gradient centrifugation) [3, 31]

37000 (E. coli MRE 600, gel filtration) [3]

40000 (Lupinus luteus, gel filtration) [3, 12]

44000–48000·(rabbit, gel filtration, equilibrium ultracentrifugation) [3, 5]

45000 (E. coli, sucrose density gradient centrifugation [33], E. coli B, gel filtration [3]) [3, 33]

46408 (E. coli, calculation from nucleotide sequence) [4]

50000 (rabbit, sedimentation analysis) [20]

53000 (E. coli B, gel filtration) [15]

59000 (Saccharomyces cerevisiae, gel filtration) [19]

62000 (Neurospora crassa, gel filtration) [18]

63000 (Ceratodon purpureus, gel filtration) [1]

71000 (Saccharomyces cerevisiae, equilibrium ultracentrifugation) [3]

Subunits

Monomer (1 × 31000, Musca domestica, SDS-PAGE [3, 31], 1 × 47000, rabbit, SDS-PAGE [3], 1 × 51500, E. coli B, SDS-PAGE [3, 15], 1 × 70000, Saccharomyces cerevisiae, SDS-PAGE [3], 1 × 45000–49000, rabbit, SDS-PAGE [5], 1 × 59000, Saccharomyces cerevisiae, SDS-PAGE [19], 1 × 45000–49000, rabbit, SDS-PAGE [20]) [3, 5, 15, 19, 20, 31]
? (x × 43000, Lactobacillus acidophilus, SDS-PAGE [3, 6], x × 45000, E. coli MRE 600, SDS-PAGE [3, 6], x × 50000, E. coli A19, SDS-PAGE [3]) [3, 6]
More (aggregation at elevated protein levels prevented by high concentration of phosphate) [20]

Glycoprotein/Lipoprotein

–

---

4 ISOLATION/PREPARATION

Source organism

Ceratodon purpureus [1]; Mouse (Ehrlich tumor cells) [2, 30]; Avian RNA tumor viruses [3, 6, 7]; Sendai virus [3, 6]; E. coli (strain UT481 (pEC4) [9], MRE 600 [3, 6], B [3, 15, 16, 35], A19 [3], overexpression (high level [9]) [4, 9], mutant with decreased AMP incorporation but normal CMP incorporation [17]) [3, 4, 6, 9, 15–17, 26, 32, 33, 35, 36]; Rat [3, 5, 20–23, 27, 28]; Rabbit [3, 5, 20–22, 27, 28]; Lactobacillus acidophilus (ATCC 4963 [6]) [3, 6]; Saccharomyces cerevisiae (wild type and overproducing [19]) [3, 8, 10, 13, 14, 19, 34]; Lupinus luteus [3, 11, 12]; Musca domestica [3, 31]; Neurospora crassa 74A [18]; Wheat [24]; Avian reticuloendotheliosis virus [29]; Bovine (calf) [25]

Source tissue

Protonema [1]; Ehrlich ascites tumor cells [2, 30]; Muscle [3]; Liver [3, 20, 21, 23, 27, 28]; Seeds [3, 11, 12]; Germ [24]; Spleen [30]; Lymph nodes [30]; Macrophage cells [30]; Whole organism [31]; Larvae [31]; Pupae [31]; Thymus [25]; More (widespread occurence in all types of cells) [3]

Localization in source

Mitochondria (matrix [23]) [3, 5, 23]; Cytoplasm [30]; Nuclei [25]; More (microsomes and nuclei are devoid of activity) [3, 23]

Purification

Ceratodon purpureus [1]; Lupinus luteus [11, 12]; E. coli (MRE 600 [6], UT481 (pEC4) [9], large scale [35], B [16, 35], mutant with decreased AMP incorporation but normal CMP incorporation [17]) [3, 4, 6, 9, 15–17, 35]; Rat [3]; Rabbit [3, 5, 20]; Lactobacillus acidophilus [6]; Saccharomyces cerevisiae (wild type and overproducing [19]) [8, 14, 19]; Neurospora crassa 74A [18]; Musca domestica [31]; Bovine (calf) [25]

Crystallization

–

Cloned

–

Renatured

–

5 STABILITY

pH

Temperature (°C)
   45 (10 min, 20% (AMP incorporation), 40% (CMP incorporation) loss of activity [12], 5 min, about 50% loss of activity [20]) [12, 20]; 50 (20 min, 50%
   loss of activity in absence of tRNA, 1 h, less than 20% loss of activity in
   presence of tRNA) [31]; More (tRNA protects against thermal inactivation
   [3], ATP and CTP protect against heat inactivation [14], high concentrations
   of tRNA without the CCA terminus (tRNA-X), but not intact tRNA protects
   against heat inactivation [14], Mg^{2+}, ATP and tRNA have little effect on heat
   inactivation [20], heating in presence of 0.15 M potassium phosphate, pH
   7.5, instead of 0.01 M buffer stabilizes [20], tRNA and to a lesser extent ATP
   and CTP protect against heat inactivation [31]) [3, 14, 20, 31]

Oxidation

Organic solvent

General stability information
   tRNA protects against thermal inactivation [3]; 2-Mercaptoethanol stabilizes
   [35]; tRNA and CTP but not ATP stabilize [1]; ATP and CTP protect against
   heat inactivation [14]; High concentrations of tRNA without the CCA terminus (tRNA-X), but not intact tRNA protects against heat inactivation [14];
   Mg^{2+}, ATP and tRNA have little effect on heat inactivation [20]; Heating in
   presence of 0.15 M potassium phosphate, pH 7.5, instead of 0.01 M buffer
   stabilizes [20]; tRNA and to a lesser extent ATP and CTP protect against
   heat inactivation [31]

Storage
   –20°C, in 50% glycerol or frozen with 0.2 mg/ml of commercial yeast tRNA,
   less than 25% loss of activity after 1 year [5]; –20°C, 50% glycerol, stable for
   at least 1 year [11]; –20°C, even at concentrations as low as 0.03 mg/ml or
   when frozen with 0.2 mg/ml of commercial yeast tRNA, stable for at least 3
   months [20]

6 CROSSREFERENCES TO STRUCTURE DATABANKS

PIR/MIPS code

Brookhaven code

7 LITERATURE REFERENCES

[1] Schneider, Z., Schneider, J.: Biochem. Physiol. Pflanz.,171,239–248 (1977)
[2] Girgenti, A.J., Whitford, T.W., Cory, J.G.: Enzyme,21,225–231 (1976)
[3] Deutscher, M.P. in "The Enzymes",3rd Ed. (Boyer, P.D., Ed.) 15,183–215 (1982) (Review)
[4] Deutscher, M.P.: Methods Enzymol.,181,434–439 (1990) (Review)
[5] Deutscher, M.P.: Methods Enzymol.,29E,70–81 (1974) (Review)
[6] Leineweber, M., Philipps, G.R.: Hoppe-Seyler's Z. Physiol. Chem.,359,473–480 (1978)
[7] Faras, A.J., Levinson, W.E., Bishop, J.M., Goodman, H.M.: Virology,58,126–135 (1974)
[8] Francis, T.A., Ehrenfeld, G.M., Gregory, M.R., Hecht, S.M.: J. Biol. Chem.,258, 4279–4284 (1983)
[9] Cudney, H., Deutscher, M.P.: J. Biol. Chem.,261,6450–6453 (1986)
[10] Kröger, M., Sternbach, H., Cramer, F.: Eur. J. Biochem.,95,341–348 (1979)
[11] Cudny, H., Pietrzak, M., Kaczkowsky, J.: Planta,142,23–27 (1978)
[12] Cudny, H., Pietrzak, M., Kaczkowski, J.: Planta,142,29–36 (1978)
[13] Rether, B., Gangloff, J., Ebel, J.-P.: Eur. J. Biochem.,50,289–295 (1974)
[14] Rether, B., Bonnet, J., Ebel, J.-P.: Eur. J. Biochem.,50,281–288 (1974)
[15] Schofield, P., Williams, K.R.: J. Biol. Chem.,252,5584–5588 (1977)
[16] Williams, K.R., Schofield, P.: Biochem. Biophys. Res. Commun.,64,262–267 (1975)
[17] McGann, R.G., Deutscher, M.P.: Eur. J. Biochem.,106,321–328 (1980)
[18] Hill, R., Nazario, M.: Biochemistry,12,482–485 (1973)
[19] Chen, J.-Y., Kirchner, G., Aebi, M., Martin, N.C.: J. Biol. Chem.,265,16221–16224 (1990)
[20] Deutscher, M.P.: J. Biol. Chem.,247,450–458 (1972)
[21] Deutscher, M.P.: J. Biol. Chem.,247,459–468 (1972)
[22] Deutscher, M.P.: J. Biol. Chem.,247,469–480 (1972)
[23] Mukerji, S.K., Deutscher, M.P.: J. Biol. Chem.,247,481–488 (1972)
[24] Dullin, P., Fabisz-Kijowska, A., Walerych, W.: Acta Biochim. Pol.,22,279–289 (1975)
[25] Edmonds, J.: J. Biol. Chem.,240,4621–4628 (1965)
[26] Hegg, L.A., Thurlow, D.L.: Nucleic Acids Res.,18,5975–5979 (1990)
[27] Evans, J.A., Deutscher, M.P.: J. Biol. Chem.,251,6646–6652 (1976)
[28] Deutscher, M.P.: Biochem. Biophys. Res. Commun.,52,216–222 (1973)
[29] Mizutani, S., Temin, H.M.: J. Virol.,19,610–619 (1976)
[30] Sato, N.L.: J. Biochem.,85,739–745 (1979)
[31] Poblete, P., Jedlicky, E., Litvak, S.: Biochim. Biophys. Acta,476,333–341 (1977)
[32] Carre, D.S., Chapeville, F.: Biochim. Biophys. Acta,361,176–184 (1974)
[33] Carre, D.S., Litvak, S., Chapeville, F.: Biochim. Biophys. Acta,361,185–197 (1974)
[34] Eckstein, F., Sternbach, H., von der Haar, F.: Biochemistry,16,3429–3432 (1977)
[35] Best, A.N., Novelli, G.D.: Arch. Biochem. Biophys.,142,527–538 (1971)
[36] Best, A.N., Novelli, G.D.: Arch. Biochem. Biophys.,142,539–547 (1971)

1 NOMENCLATURE

EC number
2.7.7.22

Systematic name
GDP:D-mannose-1-phosphate guanylyltransferase

Recommended name
Mannose-1-phosphate guanylyltransferase (GDP)

Synonyms
GDPmannose phosphorylase
Guanylyltransferase, mannose 1-phosphate (guanosine diphosphate)
GDP mannose phosphorylase
GDP-mannose 1-phosphate guanylyltransferase
Guanosine diphosphate-mannose 1-phosphate guanylyltransferase
Guanosine diphosphomannose phosphorylase
Mannose 1-phosphate guanylyltransferase

CAS Reg. No.
9026-31-7

---

2 REACTION AND SPECIFICITY

Catalyzed reaction
GDP + D-mannose 1-phosphate $\rightarrow$
$\rightarrow$ phosphate + GDPmannose

Reaction type
Nucleotidyl group transfer

Natural substrates

Substrate spectrum
1 GDPmannose + phosphate [1]
2 UDPglucose + phosphate (about 50% of activity compared to GDPmannose) [1]
3 UDPacetylglucosamine + phosphate (low activity) [1]
4 More (diphosphate cannot substitute for phosphate) [1]

Product spectrum
1 GDP + mannose 1-phosphate [1]
2 UDP + glucose 1-phosphate
3 UDP + N-acetylglucosaminyl phosphate
4 ?

Inhibitor(s)

Cofactor(s)/prosthetic group(s)/activating agents

Metal compounds/salts

Turnover number (min^{-1})

Specific activity (U/mg)

K_m-value (mM)

pH-optimum
    8 (assay at) [1]

pH-range

Temperature optimum (°C)
    30 (assay at) [1]

Temperature range (°C)

3 ENZYME STRUCTURE

Molecular weight

Subunits

Glycoprotein/Lipoprotein
    –

4 ISOLATION/PREPARATION

Source organism
    Yeast [1]

Source tissue
    Cell [1]

Localization in source

Purification
    Yeast (partial) [1]

Crystallization
    –

Cloned
    –

Renatured
    –

5 STABILITY

pH

Temperature (°C)

Oxidation

Organic solvent

General stability information

Storage

---

6 CROSSREFERENCES TO STRUCTURE DATABANKS

PIR/MIPS code

Brookhaven code

---

7 LITERATURE REFERENCES

[1] Carminatti, H., Cabib, E.: Biochim. Biophys. Acta,53,417–419 (1961)

1 NOMENCLATURE

EC number
2.7.7.23

Systematic name
UTP:N-acetyl-alpha-D-glucosamine-1-phosphate uridylyltransferase

Recommended name
UDP-N-acetylglucosamine pyrophosphorylase

Synonyms
Uridine diphosphoacetylglucosamine pyrophosphorylase [1]
UTP:2-acetamido-2-deoxy-alpha-D-glucose-1-phosphate uridylyltransferase [3]
UDP-GlcNAc pyrophosphorylase (bifunctional enzyme with activity of glucosamine-1-phosphate acetyltransferase and N-acetylglucosamine-1-phosphate uridyltransferase) [4]
GlmU uridylyltransferase (bifunctional enzyme with activity of: glucosamine-1-phosphate acetyltransferase and N-acetylglucosamine-1-phosphate uridyltransferase) [4]
Acetylglucosamine 1-phosphate uridylyltransferase
UDPacetylglucosamine pyrophosphorylase
Uridine diphosphate-N-acetylglucosamine pyrophosphorylase
Uridine diphosphoacetylglucosamine phosphorylase
Uridine diphosphoacetylglucosamine pyrophosphorylase
Uridylyltransferase, acetylglucosamine 1-phosphate

CAS Reg. No.
9023-06-7

---

2 REACTION AND SPECIFICITY

Catalyzed reaction
UTP + N-acetyl-alpha-D-glucosamine 1-phosphate →
→ diphosphate + UDP-N-acetyl-D-glucosamine

Reaction type
Nucleotidyl group transfer

Natural substrates
UTP + N-acetylglucosamine 1-phosphate (bifunctional enzyme with activity of glucosamine-1-phosphate acetyl transferase and N-acetylglucosamine-1-phosphate uridyltransferase, the two subsequent steps in the pathway of UDP-N-acetylglucosamine synthesis [4]) [3, 4]

Substrate spectrum

1 Diphosphate + UDP-N-acetyl-D-glucosamine (r [2, 4]) [1–6]
2 Diphosphate + uridine diphosphoacetylgalactosamine (Staphylococcal enzyme: 2.8% of the activity with UDP-N-acetylglucosamine, calf enzyme not) [2]
3 Diphosphate + UDPglucose (calf liver enzyme: 30% of the activity with UDP-N-acetylglucosamine, Staphylococcal enzyme not) [2]
4 More (UDP cannot substitute for UTP [2], inorganic phosphate cannot substitute for diphosphate, N-acetylmuraminic acid 1-phosphate cannot substitute for N-acetylglucosamine 1-phosphate [4]) [2, 4]

Product spectrum

1 UTP + N-acetyl-alpha-D-glucosamine 1-phosphate [1, 2]
2 ?
3 ?
4 ?

Inhibitor(s)

More (not: 5'-UMP, orotidine) [3]; 5-Hydroxyuridine [3]; 5,6-Dihydrouridine [3]; Pseudouridine [3]; Deoxyuridine (slight) [3]; UDP-N-acetylglucosamine (weak, 25% inhibition at 1 mM) [4]; $HgCl_2$ (DTT reactivates [5]) [5–7]; N-Ethylmaleimide [6, 7]; Iodoacetate (weak) [6]; Iodoacetamide [6]; Diphosphate (above 75 mM) [7]; Ni^{2+} [1]; EDTA [1]; PCMB (completely reversed by cysteine [1]) [1, 6, 7]; F^- (inhibition with Mg^{2+} as activating cation, no inhibition with Mn^{2+} as activating cation) [2]; Uridine (noncompetitive to UDP-N-acetylglucosamine, competitive to diphosphate) [3]

Cofactor(s)/prosthetic group(s)/activating agents

Dithiothreitol (stimulates) [6, 7]; Dithioerythritol (stimulates) [6, 7]

Metal compounds/salts

Mg^{2+} (required [1, 2, 4], divalent cation required [6, 7], Mg^{2+} most effective [6, 7], K_m: 2.4 mM [7], 0.4 mM (Staphylococcal enzyme) [3], optimal concentration: 1–3 mM [2], inhibition at higher concentration [2], maximal activity at a concentration of Mg^{2+} and inorganic diphosphate in a ratio of about 1:10 [5]) [1–7]; Zn^{2+} (18% of the activity with Mg^{2+}) [7]; Ca^{2+} (16% of the activity with Mg^{2+}) [7]; Sn^{2+} (14% of the activity with Mg^{2+}) [7]; Mn^{2+} (can partially replace Mg^{2+} in activation [1, 2, 5–7], 49% of the activity with Mg^{2+} [6, 7]) [1, 2, 5–7]; Co^{2+} (can partially replace Mg^{2+} in activation [1, 5–7], 38% [6], 48% [7] of the activity with Mg^{2+}) [1, 5–7]

Turnover number (min⁻¹)

Specific activity (U/mg)

5.02 [1]; 8.66 [6]; 15.1 [7]; More [2]

K_m-value (mM)
0.07 (N-acetylglucosamine 1-phosphate) [4]; 0.1 (UTP) [4]; 0.36 (uridine diphospho-N-acetylglucosamine) [1]; 0.79 (diphosphate) [1]; 2.2 (UDP-N-acetylglucosamine) [6]; 5.0 (diphosphate) [7]; 5.4 (diphosphate) [6]; 6.1 (UDP-N-acetylglucosamine) [7]

pH-optimum
7.2 [2]; 7.5 [6]; 7.5–8.5 [7]; 8.0 (Tris buffer) [1]; 8.2 [4]

pH-range
More [1, 7]; 5.5–9 (5.5: about 75% of activity maximum, 9: about 60% of activity maximum) [6]

Temperature optimum (°C)
25 (assay at) [6]; 35 [7]; 37 (assay at) [1, 2, 4]

Temperature range (°C)
20–45 (20°C: about 70% of activity maximum, 45°C: about 50% of activity maximum) [7]

---

3 ENZYME STRUCTURE

Molecular weight
37000 (Neurospora crassa, gel filtration [6], Baker's yeast, gel filtration [7]) [6, 7]

Subunits
Monomer (1 × 40000, Baker's yeast, SDS-PAGE) [7]

Glycoprotein/Lipoprotein
–

---

4 ISOLATION/PREPARATION

Source organism
Sheep [1]; Bovine (calf) [2]; Staphylococcus aureus [2]; Baker's yeast [3, 5, 7]; Neurospora crassa (IFO 6178 [3]) [3, 6]; E. coli (bifunctional enzyme with activity of glucosamine-1-phosphate acetyl transferase and N-acetylglucosamine-1-phosphate uridylyltransferase) [4]

Source tissue
Brain [1]; Liver [2]

Localization in source

Purification
Sheep (partial) [1]; Neurospora crassa [6]; Bovine (calf, partial) [2]; Staphylococcus aureus (partial) [2]; Baker's yeast [7]

Crystallization

–

Cloned

–

Renatured

–

---

5 STABILITY

pH
6.0–7.5 (4°C, stable) [6]; 7.5–8.0 (4°C, relatively stable) [7]

Temperature (°C)
40 (5 min, in presence of DTT, 20% loss of activity, in absence of DTT, 50% loss of activity) [6]; More (DTT increases thermal stability) [7]

Oxidation

Organic solvent

General stability information
Rapid decrease in activity without DTT within a few days [7]; DTT increases thermal stability [7]; Calf liver enzyme is unstable during fractionation with ammonium sulfate at several pH values [2]; Staphylococcal enzyme is labile to fractionation with organic solvents [2]; Quite stable towards ammonium sulfate [1]; Loss of activity after prolonged dialysis [1]; Uridyltransferase activity of the bifunctional enzyme is completely insensitive to millimolar concentration of thiol reagents [4]

Storage
0°C, 2 weeks, 50% loss of activity [1]; Frozen, 1 month, 50% loss of activity [1]; –20°C, about 10% loss of activity after 2 months, crude extract [2]; 4°C, 0.1 M potassium phosphate, pH 7.5, 0.01 mM DTT, stable for at least 2 weeks [6]; –20°C, unstable [7]; 4°C, 0.01 M potassium phosphate buffer, pH 7.5, 0.1 mM DTT, little loss of activity after 1 week [7]

---

6 CROSSREFERENCES TO STRUCTURE DATABANKS

PIR/MIPS code

Brookhaven code

7 LITERATURE REFERENCES

[1] Pattabiraman, T.N., Bachhawat, B.K.: Biochim. Biophys. Acta,50,129–134 (1961)
[2] Strominger, J.L., Smith, M.S.: J. Biol. Chem.,234,1822–1827 (1959)
[3] Yamamoto, K., Moriguchi, M., Kawai, H., Tochikura, T.: Biochim. Biophys. Acta,614, 367–372 (1980)
[4] Mengin-Lecreulx, D., van Heijenoort, J.: J. Bacteriol.,176,5788–5795 (1994)
[5] Yamamoto, K., Kawai, H., Moriguchi, M., Tochikura, T.: J. Ferment. Technol.,56,57–58 (1978)
[6] Yamammoto, K., Moriguchi, M., Kawai, H., Tochikura, T.: Can. J. Microbiol.,25, 1381–1386 (1979)
[7] Yamamoto, K., Kawai, H., Moriguchi, M., Tochikura, T.: Agric. Biol. Chem.,40, 2275–2281 (1976)

1 NOMENCLATURE

EC number
2.7.7.24

Systematic name
dTTP:alpha-D-glucose-1-phosphate thymidylyltransferase

Recommended name
Glucose-1-phosphate thymidylyltransferase

Synonyms
Thymidylyltransferase, glucose 1-phosphate
dTDP-glucose synthase
dTDPglucose pyrophosphorylase
Glucose 1-phosphate thymidylyltransferase
Thymidine diphosphoglucose pyrophosphorylase
Thymidine diphosphate glucose pyrophosphorylase [1]
dTDP-glucose-pyrophosphorylase [1]
TDP-glucose pyrophosphorylase [2]

CAS Reg. No.
9026-03-3

2 REACTION AND SPECIFICITY

Catalyzed reaction
dTTP + alpha-D-glucose 1-phosphate →
→ diphosphate + dTDPglucose

Reaction type
Nucleotidyl group transfer

Natural substrates
More (probably involved in conversion of D-glucose to L-rhamnose in the
biosynthesis of deoxy compound in Streptococcus faecalis) [1]

Substrate spectrum
1 dTTP + alpha-D-glucose 1-phosphate (r [2]) [1, 2]

Product spectrum
1 dTDPglucose + diphosphate [1, 2]

Inhibitor(s)

Cofactor(s)/prosthetic group(s)/activating agents

Metal compounds/salts
Mg^{2+} (absolute requirement, 2 mM: optimal) [2]

Turnover number (min^{-1})

Specific activity (U/mg)
44.8 (backward reaction) [2]

K_m-value (mM)
0.05 (TDPglucose) [2]; 0.1 (TTP) [2]

pH-optimum
8.0 (broad) [2]

pH-range

Temperature optimum (°C)
25 (assay at, backward reaction [2], assay at [1]) [1, 2]; 37 (assay at, forward reaction) [2]

Temperature range (°C)

---

3 ENZYME STRUCTURE

Molecular weight

Subunits

Glycoprotein/Lipoprotein
–

---

4 ISOLATION/PREPARATION

Source organism
Streptococcus faecalis [1]; E. coli (ATCC 13027) [1]; Leuconostoc mesenteroides (strain P-60) [1]; Lactobacillus arabinosus (ATCC 17–5) [1]; Streptococcus zymogenes (ATCC 10100) [1]; Bacillus subtilis (ATCC 6633) [1]; Sarcina lutea [1]; Alfalfa (germinated) [1]; Soybean (germinated) [1]; Pseudomonas aeruginosa (ATCC 7700) [2]

Source tissue
Cell [1, 2]

Localization in source

Purification
   Streptococcus faecalis (partial) [1]; E. coli (ATCC 13027, partial) [1]; Leuconostoc mesenteroides (strain P-60, partial) [1]; Lactobacillus arabinosus
   (ATCC 17–5, partial) [1]; Streptococcus zymogenes (ATCC 10100, partial)
   [1]; Bacillus subtilis (ATCC 6633, partial) [1]; Sarcina lutea (partial) [1]; Me-
   dicago sativa (germinated, partial) [1]; Soja max (germinated, partial) [1];
   Pseudomonas aeruginosa (ATCC 7700, partial) [2]

Crystallization
   –

Cloned
   –

Renatured
   –

---

5 STABILITY

pH

Temperature (°C)

Oxidation

Organic solvent

General stability information

Storage
   Stable for several weeks at any step of purification if kept frozen [2]

---

6 CROSSREFERENCES TO STRUCTURE DATABANKS

PIR/MIPS code
   PIR2:D64437 (Methanococcus jannaschii)

Brookhaven code

---

7 LITERATURE REFERENCES

[1] Pazur, J.H., Shuey, E.W.: J. Biol. Chem.,236,1780–1785 (1961)
[2] Kornfeld, S., Glaser, L.: J. Biol. Chem.,236,1791–1794 (1961)

1 NOMENCLATURE

EC number
2.7.7.25

Systematic name
ATP:tRNA adenylyltransferase

Recommended name
tRNA adenylyltransferase

Synonyms
tRNA CCA-pyrophosphorylase
tRNA-nucleotidyltransferase [1]
transfer-RNA nucleotidyltransferase [2]
Transfer ribonucleic acid nucleotidyl transferase [2]
CTP(ATP):tRNA nucleotidyltransferase [8]
Adenylyltransferase, transfer ribonucleate
Transfer ribonucleate adenyltransferase
transfer RNA adenylyltransferase
Nucleotidyltransferase, transfer ribonucleate
ATP (CTP):tRNA nucleotidyltransferase
Ribonucleic cytidylic cytidylic adenylic pyrophosphorylase
Transfer ribonucleate nucleotidyltransferase
Transfer ribonucleic adenylyl (cytidylyl) transferase
Transfer ribonucleic-terminal trinucleotide nucleotidyltransferase
Cytidylyltransferase, transfer ribonucleate
Ribonucleic cytidylyltransferase
Transfer ribonucleate cytidylyltransferase
-C-C-A pyrophosphorylase [3]
tRNA cytidylyltransferase [3]
ATP(CTP)-tRNA nucleotidyltransferase [3]
tRNA adenylyl(cytidylyl)transferase [16]
ATP(CTP):tRNA nucleotidyltransferase [19]
CTP:tRNA cytidylyltransferase
EC 2.7.7.21 (EC 2.7.7.21 is identical with EC 2.7.7.25)

CAS Reg. No.
52523-59-8; 9026-11-3; 9026-32-8

2 REACTION AND SPECIFICITY

Catalyzed reaction

$ATP + tRNA_n \rightarrow$
$\rightarrow$ diphosphate $+ tRNA_{n+1}$ (rabbit liver enzyme: rapid equilibrium mechanism [3], stereochemistry [34]);
$CTP + tRNA_n \rightarrow$
$\rightarrow$ diphosphate $+ tRNA_{n+1}$

Reaction type
Nucleotidyl group transfer

Natural substrates
More (mitochondria may contain a system for the maturation of tRNA molecules [23], in most organisms tRNA nucleotidyltransferase plays a role both
in tRNA biosynthesis and in tRNA repair [3], role in repair of tRNA deprived
of terminal CCA [24], regulation of AMP and CMP incorporation into
tRNAACC and tRNAC [36]) [3, 23, 24, 36]

Substrate spectrum
1 ATP + tRNA [3]
2 tRNA-N + CTP [3]
3 tRNA-C + CTP [3]
4 tRNA-C-C + ATP [3]
5 UTP + $tRNA_n$ (at 2–10% of the rate of CMP incorporation [3], not [18]) [3]
6 tRNA-C-C-A + diphosphate (r) [8]
7 tRNA-C-C + dATP (r) [8]
8 tRNA-C-A + diphosphate (r) [8]
9 tRNA-C + dATP (r) [8]
10 2-Thiocytidine 5'-triphosphate + $tRNA_n$ [10]
11 ATP + $tRNA_n$ (r [3, 8], enzyme catalyzes the incorporation of AMP and
   CMP residues into tRNA molecules from which all or part of 3'-terminal
   trinucleotide sequence-C-C-A has been removed [3], tRNA-C-C, tRNA-C
   and tRNA-N from liver, yeast or E. coli are equally active, all tRNA molecules in a mixed population are active as acceptors, reactions with 5S
   RNA, rRNA and modified tRNA's such as tRNA-C-A, tRNA-C-U and
   tRNA-C-C-C occur at much slower rates [5], overview: RNA acceptor
   specificity [3, 5], (untreated [18]) yeast tRNA [18, 25], tRNA-C-C [3, 4])
   [1–36]
12 CTP + $tRNA_n$ (r [3], snake venom phosphodiesterase reated RNA from
   Neurospora crassa, E. coli or yeast [3]) [1–36]
13 tRNA-A-C-C + ATP (not CTP or UTP) [21]
14 tRNA-A-C + CMP [21]
15 tRNA-C + UMP [21]

16 More (cytidines in tRNA that are required for activity [26], misincorporations: synthesis of sequences other than C-C-A occur when either ATP or
CTP is ommited from reaction mixture: 1. tRNA-N + ATP→ tRNA-A,
tRNA-A-A-A, tRNA-A-A-A-A, 2. tRNA-C + ATP→ tRNA-C-A, tRNA-C-A-A, 3.
tRNA-C-C + CTP→ tRNA-C-C-C, tRNA-C-C-(C)$_n$, 4. tRNA-C-C + CTP, then
ATP→ tRNA-C-C-A-A, 5. tRNA-C-C-A + CTP→ tRNA-C-C-A-C-C, 6.
tRNA-C-C-A + CTP, then ATP→ tRNA-C-C-A-C-C-A [5], enzyme also contains nucleolytic activity which removes terminal CMP residues from
tRNA-C-C and tRNA-C-C-C, other tRNA molecules (tRNA-C-C-A, tRNA-C-A,
tRNA-C-U, and tRNA-C) are not substrates [28], removal of AMP from the
terminus of tRNA proceeds optimally at 1.0 mM diphosphate, incorporation of 2'- or 3'-dAMP proceeds optimally at 6.0 mM concentration of
deoxynucleoside triphosphate [8], replacement of the terminal CCA sequence in yeast tRNA[Phe] by several unusual sequences [13], extent of
normal and anomalous nucleotide incorporation [22], low activity of incorporation of CMP into rRNA partially degraded by phosphodiesterase
[33], not: DNA [18], rRNA [18, 23], GMP [18, 21], dATP [21]) [5, 8, 13,
18, 21–23, 26, 28, 33]

Product spectrum
1 tRNA-C-C-A + diphosphate [3]
2 tRNA-C or tRNA-C-C + diphosphate [3]
3 tRNA-C-C + diphosphate [3]
4 tRNA-C-C-A + diphosphate [3]
5 Diphosphate + tRNA$_{n+1}$
6 Diphosphate + tRNA$_{n+1}$ [3, 4]
7 diphosphate + tRNA$_{n+1}$ [3]
8 ATP + tRNA-C-C [8]
9 tRNA-C-C-dA + diphosphate [8]
10 ?
11 ?
12 ?
13 ?
14 ?
15 ?
16 ?

Inhibitor(s)
Proflavine sulfate (complete inhibition at 1 mM, CMP incorporation more
sensitive than AMP incorporation) [1]; Ethidium bromide (complete inhibition
at 2 mM, CMP incorporation more sensitive than AMP incorporation) [1];
Diphosphate (nucleolytic activity [28]) [3, 28, 35]; DTNB [20]; tRNA-N (inhibition of AMP incorporation into tRNA-C-C) [3]; tRNA-C (inhibition of AMP incorporation into tRNA-C-C) [3]; tRNA-C-C (inhibition of CMP incorporation) [3];
1,10-Phenanthroline (inhibition of AMP incorporation [15, 16], no inhibition of

CMP incorporation [16], E. coli: effects AMP incorporation, no effect in CMP incorporation, no inhibition of yeast, Rous sarcoma virus and rabbit liver enzyme [3], no effect [5]) [3, 15, 16]; Bathophenanthroline [15]; CTP (AMP incorporation [3, 9, 14, 18, 24, 28, 35], competitive [14, 24], CMP incorporation [12], nucleolytic activity [28]) [3, 9, 12, 14, 18, 24, 28, 35]; ATP (CMP incorporation [3, 9, 14, 18, 24, 28, 35], AMP incorporation [12], competitive [14, 24], nucleolytic acitivity [28]) [3, 9, 12, 14, 18, 24, 28, 35]; p-Substituted mercuribenzoate (inhibition of AMP incorporation) [5]; Mersalyl (inhibition of AMP incorporation) [5, 20]; 5,5'-Dithiobis(2-nitrobenzoic acid) (inhibition of AMP incorporation [5], inhibits incorporation of AMP into tRNA-X-C-C but is without effect on the incorporation of CMP or UMP into tRNA-X [32]) [5, 32]; Hg^{2+} [5]; $HgCl_2$ [20]; PCMB (reversal by 2-mercaptoethanol [12]) [12, 20]; UTP [12]; N-Ethylmaleimide (reversal by 2-mercaptoethanol [12], inhibits incorporation of AMP into tRNA^{-X}-C-C but is without effect on incorporation of CMP or UMP into tRNA^{-X} [32]) [12, 32]; DTT [5, 20]; $(NH_4)_2SO_4$ (AMP incorporation: 50% inhibition at 0.2 ionic strength, CMP incorporation stimulated) [5]; KCl (AMP incorporation: 50% inhibition at 0.2 ionic strength, CMP incorporation stimulated) [5]; NaCl (AMP incorporation: 50% inhibition at 0.2 ionic strength, CMP incorporation stimulated) [5]; KCl (0.2 M: 50% inhibition of AMP incorporation, 5% inhibition of CMP incorporation [9], nucleolytic activity [28]) [9, 28]; tRNA^{Phe} containing iodoacetamide-alkylated 2-thiocytidine [10]; alpha,alpha-Dipyridyl [15]; 2,2,2-Terpyridyl [15]; Spermine (inhibition of AMP incorporation, stimulation of CMP incorporation) [5]; Spermidine (inhibition of AMP incorporation, stimulation of CMP incorporation) [5]; Putrescine (inhibition of AMP incorporation, stimulation of CMP incorporation) [5]; Cadaverine (inhibition of AMP incorporation, stimulation of CMP incorporation) [5]; EDTA (not [5, 15]) [12]; tRNA-X (competitive in AMP attachment to tRNA-X-C-C) [14]; tRNA-X-C-C-A (noncompetitive in CMP attachment to tRNA-X and AMP attachment to tRNA-X-C-C) [14]

Cofactor(s)/prosthetic group(s)/activating agents

2-Mercaptoethanol (activates) [35]; Polyamines (stimulates activity of rabbit liver enzyme [3], decreases requirement for Mg^{2+} (from 10 to 1 mM) [3], absolute requirement for a divalent cation which can be satisfied only by Mg^{2+}, Mn^{2+} or Co^{2+}, in addition a second function for cations has been identified which is carried out most efficiently by polyamines, although additional Mg^{2+} or monovalent cations are also effective [27], neither spermine, nor spermidine (0.1–10 mM) can replace divalent cations for the tRNA nucleotidyltransferase activity [32]) [3, 27, 32]

Metal compounds/salts

Mg^{2+} (required [3, 4, 17, 18, 21, 31, 35], Mg^{2+}, Co^{2+} or Mn^{2+} required [25], Mg^{2+} or Mn^{2+} required (for nucleolytic activity [28]) [12, 24, 28], highest rate of incorporation of AMP into tRNA-X-C-C or tRNA-X and of CMP into tRNA-X are observed in the presence of Mg^{2+} [32], optimum concentration: 1 mM

(Lactobacillus luteus) [3], 5 mM (E. coli A19 [3], inhibition above [25]) [3, 25], 5–10 mM (rabbit liver) [3], 10 mM (E. coli MRE 600) [3], 10–15 mM (mutant enzyme) [17], 5 mM (wild type enzyme) [17], 15–20 mM (higher concentrations inhibit) [18], K_m: 5 mM [18]) [3, 4, 12, 17, 18, 21, 24, 25, 28, 31, 32, 35]; Mn^{2+} (Mg^{2+}, Co^{2+} or Mn^{2+} required [25], Mg^{2+} or Mn^{2+} required (for nucleolytic acivity [28]) [12, 24, 25, 28], can partially replace Mg^{2+} in activation [3, 5, 17, 21], 30% (ATP incorporation), 40–50% (CTP incorporation) of the activity with Mg^{2+} [5], optimal concentration is lower than the level of triphosphate present and higher concentrations strongly inhibit [5], in presence of Mn^{2+} a variety of anomalous reactions catalyzed by tRNA nucleotidyltransferase are stimulated whereas normal reactions are inhibited [3], can partially replace Mg^{2+} in AMP incorporation, inefficient for CMP incorporation, in its presence UMP is incorporated instead of CMP, in presence of optimal Mg^{2+}-concentrations Mn^{2+} decreases the rate of CMP incorporation and to a lower extent of AMP, but increases the rate of UMP incorporation [32], optimal concentration: 4 mM (ATP incorporation) [5], 0.5–1 mM (CTP incorporation) [5]) [3, 5, 12, 17, 21, 24, 28, 32]; Co^{2+} (Mg^{2+}, Co^{2+} or Mn^{2+} required [25], can partially replace Mg^{2+} in activation [3, 5, 21], 15% (ATP incorporation), 20–30% (CTP incorporation) of the activity with Mg^{2+} [5], optimal concentration: 7.5–10 mM (ATP incorporation) [5], 5 mM (CTP incorporation) [5], low efficiency [32]) [3, 5, 15, 21, 25, 32]; $(NH_4)_2SO_4$ (AMP incorporation: 50% inhibition at 0.2 ionic strength, CMP incorporation stimulated) [5]; KCl (AMP incorporation: 50% inhibition at 0.2 ionic strength, CMP incorporation stimulated) [5]; NaCl (AMP incorporation: 50% inhibition at 0.2 ionic strength, CMP incorporation stimulated) [5]; More (E. coli enzyme is a metalloenzyme [16], absolute requirement for a divalent cation which can be satisfied only by Mg^{2+}, Mn^{2+} or Co^{2+}, in addition a second function for cations has been identified which is carried out most efficiently by polyamines, although additional Mg^{2+} or monovalent cations are also effective [27]) [16, 27]

Turnover number (min^{-1})
600 (Lupinus luteus, AMP incorporation) [3]; 3600 (rabbit, AMP incorporation) [3]; 7200 (Saccharomyces cerevisiae, AMP incorporation) [3]; 15600 (E. coli A19, AMP incorporation) [3]; 21000 (E. coli B, AMP incorporation) [3, 15]

Specific activity (U/mg)
280 [15]; 0.252 [35]; 58.3 [4]; 401.67 [18]; 33.33 [20]; 13.2 [19]; More [3, 5, 11, 14, 25]

K_m-value (mM)
0.000238 (tRNA) [13]; 0.015 (tRNA-C-C) [9]; 0.017 (CTP, E. coli MRE 600) [3]; 0.020 (tRNA-C) [9]; 0.028 (CTP) [1]; 0.029 (CTP, Lactobacillus acidophilus) [3]; 0.03 (CTP, E. coli A19, rabbit [3], CTP, E. coli UT481 (pEC 4) [9]) [3, 9];

0.033 (ATP [1], CTP, Musca domestica [3], ATP, E. coli UT 481(pEC 4) [9])
[1, 3, 9]; 0.07 (CTP, Lupinus luteus) [3]; 0.2 (ATP, Musca domestica, CTP,
Saccharomyces cerevisiae) [3]; 0.25 (ATP, Lupinus luteus) [3]; 0.31 (ATP, E.
coli A19) [3]; 0.4 (tRNA-C-C, Lactobacillus acidophilus) [3]; 0.71 (ATP, Lactobacillus acidophilus) [3]; 1.5–1.7 (tRNA-C-C, E. coli MRE 600) [3]; 4 (tRNA-N,
rabbit) [3]; 5.5 (tRNA-C-C, Saccharomyces cerevisiae) [3]; 6 (tRNA-C, rabbit)
[3]; 7.8 (tRNA-N, Saccharomyces cerevisiae) [3]; 11 (tRNA-C, Saccharomyces
cerevisiae) [3]; 12 (tRNA-C-C, rabbit) [3]; 13 (tRNA-C, E. coli A19) [3]; More
[3, 4, 6, 12–14, 17–19, 21, 31, 35]

pH-optimum

9.0–9.4 (E. coli A19 [3], E. coli, AMP incorporation [17]) [3, 17]; 9–10 [4]; 9
(above) [31]; 9.3–10 (rabbit) [3, 5]; 9.4 (AMP incorporation) [9]; 9.5 (Lupinus
luteus, Saccharomyces cerevisiae, E. coli MRE 600 [3], AMP incorporation
[25]) [3, 12, 19, 25]; 10 (CMP incorporation [9, 17], nucleolytic activity [28])
[9, 17, 28]

pH-range

7.1–10 (7.1: 20–30% of activity maximum, 9.3–10: activity maximum, rabbit)
[5]; More [12, 25]

Temperature optimum (°C)

30 (assay at) [11]; 35 (assay at) [17]; 37 (assay at) [15, 16, 28, 30]; 45 [12]

Temperature range (°C)

---

3 ENZYME STRUCTURE

Molecular weight

30000 (Musca domestica, gel filtration, sucrose density gradient centrifugation) [3, 31]

37000 (E. coli MRE 600, gel filtration) [3]

40000 (Lupinus luteus, gel filtration) [3, 12]

44000–48000 (rabbit, gel filtration, equilibrium ultracentrifugation) [3, 5]

45000 (E. coli, sucrose density centrifugation [33], E. coli B, gel filtration [3])
[3, 33]

46408 (E. coli, determination of nucleotide sequence) [4]

50000 (rabbit, sedimentation analysis) [20]

53000 (E. coli B, gel filtration) [15]

59000 (Saccharomyces cerevisiae, gel filtration) [19]

62000 (Neurospora crassa, gel filtration) [18]

63000 (Ceratodon purpureus, gel filtration) [1]

71000 (Saccharomyces cerevisiae, equilibrium ultracentrifugation) [3]

Subunits

Monomer (1 × 31000, Musca domestica, SDS-PAGE [3, 31], 1 × 47000, rabbit, SDS-PAGE [3, 20], 1 × 51500, E. coli B, SDS-PAGE [3, 15], 1 × 70000,
Saccharomyces cerevisiae, SDS-PAGE [3], 1 × 45000–49000, rabbit,
SDS-PAGE [5], 1 × 59000, Saccharomyces cerevisiae, SDS-PAGE [19]) [3, 5,
15, 19, 20, 31]
? (x × 43000, Lactobacillus acidophilus, SDS-PAGE [3, 6], x × 45000, E. coli
MRE 600, SDS-PAGE [3, 6], x × 50000, E. coli A19, SDS-PAGE [3]) [3, 6]
More (aggregation at elevated protein levels prevented by high concentration of phosphate) [20]

Glycoprotein/Lipoprotein

–

4 ISOLATION/PREPARATION

Source organism

Ceratodon purpureus [1]; Mouse (Ehrlich tumor cells) [2, 30]; Avian RNA tumor viruses [3, 6, 7]; Sendai virus [3, 6]; E. coli (strain UT481 (pEC4) [9],
MRE 600 [3, 6], B [3, 15, 16, 35], A19 [3], overexpression (high level) [9]) [4,
9], mutant with decreased AMP incorporation but normal CMP incorporation
[17]) [3, 4, 6, 9, 15–17, 26, 32, 33, 35, 36]; Rat [3, 5, 20–23, 27, 28]; Lactobacillus acidophilus (ATCC 4963 [6]) [3, 6]; Saccharomyces cerevisiae (wild
type and overproducing [19]) [3, 8, 10, 13, 14, 19, 34]; Lupinus luteus [3,
11, 12]; Musca domestica [3, 31]; Neurospora crassa 74A [18]; Wheat [24];
Avian reticuloendotheliosis virus [29]; Rabbit [3, 5, 20–22, 25, 27, 28]; Bovine (calf) [25]

Source tissue

Protonema [1]; Ehrlich ascites tumor cells [2, 30]; Muscle [3]; Liver [3, 20,
21, 23, 27, 28]; Seeds [3, 11, 12]; Germ [24]; Spleen [30]; Lymph nodes
[30]; Macrophage cells [30]; Whole organism [31]; Larvae [31]; Pupae [31];
Muscle [25]; More (widespread occurence in all types of cells) [3]

Localization in source

Mitochondria (matrix [23]) [3, 5, 23]; Cytoplasm [30]; More (microsomes and
nuclei are devoid of activity) [3, 23]

Purification

Ceratodon purpureus [1]; Lupinus luteus [11, 12]; E. coli (MRE 600 [6],
UT481 (pEC4) [9], large scale [35], B [16, 35], mutant with decreased AMP
incorporation but normal CMP incorporation [17]) [3, 4, 6, 9, 15–17, 35]; Rat
[3]; Rabbit [3, 5, 20, 25]; Lactobacillus acidophilus [6]; Saccharomyces cerevisiae (wild type and overproducing [19]) [8, 14, 19]; Neurospora crassa
74A [18]; Musca domestica [31]

Crystallization
–

Cloned
–

Renatured
–

---

5 STABILITY

pH

Temperature (°C)
45 (10 min, 20% (AMP incorporation), 40% (CMP incorporation) loss of activity [12], 5 min, about 50% loss of activity [20]) [12, 20]; 50 (20 min, 50%
loss of activity in absence of tRNA, 1 h, less than 20% loss of activity in presence of tRNA) [31]; More (tRNA protects against thermal inactivation [3], ATP
and CTP protect against heat inactivation [14], high concentrations of tRNA
without the CCA terminus (tRNA-X), but not intact tRNA protects against
heat inactivation [14], Mg^{2+}, ATP and tRNA have little effect on heat inactivation [20], heating in presence of 0.15 M potassium phosphate, pH 7.5, instead of 0.01 M buffer stabilizes [20], tRNA and to a lesser extent ATP and
CTP protect against heat inactivation [31]) [3, 14, 20, 31]

Oxidation

Organic solvent

General stability information
tRNA protects against thermal inactivation [3]; 2-Mercaptoethanol stabilizes
[35]; tRNA and CTP but not ATP stabilize [1]; ATP and CTP protect against
heat inactivation [14]; High concentrations of tRNA without the CCA terminus (tRNA-X), but not intact tRNA protects against heat inactivation [14];
Mg^{2+}, ATP and tRNA have little effect on heat inactivation [20]; Heating in
presence of 0.15 M potassium phosphate, pH 7.5, instead of 0.01 M buffer
stabilizes [20]; tRNA and to a lesser extent ATP and CTP protect against
heat inactivation [31]

Storage
–20°C, in 50% glycerol or frozen with 0.2 mg/ml of commercial yeast tRNA,
less than 25% loss of activity after 1 year [5]; –20°C, 50% glycerol, stable for
at least 1 year [11]; –20°C, even at concentrations as low as 0.03 mg/ml or
when frozen with 0.2 mg/ml of commercial yeast tRNA, stable for at least 3
months [20]

6 CROSSREFERENCES TO STRUCTURE DATABANKS

PIR/MIPS code
PIR1:RNECTA (Escherichia coli)

Brookhaven code

---

7 LITERATURE REFERENCES

[1] Schneider, Z., Schneider, J.: Biochem. Physiol. Pflanz.,171,239–248 (1977)
[2] Girgenti, A.J., Whitford, T.W., Cory, J.G.: Enzyme,21,225–231 (1976)
[3] Deutscher, M.P. in "The Enzymes",3rd Ed. (Boyer, P.D., Ed.) 15,183–215 (1982) (Review)
[4] Deutscher, M.P.: Methods Enzymol.,181,434–439 (1990) (Review)
[5] Deutscher, M.P.: Methods Enzymol.,29E,70–81 (1974) (Review)
[6] Leineweber, M., Philipps, G.R.: Hoppe-Seyler's Z. Physiol. Chem.,359,473–480 (1978)
[7] Faras, A.J., Levinson, W.E., Bishop, J.M., Goodman, H.M.: Virology,58,126–135 (1974)
[8] Francis, T.A., Ehrenfeld, G.M., Gregory, M.R., Hecht, S.M.: J. Biol. Chem.,258, 4279–4284 (1983)
[9] Cudney, H., Deutscher, M.P.: J. Biol. Chem.,261,6450–6453 (1986)
[10] Kröger, M., Sternbach, H., Cramer, F.: Eur. J. Biochem.,95,341–348 (1979)
[11] Cudny, H., Pietrzak, M., Kaczkowsky, J.: Planta,142,23–27 (1978)
[12] Cudny, H., Pietrzak, M., Kaczkowski, J.: Planta,142,29–36 (1978)
[13] Rether, B., Gangloff, J., Ebel, J.-P.: Eur. J. Biochem.,50,289–295 (1974)
[14] Rether, B., Bonnet, J., Ebel, J.-P.: Eur. J. Biochem.,50,281–288 (1974)
[15] Schofield, P., Williams, K.R.: J. Biol. Chem.,252,5584–5588 (1977)
[16] Williams, K.R., Schofield, P.: Biochem. Biophys. Res. Commun.,64,262–267 (1975)
[17] McGann, R.G., Deutscher, M.P.: Eur. J. Biochem.,106,321–328 (1980)
[18] Hill, R., Nazario, M.: Biochemistry,12,482–485 (1973)
[19] Chen, J.-Y., Kirchner, G., Aebi, M., Martin, N.C.: J. Biol. Chem.,265,16221–16224 (1990)
[20] Deutscher, M.P.: J. Biol. Chem.,247,450–458 (1972)
[21] Deutscher, M.P.: J. Biol. Chem.,247,459–468 (1972)
[22] Deutscher, M.P.: J. Biol. Chem.,247,469–480 (1972)
[23] Mukerji, S.K., Deutscher, M.P.: J. Biol. Chem.,247,481–488 (1972)
[24] Dullin, P., Fabisz-Kijowska, A., Walerych, W.: Acta Biochim. Pol.,22,279–289 (1975)
[25] Starr, J.L., Goldthwait, D.A.: J. Biol. Chem.,238,682–689 (1963)
[26] Hegg, L.A., Thurlow, D.L.: Nucleic Acids Res.,18,5975–5979 (1990)
[27] Evans, J.A., Deutscher, M.P.: J. Biol. Chem.,251,6646–6652 (1976)
[28] Deutscher, M.P.: Biochem. Biophys. Res. Commun.,52,216–222 (1973)
[29] Mizutani, S., Temin, H.M.: J. Virol.,19,610–619 (1976)
[30] Sato, N.L.: J. Biochem.,85,739–745 (1979)
[31] Poblete, P., Jedlicky, E., Litvak, S.: Biochim. Biophys. Acta,476,333–341 (1977)
[32] Carre, D.S., Chapeville, F.: Biochim. Biophys. Acta,361,176–184 (1974)
[33] Carre, D.S., Litvak, S., Chapeville, F.: Biochim. Biophys. Acta,361,185–197 (1974)
[34] Eckstein, F., Sternbach, H., von der Haar, F.: Biochemistry,16,3429–3432 (1977)
[35] Best, A.N., Novelli, G.D.: Arch. Biochem. Biophys.,142,527–538 (1971)
[36] Best, A.N., Novelli, G.D.: Arch. Biochem. Biophys.,142,539–547 (1971)

---

1 NOMENCLATURE

EC number
2.7.7.27

Systematic name
ATP:alpha-D-glucose-1-phosphate adenylyltransferase

Recommended name
Glucose-1-phosphate adenylyltransferase

Synonyms
ADPglucose pyrophosphorylase
Adenylyltransferase, glucose 1-phosphate
Adenosine diphosphate glucose pyrophosphorylase
Adenosine diphosphoglucose pyrophosphorylase
ADP-glucose pyrophosphorylase
ADP-glucose synthase
ADP-glucose synthetase
ADPG pyrophosphorylase

CAS Reg. No.
9027-71-8

---

2 REACTION AND SPECIFICITY

Catalyzed reaction
ATP + alpha-D-glucose 1-phosphate $\rightarrow$
$\rightarrow$ diphosphate + ADPglucose (mechanism (Rhodospirillum rubrum, E. coli) [32])

Reaction type
Nucleotidyl group transfer

Natural substrates
ATP + glucose 1-phosphate (first step of starch biosynthesis [1], key regulatory enzyme of starch biosynthesis [2, 10], one of the main regulatory steps in starch biosynthesis in plants [6, 32] and glycogen in bacteria [13, 32], first unique reaction in synthesis of alpha-1,4-glucosidic linkage [32]) [1, 2, 6, 10, 13, 32]

Substrate spectrum
1 ATP + glucose 1-phosphate (r [3, 5–8, 13, 15, 18–23, 29, 32], specific for ATP [7, 13, 18, 23], specific for ADPglucose in the reverse direction (Mycobacterium smegmatis) [32], ADPglucose synthesis at 50% (strain 274)

or 75% (strain 15365) the rate of pyrophosphorolysis [23], substrate binding studies (E. coli): 1 mol glucose 1-phosphate per mol subunit, 4 mol
ADPglucose per mol tetrameric enzyme [32]. No substrates are UTP [7,
18], TTP [7], GTP, dATP, CTP [7, 13, 23], ITP, dTTP, XTP [13, 23],
alpha-D-mannose 1-phosphate or alpha-D-galactose 1-phosphate [24])
[1–32]
2 GTP + glucose 1-phosphate (can somewhat replace ATP) [18]

Product spectrum
1 Diphosphate + ADPglucose [3, 5–8, 13, 15, 18–23, 29, 32]
2 Diphosphate + GDPglucose

Inhibitor(s)
Phosphate (allosteric inhibitor [6–8, 10, 11, 15, 32], plants, algae, cyano-
bacteriae [32], in vitro, no physiological relevance [14], kinetics (Rhodopseudomonas capsulata or sphaeroides [32]) [3, 5, 32], strong [3, 4], most
effective (Rhodopseudomonas sphaeroides, fructose 6-phosphate reverses)
[32], endospermal enzyme: weak [1], 3-phosphoglycerate reverses, increasing concentration of Mg^{2+} does not overcome the inhibition [5], not
(Enterobacter hafniae [13], up to 10 mM [23], aeromonads, Mycobacterium
smegmatis, Rhodospirillum rubrum, Rhodospirillum molischianum [32]) [13,
23, 32]) [1, 3–8, 10, 11, 13–15, 18, 29, 32]; Chromium adenosine triphosphate (strong, E. coli) [32]; Trehalose phosphate (at high concentrations
[18], Mycobacterium smegmatis, feed-back inhibition [32]) [18, 32]; ADP
(allosteric inhibitor [22, 32], kinetics (Rhodopseudomonas capsulata [32])
[5, 32], 1 mM [10], most effective inhibitor of aeromonads [32], less effective
than phosphate [3] or AMP (Serratia marcescens [23, 32], Rhodospirillum
tenue [32]) [22, 23, 32], plants, algae, cyanobacteria [32], Aeromonas
hydrophila: most effective, fructose 6-phosphate or fructose 1,6-diphosphate
partially reverses, Enterobacter hafniae: strong, phosphoenolpyruvate reverses [13], phosphoenolpyruvate, fructose diphosphate, pyridoxal phosphate, NADPH, 3-phosphoglycerate, fructose 6-phosphate or pyruvate does
not reverse [23], glycolytic metabolites do not affect inhibition (Serratia
marcescens) [32], not (Mycobacterium smegmatis, Rhodospirillum rubrum,
Rhodospirillum molischianum [32]) [11, 32]) [3–5, 7, 10, 13, 22, 23, 32];
AMP (allosteric inhibitor [22, 32], kinetics [32], less effective than phosphate
[3], Serratia marcescens (ATCC 15365: very strong, ATP reverses [23]) [23,
32], genus Serratia, E. coli B and mutants [32], Enterobacter hafniae:
strong, phosphoenolpyruvate partially reverses [13], glycolytic metabolites
do not affect inhibition (Serratia marcescens) [32], not (Mycobacterium
smegmatis, aeromonads, Rhodospirillum rubrum, Rhodospirillum molischianum [32]) [4, 11, 32]) [3, 7, 13, 14, 22, 23, 32]; Cyclic-3',5'-AMP (less effective than phosphate) [3]; ATP (at high concentrations, Mg^{2+} does not reverse) [18]; Pyruvate (not [18]) [7]; $NADP^+$ (less effective than phosphate
[3]) [3, 4]; NAD^+ [23]; Phosphoenolpyruvate (Rhodopseudomonas capsu-

lata, Agrobacterium tumefaciens, Arthrobacter viscosus, most effective in vitro, fructose 6-phosphate protects [32]) [7, 32]; Rabbit antiserum raised
against spinach enzyme (not against E. coli enzyme [10, 11]) [8, 10, 11];
Diphosphate (substrate inhibition, 1 mM [10]) [7, 10]; GTP (physiological
relevant inhibition) [14]; SO_4^{2-} (5 mM, weak) [7]; NO_3^- (not at 5 mM, only at
10 mM) [7]; Cl^- (not at 5 mM, only at 10 mM) [7]; F^- (NaF [8], not at 5 mM,
only at 10 mM [7]) [7, 8]; Phenylglyoxal (in the presence of 3-phosphoglyce-
rate) [11]; UDPglucose [7]; Glucose 6-phosphate [18]; Citrate [23]; Isocitrate [23]; FAD^+ [23]; More (activator/inhibitor interaction in vivo and in vitro
(E. coli) [32], no inhibition by trehalose, trehalose diphosphate, fructose
6-phosphate, ribose 5-phosphate or 3-phosphoglyceraldehyde [18]) [18, 32]

Cofactor(s)/prosthetic group(s)/activating agents

3-Phosphoglycerate (activation (slight [1, 23]) [1–8, 10, 11, 13, 23–25, 27,
29, 32], allosteric activation [3–8, 10, 11, 24], strong [29], pH-dependent
(maize endosperm [32]) [5, 32], kinetics [11, 32], Enterobacter hafniae [13],
endospermal enzyme [1], plant tissues, green algae, cyanobacteria [32],
not (Aeromonas hydrophila [13]) [13, 22]) [1–8, 10, 11, 13, 23–25, 27, 29,
32]; 2-Phospho-D-glycerate (activation (slight [4, 22]) [4, 11, 22], pyrophosphorolysis [4], not (Synechocystis sp. [11], Enterobacter hafniae, Aeromonas hydrophila [13]) [8, 11, 13]) [4, 11, 22]; 2,3-Diphosphoglycerate (activation [5, 22], allosteric activator [22], ADPglucose synthesis [5], not [4, 8,
11]) [5, 22]; Phosphoglycolate (slight activation, ADPglucose synthesis) [5];
6-Phosphogluconate (slight activation) [22]; D-Fructose 6-phosphate (activation (slight [10, 11, 23]) [4–8, 10, 11, 13, 18, 19, 23, 25, 29, 32], most effective allosteric activator of aeromonad enzyme [13], less effective than
3-phosphoglycerate [7, 8, 25], pyrophosphorolysis [4], ADPglucose synthesis [5, 6], Mycobacterium smegmatis, Rhodopseudomonas capsulata,
Rhodopseudomonas palustris, Rhodomicrobium vanniellii, Chromatium vinosum, Chlorobium limicola f. thiosulfatophilum, Arthrobacter viscosus,
Agrobacterium tumefaciens, Rhodopseudomonas gelatinosa, Rhodopseudomonas sphaeroides, Micrococcus luteus, Aeromonas formicans, Aeromonas hydrophila, Aeromonas liquefaciens [32], Serratia marcescens [23],
plants [29], not Enterobacter hafniae [13], Serratia marcescens, Serratia liquefaciens, Clostridium pasteurianum [32]) [4–8, 10, 11, 13, 18, 19, 23, 25,
29, 32]; Fructose 1,6-diphosphate (activation (slight [3, 4, 11, 23]), pyrophosphorolysis [4–6], ADPglucose synthesis [3, 5, 6], kinetics [5], Aeromonas hydrophila [13], one of the most effective activators for Enterobacter tarda, Klebsiella pneumoniae, Salmonella enteritidis and Shigella dysenteriae
enzyme [13], less effective than 3-phosphoglycerate [25], E. coli [29], not
[8], allosteric activator [15, 22, 32], physiological modulator [22], E. coli, Mycobacterium smegmatis, Rhodopseudomonas gelatinosa, Rhodopseudomonas sphaeroides, Micrococcus luteus, Aeromonas formicans, Aeromonas
hydrophila, Aeromonas liquefaciens, Enterobacteriaceae [32], most important in vivo (E. coli) [32]) [3–6, 10, 11, 13, 15, 18, 22, 23, 25, 29, 32]; Fruc-

tose 1-phosphate (activation, less effective than fructose 6-phosphate Aeromonas hydrophila) [13]; D-Glucose 6-phosphate (activation (slight [10]) [8,
10, 29], less effective than 3-phosphoglycerate [8], not (Enterobacter hafniae, Aeromonas hydrophila [13]) [13, 23]) [8, 10, 29]; Glucose 1,6-diphos-
phate (slight activation [10, 22], not [8, 11, 23]) [10, 22]; 1,6-Hexanediol bis-
phosphate (activation [13, 32], fructose 1,6-diphosphate analog, E. coli B:
most effective [32], Aeromonas hydrophila [13]) [13, 32]; Mannose 6-phos-
phate (activation, less effective than fructose 6-phosphate, Aeromonas
hydrophila) [13]; D-Ribose 5-phosphate (activation (slight [5]) [3–5, 7, 23],
fructose 6-phosphate analog [32], pyrophosphorolysis [4], ADPglucose synthesis [3, 5], less effective than 3-phosphoglycerate [7], not [22]) [3–5, 7,
23, 32]; Deoxyribose 5-phosphate (activation, fructose 6-phosphate analog)
[32]; Sedoheptulose 1,7-diphosphate (activation [13, 32], isosteric analog of
fructose diphosphate, E. coli [32], Aeromonas hydrophila, Enterobacter hafniae [13]) [13, 32]; D-Arabinitol 1,5-diphosphate (activation, E. coli) [32];
Glyceraldehyde 3-phosphate (slight activation [11, 22], less effective than
fructose diphosphate (E. coli) [32], not [8, 23]) [11, 22, 32]; 2-Deoxy-D-ribo-
se 5-phosphate (activation, pyrophosphorolysis [4], not [23]) [4]; Dihydroxy-
acetone phosphate (activation, less effective than 3-phosphoglycerate [8],
not (Enterobacter hafniae, Aeromonas hydrophila [13]) [11, 13, 23]) [8];
2-Keto-3-deoxy phosphogluconate (activation [23, 32], fructose 6-phosphate
and pyruvate analog, less effective than fructose diphosphate (E. coli B)
[32]) [23, 32]; 2-Oxobutyrate (slight activation, pyruvate analog) [32];
Hydroxypyruvate (slight activation, pyruvate analog) [32]; Pyridoxal phosphate (activation (slight [23]) [11, 13, 22, 23, 29, 32], allosteric activator [22,
32], E. coli B and mutant SG14 [32], Enterobacteriaceae [32], Serratia marcescens [23], one of the most effective activators for Enterobacter tarda,
Klebsiella pneumoniae, Salmonella enteritidis and Shigella dysenteriae enzyme, less effective for Enterobacter hafniae [13], not Aeromonas sp. [32])
[11, 13, 22, 23, 29, 32]; 4-Pyridoxic acid 5-phosphate (activation, E. coli)
[32]; Glycerol 1,3-diphosphate (activation, E. coli) [32]; Pyrophosphorylribo-
se 5-phosphate (activation, E. coli) [32]; NADPH (activation (slight [10]) [10,
13, 22], one of the most effective activators for Enterobacter tarda, Klebsiella pneumoniae, Salmonella enteritidis and Shigella dysenteriae enzyme
[13], Enterobacteriaceae, E. coli B [32], not (Enterobacter hafniae, Aeromonas hydrophila [13], E. coli SG14, Aeromonas sp. [32]) [13, 32]) [10, 13, 22,
32]; NADP+ (slight activation [22], less effective than fructose diphosphate
(E. coli B) [32], not [23]) [22, 32]; Pyruvate (activation, kinetics, Rhodospirillum rubrum, Rhodospirillum molischianum, Rhodospirillum tenue,
Rhodopseudomonas palustris, Rhodopseudomonas capsulata, Rhodomicrobium vanniellii, Chromatium vinosum, Chlorobium limicola f. thiosulfatophilum, Arthrobacter viscosus, Agrobacterium tumefaciens, Rhodopseudomonas gelatinosa, Rhodopseudomonas sphaeroides) [32]; Phosphoenolpyruvate (activation (slight [11]) [3–6, 8, 10, 11, 13, 18, 23, 32], ADPglucose

synthesis [3, 5, 6], pyrophosphorolysis [3–6], kinetics [23], most effective for
Enterobacter hafniae [13], less effective than 3-phosphoglycerate [8] or
fructose diphosphate (E. coli B) [32]) [3–6, 8, 10, 11, 13, 18, 23, 32]; Sodium oxamate (activation, structural analog of pyruvate) [20]; alpha-Glycerol
phosphate (activation, pyrophosphorolysis [4], not [3]) [4]; DTT (activation [3,
4], kinetics [4], only slight activation in the presence of 3-phosphoglycerate
[4]) [3, 4]; 2-Mercaptoethanol (activation, at high concentration) [4]; GSH
(activation, at high concentration) [4]; Cysteine (slight activation, at high
concentration) [4]; AMP (activation, less effective than 3-phosphoglycerate
[29], not (Enterobacter hafniae or Aeromonas hydrophila [13]) [11, 13]) [29];
ADP (activation, less effective than 3-phosphoglycerate [29], not (Enterobacter hafniae or Aeromonas hydrophila [13]) [11, 13]) [29]; More (activa-
tor/inhibitor interaction in vivo and in vitro (E. coli) [32], effector binding
studies (E. coli) and structural requirements for an activator of E. coli enzyme [32], genus Serratia: no significant (below 20%) activation by glycolytic metabolites [32], Enterobacteriaceae with a nonspecific activator site
[32], no activation of pyrophosphorolysis [4] or ADPglucose synthesis [3] by
pyruvate [3, 4, 8, 11, 23], fructose 2,6-bisphosphate, Ca^{2+}/calmodulin [8],
L-lactate, citrate [3, 4], oxalacetate, L-malate [3, 4, 23], 2-oxoglutarate [3,
23], NADH, fumarate, succinate, acetyl-CoA, Glu, Ala, Asp, riboflavine
5'-phosphate [23], no activation of Enterobacter hafniae or Aeromonas
hydrophila enzyme by NADH, pyruvate, NAD^+, $NADP^+$, phosphate, glycerol
3-phosphate, erythrose 4-phosphate, malate, cAMP [13]) [3, 4, 8, 11, 13, 23,
32]

Metal compounds/salts
Mg^{2+} (requirement [3, 7, 10, 11, 15, 18–25, 29, 32], ADPglucose synthesis
[3, 23], 6–8 mM [7], E. coli: 2 mol MgATP per mol tetrameric enzyme [32])
[3, 7, 10, 11, 15, 18–25, 29, 32]; More (no activation by Mn^{2+}, Zn^{2+}, NH_4^+,
Na^+) [7]

Turnover number (min^{-1})

Specific activity (U/mg)
0.125–0.287 (Serratia marcescens) [23]; 1.43 [9]; 2.5 (Enterobacter hafniae)
[13]; 2.8 [3]; 4.36 [7]; 24 [5]; 28.2 (Aeromonas hydrophila) [13]; 34 [8]; 64
(Rhodospirillum rubrum) [20]; 81 [10]; 90–110 [15]; 106 [16, 22]; 111 (Anabaena sp.) [11]; 114 (Rhodospirillum tenue) [21]; 134 (Synechocystis sp.)
[11]; 156 [6]; 168 (Rhodopseudomonas sphaeroides) [19]

K_m-value (mM)
More (effect of activators on substrate kinetic parameters of bacterial enzymes [32], kinetic parameters (Serratia marcescens [23]) [4, 6, 8, 10, 11,
13, 23, 24, 32], of proteolyzed and non-proteolyzed enzyme [8], the allosteric properties of Enterobacter hafniae are distinctly different from other
bacteria of the genus Enterobacter [13]) [4, 6, 8, 10, 11, 13, 23, 24, 32];

0.032 (ATP) [7]; 0.033 (diphosphate) [7]; 0.038 (glucose 1-phosphate) [9]; 0.14 (glucose 1-phosphate) [3]; 0.18 (ATP) [9]; 0.19 (ATP [3], glucose 1-phosphate) [7]; 0.24 (ADPglucose) [3, 4]; 0.6 (ATP) [18]; 0.62 (ADPglucose) [7]; 0.8 (glucose 1-phosphate) [18]

pH-optimum

More (pI: 7.02, ADPglucose pyrophosphorylase B [9], Arthrobacter viscosus: shift of pH-optimum in the presence of activator, not Rhodopseudomonas capsulata or Agrobacterium tumefaciens [32]) [9, 32]; 7 (Aeromonas hydrophila, ADPglucose synthesis, HEPES buffer) [13]; 7–7.5 (ADPglucose synthesis, Tris buffer) [7]; 7–8 (ADPglucose synthesis, HEPES buffer) [7]; 7.4 (broad) [8]; 7.5 (Serratia marcescens ATCC 274, HEPES buffer) [23]; 8 (ADPglucose synthesis [3, 13], HEPES, bicine and glycylglycine buffer [3], pyrophosphorolysis [4], Enterobacter hafniae, bicine buffer [13], Serratia marcescens ATCC 15365, HEPES buffer [23], Arthrobacter viscosus, in the presence of activator [32]) [3, 4, 13, 32]; 8.5 (Tris buffer [3, 23], ADPglucose synthesis [3]) [3, 18, 23]; 10 (Arthrobacter viscosus, in the absence of activator) [32]

pH-range

6–8.3 (about half-maximal activity at pH 6 and 8.3, Tris buffer) [7]; 6.1–8.5 (about half-maximal activity at pH 6.1 and about 80% of maximal activity at pH 8.5, HEPES buffer) [7]; 6.8–9.5 (about half-maximal activity at pH 6.8 and about 70% of maximal activity at pH 9.5) [4]; 7.2–9 (about half-maximal activity at pH 7.2 and about 90% of maximal activity at pH 9) [18]

Temperature optimum (°C)

37 (assay at) [3–10, 13, 15, 18, 20–23]

Temperature range (°C)

---

3 ENZYME STRUCTURE

Molecular weight

More (amino acid composition (E. coli [32]) [15, 19–21, 32] compared to that of Rhodospirillum tenue [19, 20], E. coli [19, 20, 27], Salmonella typhimurium [19, 21], rice endosperm small subunit or maize endosperm large subunit [27] and N-terminal sequence [19–21] compared to those of Rhodospirillum tenue [19, 20], E. coli, Salmonella typhimurium [19, 21], amino- and carboxy terminal sequence studies (E. coli) [32]) [15, 19–21, 27, 32]
96000 (Serratia marcescens, dimer, dimer-tetramer equilibrium presumed) [32]
185000 (E. coli mutant AC70R1, sedimentation velocity measurement [15], E. coli B, sedimentation equilibrium centrifugation [32]) [15, 32]
186000 (Serratia marcescens, tetramer, dimer-tetramer equilibrium presumed) [32]

190000 (Salmonella typhimurium, sucrose density gradient centrifugation) [22]
195000 (Salmonella typhimurium, PAGE) [22]
200000 (Solanum tuberosum, sucrose density gradient centrifugation [3],
Enterobacter hafniae, Aeromonas hydrophila, sucrose density gradient
method [13], Rhodospirillum rubrum, sucrose density gradient centrifugation [20]) [3, 13, 20]
204000 (Rhodopseudomonas sphaeroides, sucrose density gradient centrifugation) [19]
206000 (Spinacia oleracea, sucrose density gradient ultracentrifugation) [6]
210000 (Chlamydomonas reinhardtii [10], E. coli, sucrose density gradient
centrifugation [16], Arabidopsis thaliana, gel filtration [24]) [10, 16, 24]
215000 (Rhodospirillum tenue, sucrose density gradient centrifugation) [21]
216000 (Synechocystis sp., gel filtration) [11]
225000 (Anabaena sp., gel filtration) [11]
230000 (Zea mays, gel filtration) [8]
235000 (Zea mays, ATPG pyrophosphorylase B, gel filtration) [9]
237000–253000 (Zea mays, sucrose density gradient centrifugation) [26]
375000 (Zea mays, ATPG pyrophosphorylase A, gel filtration) [9]
400000 (Zea mays, ATPG pyrophosphorylase A, PAGE) [9]

Subunits

Tetramer ($x \times 44000 + x \times 48000$, Spinacia oleracea, SDS-PAGE [6],
$x \times 48000 + x \times 54000$, Arabidopsis thaliana, wild-type, SDS-PAGE [24],
$x \times 50000 + x \times 53000$, Chlamydomonas reinhardtii, SDS-PAGE [10],
4×46000, Rhodopseudomonas sphaeroides, SDS-PAGE [19], 4×48000,
Salmonella typhimurium, SDS-PAGE [22], Arabidopsis thaliana, mutant,
SDS-PAGE [24], 4×48180, Synechocystis sp., deduced from nucleotide se-
quence [12], 4×48347, Anabaena sp., calculated from nucleotide se-
quence [27], 4×50000, Solanum tuberosum, SDS-PAGE [3], Anabaena sp.,
SDS-PAGE [11], E. coli AC70R1, SDS-PAGE [15], E. coli B, SDS-PAGE [32],
Rhodospirillum rubrum, SDS-PAGE [20], 4×51000, E. coli, SDS-PAGE [16],
Rhodospirillum tenue, SDS-PAGE [21], 4×53000, Synechocystis sp.,
SDS-PAGE [11], 4×54000, Zea mays, SDS-PAGE [8], 4×96000, Zea mays,
SDS-PAGE [9]) [3, 6, 8–12, 15, 16, 19–22, 24, 27, 32]
More (plant enzymes exist as heterotetramers composed of two large and
two small subunits [29], E. coli SG5: oligomer formation with several times
the tetramer MW in the presence of 1 mM fructose diphosphate, not E. coli
B wild-type [32]) [29, 32]

Glycoprotein/Lipoprotein
–

4 ISOLATION/PREPARATION

Source organism

Arabidopsis thaliana (wild-type and starch-deficient mutant) [24]; Avocado [32]; Carrot [32]; Hordeum vulgare (barley, var. disticum cv. Bomi [1]) [1, 28, 29, 32]; Kidney bean [32]; Lettuce [32]; Mung bean [32]; Oryza sativa (rice, cv. Biggs M201 [2]) [2, 29, 32]; Peanut [32]; Pisum sativum (pea) [32]; Solanum tuberosum (potato, var. Norchip [3, 4]) [3, 4, 29, 32]; Sorghum [32]; Spinacia oleracea (spinach) [5, 6, 19, 21, 29, 32]; Sugar beet [32]; Tobacco [32]; Tomato [32]; Wheat [32]; Zea mays (sweet corn, var. Golden Beauty [7], starchy maize [8, 9], var. TX-40, TX-601 [9]) [7–9, 26, 29, 32]; Chlamydomonas reinhardtii (strain 137C [10], green alga) [10, 32]; Chlorella pyrenoidosa (green alga) [32]; Chlorella vulgaris (green alga) [32]; Scenedesmus obliquus (green alga) [32]; Anabaena sp. (PCC 7120, Cyanobacterium) [11, 27]; Aphanocapsa sp. (6308, Cyanobacterium) [32]; Synechocystis sp. (PCC 6803, Cyanobacterium) [11, 12]; Synechococcus sp. (6301, formerly Anacystis nidulans, cyanobacterium) [25, 32]; Aeromonas formicans [32]; Aeromonas hydrophila [13, 32]; Aeromonas liquefaciens [32]; Agrobacterium tumefaciens [32]; Arthrobacter viscosus [32]; Chlorobium limicola f. thiosulfatophilum [32]; Chromatium vinosum [32]; Citrobacter freundii [30, 32]; Clostridium pasteurianum [32]; Enterobacter aerogenes [30, 32]; Enterobacter cloacae [30, 32]; Enterobacter hafniae [13]; Enterobacter tarda [13]; Escherichia aurescens [30, 32]; E. coli (strains K-12 [32], B [14, 15, 32] and mutant strains AC70R1 [15, 16, 32], SG14 [32], SG5 and CL1136 [14, 32]) [14–17, 19, 21, 29, 32]; Klebsiella pneumoniae [13]; Micrococcus luteus [32]; Mycobacterium smegmatis [18, 32]; Rhodomicrobium vanniellii [19, 21, 32]; Rhodopseudomonas sphaeroides (strains 3244 and JT [19, 21]) [19, 21, 32]; Rhodopseudomonas capsulata [19, 21, 32]; Rhodopseudomonas acidophila [19, 21]; Rhodopseudomonas gelatinosa [19, 21, 32]; Rhodopseudomonas globiformis [19, 21]; Rhodopseudomonas palustris [19, 21, 32]; Rhodopseudomonas viridis [19, 21]; Rhodospirillum rubrum (ICPB 2204 [20]) [19–21, 32]; Rhodospirillum tenue [19, 21, 32]; Rhodospirillum fulvum [19, 21]; Rhodospirillum molischianum [19, 21, 32]; Salmonella enteritidis [13]; Salmonella typhimurium (strains LT-2 [31] and LT-2 mutant JP102 [22]) [13, 22, 31, 32]; Serratia marcescens (ATCC 274, ATCC 15365 and wild-type strain HY [23]) [13, 23, 32]; Serratia liquefaciens [13, 23, 32]; Shigella dysenteriae [13, 32]; More (not in Proteus vulgaris or Erwinia carotovora [30], distribution in genus Serratia [23], immunological relationship between enzymes from Rhodopseudomonas sphaeroides [19] or Rhodospirillum tenue [21] and other Rhodospirillaceae [19, 21]) [19, 21, 23, 30]

Source tissue

Endosperm (22 days old [26], wheat, rice, barley [29], maize [29, 32]) [1, 2, 8, 9, 26, 28, 29, 32]; Embryo (maize) [32]; Germ (wheat) [32]; Leaf (barley, lettuce, kidney bean, peanut, rice, sorghum, sugar beet, tobacco, tomato [32], wheat, spinach [29, 32]) [5, 6, 24, 28, 29, 32]; Mesocarp (avocado) [32]; Seedlings (mung bean) [32]; Seeds (etiolated peas [32]) [7, 32]; Tuber (growing [3, 4], potato [3, 4, 29, 32]) [3, 4, 29, 32]; Cell [10–23, 25, 27, 30–32]

Localization in source

Amyloplast (potato leaf [29]) [1, 2, 29]; Soluble [5, 6]

Purification

Solanum tuberosum (partial) [3, 4]; Spinacia oleracea [5, 6, 29, 32]; Zea mays (partial [7–9, 26, 32], endosperm [8, 32], whole kernels [8], ATPglucose pyrophosphorylase A [9]) [7–9, 26, 32]; Chlamydomonas reinhardtii [10]; Anabaena sp. (to near homogeneity) [11]; Synechocystis sp. (to near homogeneity) [11]; Aeromonas hydrophila (partial) [13, 32]; Enterobacter hafniae (partial) [13]; E. coli (mutant AC70R1, affinity chromatography) [15, 16, 32]; Mycobacterium smegmatis (partial) [18]; Rhodopseudomonas sphaeroides (partial [32], to near homogeneity [19]) [19, 32]; Rhodospirillum molischianum (partial) [32]; Rhodospirillum rubrum (affinity chromatography) [20, 32]; Rhodospirillum tenue (partial [32]) [21, 32]; Salmonella typhimurium [22]; Serratia marcescens (partial [23, 32], strains ATCC 274 and ATCC 15365 [23]) [23, 32]; Arabidopsis thaliana (partial) [24]; Synechococcus sp. (partial) [25]

Crystallization

(E. coli) [17]

Cloned

(Hordeum vulgare, large subunit [1, 28], Oryza sativa [2], Anabaena sp., cloned and expressed in E. coli B mutant strain AC70R1–504 [27]) [1, 2, 27, 28]; More (survey of plant species and tissues from which cDNA sequences encoding the small and large subunit of ADPG pyrophosphorylase are available) [29]

Renatured

–

5 STABILITY

pH

Temperature (°C)
More (heat stable in the presence of phosphate [13], low heat stability [10])
[10, 13]; 25 (4 days, Rhodospirillum tenue, stable in the presence of 30 mM
potassium phosphate, pH 7) [21]; 37 (4 days, Rhodospirillum tenue, stable
in the presence of 30 mM potassium phosphate, pH 7) [21]; 60 (5 min, inactivation) [10]

Oxidation

Organic solvent

General stability information
Freeze-thawing reduces activity by 70%, high ionic strength, ATP and Mg^{2+}
stabilize [7]; Polyvinylpyrrolidone, i.e. PVP fractionation stabilizes during purification, 20% sucrose and 30 mM phosphate stabilize during purification
[6]; PMSF and chymostatin prevent proteolysis during purification, not
p-aminobenzamidine, N^{alpha}-p-tosyl-L-lysine chloromethylketone, NEM,
PCMB, benzamidine, leupeptin, pepstatin or EDTA [8]; Irreversibly insoluble
when dialyzed against deionized water [15]; Phosphate stabilizes [21, 22];
No reducing reagent required during purification of Rhodospirillum tenue
enzyme [21]; Glycerol, 20% stabilizes [22]

Storage
−70°C, in 50 mM Tris-HCl buffer, pH 7.2, 10 mM potassium phosphate, 0.5
mM dithioerythritol, 20% glycerol, stable [22]; −70°C, 50 mM Tris-HCl, pH
7.2, 2.5 mM dithioerythritol, at least 1 year [15]; −20°C, in 60% $(NH_4)_2SO_4$,
15 weeks [7]; Frozen or 4°C, most stable in buffer containing 20% sucrose
and 30 mM phosphate [6]; Frozen, partially purified preparation, up to 2
weeks [18]; 0°C, partially purified, at least several weeks [18]; 0°C, Rhodospirillum tenue, in 30 mM potassium phosphate buffer, pH 7, 4 days [21];
4°C, in 50 mM HEPES buffer, pH 7.5, 20% sucrose, 1 mM EDTA, 2 mM GSH,
at least 6 weeks [3]; 4°C, Rhodospirillum tenue, in 50 mM HEPES buffer, pH
7, 20% sucrose, 0.2 mM dithioerythritol, 1 mM EDTA, 2 weeks, about 60%
loss of activity within 1 month [6]; 4°C, 30–50% loss of activity overnight [21]

---

6 CROSSREFERENCES TO STRUCTURE DATABANKS

PIR/MIPS code
PIR2:S05079 ((clone AGA.1) wheat (fragment)); PIR2:S05078 ((clone AGA.3)
wheat (fragment)); PIR2:S24991 (Anabaena sp. (PCC 7120)); PIR2:C56639
(Bacillus caldolyticus (fragment)); PIR2:S40049 (Bacillus subtilis);
PIR2:S24984 (barley); PIR1:YUEC (Escherichia coli); PIR2:I41233 (Escheri-
chia coli (fragment)); PIR2:S41292 (fava bean); PIR2:S41293 (fava bean);

PIR2:S49439 (maize); PIR2:JQ1005 (maize (fragment)); PIR2:S38872 (potato); PIR2:S37147 (potato); PIR2:S13380 (potato (fragment)); PIR2:JU0444 (rice); PIR2:S39504 (wheat); PIR2:S53991 (isoform 2 potato); PIR2:S22525 (large chain barley (fragment)); PIR2:S22526 (large chain barley (fragment)); PIR2:S18237 (large chain potato (fragment)); PIR3:S60572 (large chain wheat); PIR2:S42548 (large chain 1 Arabidopsis thaliana (fragment)); PIR2:S42547 (large chain 2 Arabidopsis thaliana (fragment)); PIR2:S42545 (large chain 3 Arabidopsis thaliana (fragment)); PIR2:S51944 (large chain S1 precursor beet); PIR2:S05077 (precursor (clone AGA.7) wheat (fragment)); PIR2:A34318 (precursor rice); PIR2:S42546 (small chain Arabidopsis thaliana (fragment)); PIR2:S22524 (small chain barley (fragment)); PIR2:A55317 (small chain potato); PIR3:S61478 (small chain A barley); PIR3:S61479 (small chain B barley); PIR3:S61481 (small chain B barley); PIR2:S51943 (small chain B1 precursor beet (fragment)); PIR3:S61480 (chain A barley)

Brookhaven code

---

7 LITERATURE REFERENCES

[1] Villand, P., Olsen, O.-A., Kilian, A., Kleczkowski, L.A.: Plant Physiol.,100,1617–1618 (1992)
[2] Anderson, J.M., Hnilo, J., Larson, R., Okita, T.W., Morell, M., Preiss, J.: J. Biol. Chem., 264,12238–12242 (1989)
[3] Sowokinos, J.R., Preiss, J.: Plant Physiol.,69,1459–1466 (1982)
[4] Sowokinos, J.R.: Plant Physiol.,68,924–929 (1981)
[5] Ghosh, H.P., Preiss, J.: J. Biol. Chem.,241,4491–4504 (1966)
[6] Copeland, L., Preiss, J.: Plant Physiol.,68,996–1001 (1981)
[7] Amir, J., Cherry, J.H.: Plant Physiol.,49,893–897 (1972)
[8] Plaxton, W.C., Preiss, J.: Plant Physiol.,83,105–112 (1987)
[9] Fuchs, R.L., Smith, J.D.: Biochim. Biophys. Acta,566,40–48 (1979)
[10] Iglesias, A.A., Charng, Y.-Y., Ball, S., Preiss, J.: Plant Physiol.,104,1287–1294 (1994)
[11] Iglesias, A.A., Kakefuda, G., Preiss, J.: Plant Physiol.,97,1187–1195 (1991)
[12] Kakefuda, G., Charng, Y.-Y., Iglesias, A.A., McIntosh, L., Preiss, J.: Plant Physiol., 99,359–361 (1992)
[13] Yung, S.-G., Paule, M., Beggs, R., Greenberg, E., Preiss, J.: Arch. Microbiol.,138, 1–8 (1984)
[14] Dietzler, D.N., Porter, S.E., Roth, W.G., Leckie, M.P.: Biochem. Biophys. Res. Commun.,122,289–296 (1984)
[15] Haugen, T.H., Ishaque, A., Preiss, J.: J. Biol. Chem.,251,7880–7885 (1976)
[16] Haugen, T.H., Ishaque, A., Chatterjee, A.K., Preiss, J.: FEBS Lett.,42,205–208 (1974)
[17] Mulichak, A.M., Skrzypczak-Jankun, E., Rydel, T.J., Tulinsky, A., Preiss, J.: J. Biol. Chem.,263,17237–17238 (1988)
[18] Lapp, D., Elbein, A.D.: J. Bacteriol.,112,327–336 (1972)
[19] Yung, S.-G., Preiss, J.: J. Bacteriol.,151,742–749 (1982)
[20] Preiss, J., Huebner, J., Greenberg, E.: Curr. Microbiol.,7,257–262 (1982)

---

[21] Yung, S.-G., Preiss, J.: J. Bacteriol.,147,101–109 (1981)
[22] Lehmann, M., Preiss, J.: J. Bacteriol.,143,120–127 (1980)
[23] Preiss, J., Crawford, K., Downey, J., Lammel, C., Greenberg, E.: J. Bacteriol.,127, 193–203 (1976)
[24] Li, L., Preiss, J.: Carbohydr. Res.,227,227–239 (1992)
[25] Levi, C., Preiss, J.: Plant Physiol.,58,753–756 (1976)
[26] Hannah, L.C., Nelson, O.E.: Plant Physiol.,55,297–302 (1975)
[27] Charng, Y.Y., Kakefuda, G., Iglesias, A.A., Buikema, W.J., Preiss, J.: Plant Mol. Biol., 20,37–47 (1992)
[28] Villand, P., Aalen, R., Olsen O.-A., Lüthi, E., Lönneborg, A., Kleczkowski, L.A.: Plant Mol. Biol.,19,381–389 (1992)
[29] Kleczkowski, L.A., Villand, P., Lönneborg, A., Olsen O.-A., Lüthi, E.: Z. Naturforsch., 46c,605–612 (1991) (Review)
[30] Ribéreau-Gayon, G., Sabraw, A., Lammel, C., Preiss, J.: Arch. Biochem. Biophys., 142,675–692 (1971)
[31] Lehmann, M., Preiss, J.: J. Bacteriol.,143,120–127 (1980)
[32] Preiss, J.: Adv. Enzymol. Relat. Areas Mol. Biol.,46,317–381 (1978) (Review)

1 NOMENCLATURE

EC number
2.7.7.28

Systematic name
NTP:hexose-1-phosphate nucleotidyltransferase

Recommended name
Nucleoside-triphosphate-hexose-1-phosphate nucleotidyltransferase

Synonyms
Nucleotidyltransferase, hexose 1-phosphate
Hexose 1-phosphate nucleotidyltransferase
Hexose nucleotidylating enzyme
NDP hexose pyrophosphorylase
Nucleoside diphosphohexose pyrophosphorylase
More (presumably identical with EC 2.7.7.29 [1])

CAS Reg. No.
37278-26-5

2 REACTION AND SPECIFICITY

Catalyzed reaction
NTP + hexose 1-phosphate →
→ diphosphate + NDPhexose

Reaction type
Nucleotidyl group transfer

Natural substrates

Substrate spectrum
1 GDPglucose + diphosphate (r, best substrate) [1]
2 GDPmannose + diphosphate (r, reaction at about half the rate of GDPglucose) [1]
3 IDPmannose + diphosphate (r, reaction at 21% the rate of GDPglucose) [1]
4 IDPglucose + diphosphate (r, reaction at 7% the rate of GDPglucose) [1]

Product spectrum
1 GDP + glucose 1-phosphate [1]
2 GDP + mannose 1-phosphate [1]
3 IDP + mannose 1-phosphate [1]
4 IDP + glucose 1-phosphate [1]

Inhibitor(s)

Mannose 1-phosphate (strong) [1]; GDPmannose (strong) [1]; More (no inhibition by glucose 6-phosphate, mannose 6-phosphate, fructose 6-phosphate, ribose 1-phosphate, ribose 5-phosphate, lactose 1-phosphate, galactose 1-phosphate, galactose 6-phosphate, fructose 1-phosphate, glucose 1-phosphate (the latter with GTP and mannose 1-phosphate as substrate)) [1]

Cofactor(s)/prosthetic group(s)/activating agents

Metal compounds/salts

Mg^{2+} (requirement, 5 mM, twice the diphosphate-concentration) [1]

Turnover number (min^{-1})

Specific activity (U/mg)

K_m-value (mM)

pH-optimum

7.5–8 [1]

pH-range

6.5–9 (about 40% of maximal activity at pH 6.5 and about half-maximal activity at pH 9) [1]

Temperature optimum (°C)

25 (assay at) [1]

Temperature range (°C)

3 ENZYME STRUCTURE

Molecular weight

Subunits

Glycoprotein/Lipoprotein

–

4 ISOLATION/PREPARATION

Source organism

Rat [1]; Pig [1]; Sheep (ewe) [1]; Bovine (calf) [1]; Rabbit [1]

Source tissue

Mammary gland (lactating) [1]; Liver (rat, calf) [1]; Muscle (rat) [1]; Kidney (rat) [1]; Brain (rat) [1]

Localization in source
  Cytoplasm [1]

Purification
  Bovine (calf) [1]

Crystallization
  –

Cloned
  –

Renatured
  –

---

5 STABILITY

pH
  6–6.5 ($t_{1/2}$: 60 days at 4°C) [1]

Temperature (°C)
  50 (after 5 or 10 min: 15% or 33% loss of activity, respectively, bovine) [1];
  55 (1 min, 57% loss of activity, bovine) [1]

Oxidation

Organic solvent

General stability information

Storage
  4°C, at pH 6–6.5, $t_{1/2}$: 60 days [1]

---

6 CROSSREFERENCES TO STRUCTURE DATABANKS

PIR/MIPS code

Brookhaven code

---

7 LITERATURE REFERENCES

[1] Verachtert, H., Rodriguez, P., Bass, S.T., Hansen, R.G.: J. Biol. Chem.,241,2007–
    2013 (1966)

1 NOMENCLATURE

EC number
 2.7.7.29

Systematic name
 GTP:alpha-D-hexose-1-phosphate guanylyltransferase

Recommended name
 Hexose-1-phosphate guanylyltransferase

Synonyms
 GDPhexose pyrophosphorylase
 Guanylyltransferase, hexose 1-phosphate
 GDP hexose pyrophosphorylase
 Guanosine diphosphohexose pyrophosphorylase
 Hexose 1-phosphate guanylyltransferase
 More (may be identical with EC 2.7.7.28)

CAS Reg. No.
 9033-17-4

2 REACTION AND SPECIFICITY

Catalyzed reaction
 GTP + alpha-D-hexose 1-phosphate →
 → diphosphate + GDPhexose

Reaction type
 Nucleotidyl group transfer

Natural substrates

Substrate spectrum
 1 GDPglucose + diphosphate (r, highest activity) [1]
 2 GDPmannose + diphosphate (r) [1]
 3 IDPmannose + diphosphate [1]
 4 IDPglucose + diphosphate [1]
 5 ADPmannose + diphosphate [1]
 6 UDPglucose + diphosphate [1]
 7 UDPmannose + diphosphate [1]

Product spectrum
1 GTP + alpha-D-glucose-1-phosphate
2 GTP + mannose 1-phosphate
3 ITP + mannose 1-phosphate
4 ITP + glucose 1-phosphate
5 ATP + mannose 1-phosphate
6 UTP + glucose 1-phosphate
7 UTP + mannose 1-phosphate

Inhibitor(s)
Mannose 1-phosphate (when GDPglucose is substrate) [1]; GDPmannose (when GDPglucose is substrate) [1]; Zn^{2+} (when added with Mg^{2+} or Mn^{2+} or Co^{2+}) [1]

Cofactor(s)/prosthetic group(s)/activating agents

Metal compounds/salts
Mg^{2+} (required for optimum activity) [1]; Co^{2+} (can substitute for Mg^{2+}) [1]; Mn^{2+} (can partially substitute for Mg^{2+}) [1]

Turnover number (min^{-1})

Specific activity (U/mg)
0.38 (GDPglucose) [1]

K_m-value (mM)
0.0000001 (GDPmannose) [1]; 0.00062 (GDPglucose) [1]; 0.00082 (diphosphate) [1]; 0.0026 (glucose 1-phosphate, GTP) [1]

pH-optimum
7.5–8.0 [1]

pH-range

Temperature optimum (°C)
25 (assay at) [1]

Temperature range (°C)

3 ENZYME STRUCTURE

Molecular weight

Subunits

Glycoprotein/Lipoprotein
–

4 ISOLATION/PREPARATION

Source organism
  Bovine (calf) [1]

Source tissue
  Liver [1]

Localization in source

Purification
  Bovine (calf, partial) [1]

Crystallization
  –

Cloned
  –

Renatured
  –

---

5 STABILITY

pH

Temperature (°C)

Oxidation

Organic solvent

General stability information

Storage

---

6 CROSSREFERENCES TO STRUCTURE DATABANKS

PIR/MIPS code

Brookhaven code

---

7 LITERATURE REFERENCES

[1] Hansen, R.G., Verachtert, H., Rodriguez, P., Bass, S.T.: Methods Enzymol.,8,269–271 (1966)

1 NOMENCLATURE

EC number
2.7.7.30

Systematic name
GTP:L-fucose-1-phosphate guanylyltransferase

Recommended name
Fucose-1-phosphate guanylyltransferase

Synonyms
GDPfucose pyrophosphorylase
Guanosine diphosphate L-fucose pyrophosphorylase [1]
GDP-L-fucose pyrophosphorylase [1]
GDP-fucose pyrophosphorylase [3]

CAS Reg. No.
9033-14-1

---

2 REACTION AND SPECIFICITY

Catalyzed reaction
GTP + L-fucose 1-phosphate $\rightarrow$
$\rightarrow$ diphosphate + GDP-L-fucose

Reaction type
Nucleotidyl group transfer

Natural substrates
GTP + L-fucose 1-phosphate (pathway of fucose production, which is an important component of complex heterooligosaccharides) [3]

Substrate spectrum
1 GTP + L-fucose 1-phosphate (r [1]) [1–3]
2 Diphosphate + GDP-L-fucose (r) [1]
3 Diphosphate + alpha-D-mannose 1-phosphate (17.5% of the activity with GDP-L-fucose) [1]
4 UTP + L-fucose 1-phosphate (18.7% of the activity with GTP) [1]

Product spectrum
1 Diphosphate + GDP-L-fucose [3]
2 GTP + L-fucose 1-phosphate [1]
3 ?
4 ?

Inhibitor(s)

Cofactor(s)/prosthetic group(s)/activating agents

Metal compounds/salts
   Mg^{2+} (divalent cation required, Mg^{2+} most effective) [1]; Co^{2+} (can partially replace Mg^{2+} in activation) [1]

Turnover number (min^{-1})

Specific activity (U/mg)
   More [1]

K_m-value (mM)
   0.12 (GDP-L-fucose) [1]

pH-optimum
   7.8 [1]

pH-range
   6–9.5 (6: about 85% of activity maximum, 9.5: about 70% of activity maximum) [1]

Temperature optimum (°C)

Temperature range (°C)

3 ENZYME STRUCTURE

Molecular weight

Subunits

Glycoprotein/Lipoprotein
   –

4 ISOLATION/PREPARATION

Source organism
   Pig [1, 3]; Mouse [2]

Source tissue
   D51 clonal derivative of mouse L 929 cells [2]; Liver [1]; Submaxillary glands [3]

Localization in source
   Cytoplasm [2]

Purification
   Pig (partial) [1]

Crystallization

–

Cloned

–

Renatured

–

---

5 STABILITY

pH

Temperature (°C)

Oxidation

Organic solvent

General stability information

Storage
Unstable to storage at either 16°C or 0°C [1]

---

6 CROSSREFERENCES TO STRUCTURE DATABANKS

PIR/MIPS code

Brookhaven code

---

7 LITERATURE REFERENCES

[1] Ishihara, H., Heath, E.C.: J. Biol. Chem.,243,1110–1115 (1968)
[2] Coates, S.W., Gurney, T., Sommers, L.W., Yeh, M., Hirschberg, C.B.: J. Biol. Chem., 255,9225–9229 (1980)
[3] Stiller, R., Thiem, J.: Liebigs Ann. Chem.,461–466 (1992)

1 NOMENCLATURE

EC number
2.7.7.31

Systematic name
Nucleoside-triphosphate:DNA deoxynucleotidylexotransferase

Recommended name
DNA nucleotidylexotransferase

Synonyms
Terminal deoxyribonucleotidyltransferase
Terminal addition enzyme
Nucleotidyltransferase, terminal deoxyribo-
Addase
Deoxynucleotidyl terminal transferase
Deoxyribonucleic acid nucleotidyltransferase
Deoxyribonucleic nucleotidyltransferase
Terminal deoxynucleotide transferase
Terminal deoxynucleotidyltransferase
TdT [4, 5]

CAS Reg. No.
9027-67-2

2 REACTION AND SPECIFICITY

Catalyzed reaction
Deoxynucleoside triphosphate + $DNA_n \rightarrow$
$\rightarrow$ diphosphate + DNA_{n+1} (rapid equilibrium random mechanism [6, 30])

Reaction type
Nucleotidyl group transfer

Natural substrates
Deoxynucleoside triphosphate + DNA (hypothesis: terminal transferase acts
as a somatic mutator, diversifying the amino acid sequence in the variable
region of immunoglobulin molecules by changing one of the nucleotides
[10], enzyme may act as a random-sequence generator in highly localized
regions of the DNA [12], enzyme may be an intracellular marker for a sub-
population of cells [12]) [10, 12]

Substrate spectrum

1 Deoxynucleoside triphosphate + DNA$_n$ (ir [2, 10, 11], enzyme also catalyzes diphosphate exchange and pyrophosphorolysis [26], reaction is not template directed [1–31], primer required [1–31]: prefers non-denatured to single-stranded DNA as primer [16], heat-denatured DNA [2], DNase I-treated DNA [2], DNase II-treated DNA [2], oligodeoxynucleotides [16], polydeoxynucleotides [16], oligonucleotides containing at least 3 phosphate groups and a free 3'-hydroxyl end [10], preferred length 12–18 residues [18], as the initiator chain length is increased to 5–7 nucleotides the polymerization rate also increases [10], longer chains interact well with the enzyme, but synthesis rate decreases [10], the preferred combination is oligo(dT) initiator with dTTP substrate [18], low levels of activity with dGTP and oligo(dA), oligo(dC) or oligo(dT) [18], relative rates of polymerization with several nucleotides [11], deoxynucleoside triphosphate preference: dGTP, dCTP, dTTP, dATP (42000 MW enzyme) [13], polymerization is highly efficient if only one of the four deoxyribonucleoside 5'-triphosphates is present [16], no nucleotide specificity [2]) [1–31]
2 Ribonucleoside triphosphate + DNA$_n$ [1]

Product spectrum

1 Diphosphate + DNA$_{n+1}$ (extension of the 3'-hydroxy terminus for approximately 40 nucleotides is achieved if non-denatured DNA and dTTP are used as primer and substrate [16]) [1–31]
2 ?

Inhibitor(s)

(E)-5-(2-Bromovinyl)-2'-deoxyuridine 5'-triphosphate [20]; Diphosphate [2]; 3'-Deoxy-3'-fluorothymidine 5'-triphosphate [5]; 2',3'-Dideoxythymidine 5'-triphosphate [5]; Oligo(dA) (product inhibition) [6]; NH$_4^+$ [9]; Cl$^-$ [9]; I$^-$ [9]; Phosphate [9]; High ionic strength [11]; Metal chelators [11]; EDTA [11]; 1,10 Phenanthroline (not m-isomer) [11]; Cysteine [11]; Tris (above 50 mM) [17]; Triethylenetetramine [11]; 2,2'-Bipyridyl [11]; Histidine [11]; Mercaptoacetate [11]; NEM [13]; ATP (selective inhibitor, inactive in DNA-synthesizing systems both with DNA polymerase alpha or beta [14]) [13, 14]; Diadenosine 5',5'-diphosphate [24]; Diadenosine 5',5'-triphosphate [24]; Diadenosine 5'-5'-tetraphosphate [21, 24]; Diadenosine 5',5'-pentaphosphate [24]; Diadenosine 5',5'-hexaphosphate [24]; 9-beta-Arabinofuranosyladenine 5'-triphosphate [14]; 9-alpha-Arabinofuranosyladenine 5'-triphosphate [14]; 3'-dATP (selective inhibitor, inactive in DNA-synthesizing systems both with DNA polymerase alpha or beta) [14]; 2',3'-Dideoxyadenosine (inhibition is stronger with (dA)$_{12-18}$ as primer than with heat denatured DNA) [23]; 2',3'-Dideoxycytidine (inhibition is stronger with (dA)$_{12-18}$ as primer than with heat denatured DNA) [23]; 2',3'-Dideoxyguanosine (inhibition is stronger with (dA)$_{12-18}$ as primer than with heat denatured DNA) [23]; 2',3'-Dideoxythymidine (inhibition is stronger with (dA)$_{12-18}$ as primer than with heat denatured DNA) [23]; dATP (mechanism of Mn^{2+}-dependent inhibition) [27];

Pyrans [15]; Streptolydigin (does not significantly inhibit DNA polymerase alpha, beta, and gamma or RNA polymerase, non-competitive to $(dA)_{12-18}$ and dGTP) [19]

Cofactor(s)/prosthetic group(s)/activating agents
EDTA (stimulates) [1]; Sulfhydryl compound (required) [1]

Metal compounds/salts
Mg^{2+} (required [1], activates [10], requires both Mg^{2+} and Mn^{2+} [16], divalent metal required [4, 18, 30, 31], Mg^{2+} or Co^{2+} [11], Mn^{2+} or Mg^{2+} required [4, 18], order of specificity: $Mg^{2+} > Zn^{2+} > Co^{2+} > Mn^{2+}$ [30], order of efficiency for elongation of oligonucleotide primers with dATP: $Mg^{2+} > Zn^{2+} > Co^{2+}$ [9], dGTP also optimally added in the presence of Mg^{2+} [9], polymerization of pyrimidines is best with Co^{2+} [9], human enzyme catalyzes the polymerization reaction as well or better in the presence of Mn^{2+} or Co^{2+} than in presence of Mg^{2+}, calf thymus enzyme prefers Mg^{2+} [17], Mg^{2+} and Co^{2+} are equally effective in phosphorolysis [26], Mg^{2+} is more effective than Mn^{2+} [31]) [1, 4, 9–11, 16–18, 26, 30, 31]; Mn^{2+} (divalent metal required [30], order of specificity: $Mg^{2+} > Zn^{2+} > Co^{2+} > Mn^{2+}$ [30], Mg^{2+} or Mn^{2+} required [4, 18, 31], Mg^{2+} is more effective than Mn^{2+} [31], human enzyme from leukemic cells shows maximum activity with Mn^{2+} as divalent cation [10], requires both Mn^{2+} and Mg^{2+} [16], human enzyme catalyzes the polymerization reaction as well or better in the presence of Mn^{2+} or Co^{2+} than in presence of Mg^{2+} [17]) [4, 10, 16–18, 30, 31]; Co^{2+} (divalent metal required [30], order of specificity: $Mg^{2+} > Zn^{2+} > Co^{2+} > Mn^{2+}$ [30], order of efficiency for elongation of oligonucleotide primers with dATP: $Mg^{2+} > Zn^{2+} > Co^{2+}$ [9], Co^{2+} is the best activator in elongation of chains with dCTP or dTTP [10], human enzyme catalyzes the polymerization reaction as well or better in the presence of Mn^{2+} or Co^{2+} than in presence of Mg^{2+} [17], Mg^{2+} and Co^{2+} are equally effective in phosphorolysis [26]) [9, 10, 17, 26, 30]; Zn^{2+} (divalent metal required [30], order of specificity: $Mg^{2+} > Zn^{2+} > Co^{2+} > Mn^{2+}$ [30], order of efficiency for elongation of oligonucleotide primers with dATP: $Mg^{2+} > Zn^{2+} > Co^{2+}$ [9]) [9, 30]

Turnover number (min⁻¹)
Turnover number (min^{-1})
50 (dATP) [10, 11]

Specific activity (U/mg)
More [13, 16, 31]

K_m-value (mM)
More (overview, K_m-values for dATP polymerization with various primers [11], effect of metals on K_m-values [17]) [4, 11, 12, 17, 30, 31]; 0.0003 (poly$(dA)_{50}$, Mn^{2+}-activated enzyme) [17]; 0.001 (oligonucleotide primers) [9]; 0.0025 (poly$(dA)_{50}$, Mg^{2+}-activated) [17]; 0.01 (dATP, dGTP) [9]; 0.02 (oligo$(dA)_{12-18}$) [13]; 0.1 (dGTP) [17]; 0.5 (dTTP, dCTP) [9]; 1 (homopolymer primers) [9]

pH-optimum
6.9 [18]; 7.2 (assay at [3]) [3, 10]; 7.5–8.5 [17]

pH-range

Temperature optimum (°C)
    35 (assay at) [3, 11]; 37 (assay at) [13]

Temperature range (°C)

---

3 ENZYME STRUCTURE

Molecular weight
    32360 (bovine, equilibrium sedimentation) [3]
    40000–45000 (chicken) [7]
    42000–60000 (bovine, gel filtration, the 2 subunits of the calf thymus en-
    zyme reported earlier may be proteolytic products derived from a single po-
    lypeptide of MW 60000, which may be the native form) [28]
    45000 (2 MW forms: 45000 and 57000, pig, gel filtration) [13]
    57000 (2 MW forms: 45000 and 57000, pig, gel filtration) [13]
    60000 (various organisms [9], mouse [10, 28], rat [28]) [9, 10, 28]
    62000 (human, gel filtration) [31]
    67000 (rat, mouse, gel filtration) [28]
    79000 (bovine) [10]
    500000 (wheat) [16]
    More (2 high MW forms: 58000 and 45000 and one two subunit form of
    44000 MW [4]) [4, 12]

Subunits
    Monomer (1 × 40000–45000, chicken [7], 1 × 60000, various organism [9],
    1 × 62000, human, SDS-PAGE [31], 1 × 60000, mouse [10], 1 × 79000, bo-
    vine [10], 1 × 42000, low-MW form, pig, SDS-PAGE [13], 1 × 57000, high-MW
    form, pig, SDS-PAGE [13]) [7, 9, 10, 13, 31]
    Dimer (1 × 8000 (alpha) + 1 × 26500 (beta), bovine, SDS-PAGE) [3]

Glycoprotein/Lipoprotein
    More (may contain a small amount of carbohydrate) [31]

---

4 ISOLATION/PREPARATION

Source organism
    Bovine (calf) [1–4, 8, 10–12, 14, 17, 20–24, 26–29]; Human [6, 8, 15, 17, 19,
    29, 30, 31]; Chicken [7]; Mammalia [10]; Pig [13]; Wheat [16]; Mouse mam-
    mary tumor virus [18]; Xenopus sp. [25]; Mouse [10, 28]; Rat [28]; More
    (various organisms) [9]

Source tissue
    Thymus [1–4, 7, 8, 10, 11, 13, 14, 17, 20–24, 26–29]; Bone marrow [10, 29];
    Lymphoblasts (from leukemic patients [31], multiple MW forms from leuke-
    mic cells [6], multiple forms from patients with acute lymphoblastic leukemia
    [8, 15, 17] and with chronic myelogenous leukemia [8]) [6, 8, 15, 17, 29,

31]; Lymphocytes (enzyme activity is only found in the nuclei of pre-T and pre-B lymphocytes) [9]; Germ [16]; Molt-4 cells [19]; Leukocytes [19, 29]; Leukemic cells [30]; Commercial preparation [9]

Localization in source
Nucleus [1, 2, 9]; Chromatin [4]; Viral core [18]

Purification
Bovine (calf [2–4, 22], 2 high MW forms: 58000 and 45000 and one two sub-unit form of 44000 MW [4], single step immunoaffinity purification [22]) [2–4, 10, 14, 22, 28, 29]; Human (partial) [8, 17, 31]; Pig (2 MW forms: 45000 and 57000) [13]; Wheat [16]; Mouse [28]; Rat [28]

Crystallization
–

Cloned
[9, 25]

Renatured
–

---

5 STABILITY

pH
4.5 (stable at) [10, 11]

Temperature (°C)
40 (not stable above) [10, 11]

Oxidation

Organic solvent
Not stable in organic solvents [10, 11]

General stability information
Rapid loss of activity can be eliminated by addition of albumin to the reaction mixture [2]; Not stable in urea, SDS and organic solvents [10, 11]

Storage
–20°C, 40% loss of activity after 7 months [3]; –20°C, 50 mM potassium phosphate buffer, pH 7.0, stable for 9 years [11]

---

6 CROSSREFERENCES TO STRUCTURE DATABANKS

PIR/MIPS code
PIR2:S55786 (chicken); PIR1:WXHU (human); PIR2:B23595 (mouse); PIR2:S30235 (mouse); PIR2:A23595 (long form bovine)

Brookhaven code

---

7 LITERATURE REFERENCES

[1] Krakow, J.S., Coutsogeorgopoulos, C., Canellakis, E.S.: Biochim. Biophys. Acta, 55,639–650 (1962)

[2] Gottesman, M.E., Canellakis, E.S.: J. Biol. Chem.,241,4339–4352 (1966)

[3] Chang, L.M.S., Bollum, F.J.: J. Biol. Chem.,246,909–916 (1971)

[4] Pandey, V., Modak, M.J.: Prep. Biochem.,17,359–377 (1987)

[5] Matthes, E., Lehman, C., Drescher, B., Buettner, W., Langen, P.: Biomed. Biochim. Acta,44, K63-K73 (1985)

[6] Deibel, M.R., Coleman, M.S., Hutton, J.J.: Adv. Exp. Med. Biol.,145,37–60 (1982)

[7] Penit, C., Gelabert, M.J., Transy, C., Rouget, P.: Adv. Exp. Med. Biol.,145,61–73 (1982)

[8] Deibel, M.R., Coleman, M.S., Acree, K., Hutton, J.J.: J. Clin. Invest.,67,725–734 (1981)

[9] Grosse, F., Manns, A. in "Methods in Molecular Biology" (Burrell, M.M., Ed.) 16,95–105, Humana Press Inc., Totowa, NJ (1993)

[10] Ratliff, R.L. in "The Enzymes",3rd Ed. (Boyer, P.D., Ed.) 14,105–118 (1981) (Review)

[11] Bollum, F.J. in "The Enzymes",3rd Ed. (Boyer, P.D., Ed.) 10,145–171, Academic, New York (1974) (Review)

[12] Bollum, F.J.: Adv. Enzymol. Relat. Areas Mol. Biol.,47,347–374 (1978) (Review)

[13] Kaneda, T., Kuroda, S., Koiwai, O., Yoshida, S.: J. Biochem.,90,1421–1427 (1981)

[14] Müller, W.E.G., Zahn, R.K., Arendes, J.: FEBS Lett.,94,47–50 (1978)

[15] DiCioccio, R.A., Sahai Srivastava, B.I.S.: Biochem. J.,175,519–524 (1978)

[16] Brodniewicz-Proba, T., Buchowicz, J.: Biochem. J.,191,139–145 (1980)

[17] Coleman, M.S.: Arch. Biochem. Biophys.,182,525–532 (1977)

[18] Ashley, R.L., Cardiff, R.D., Manning, J.S.: Virology,77,367–375 (1977)

[19] DiCioccio, R.A., Srivastava, B.I.S.: Biochem. Biophys. Res. Commun.,72,1343–1349 (1976)

[20] Ono, K., Nakane, H., Colla, L., De Clercq, E.: Nucleic Acids Res.,12,123–126 (1983)

[21] Ono, K., Iwata, Y., Nakamura, H., Matsukage, A.: Biochem. Biophys. Res. Commun., 95,34–40 (1980)

[22] Fuller, S.A., Philips, A., Coleman, M.S.: Biochem. J.,231,105–113 (1985)

[23] Ono, K.: Biochim. Biophys. Acta,1049,15–20 (1990)

[24] Pandey, V.N., Amrute, S.B., Satav, J.G., Modak, M.J.: FEBS Lett.,213,205–208 (1987)

[25] Lee, A., Hsu, E.: J. Immunol.,152,4500–4507 (1994)

[26] Srivastava, A., Modak, M.J.: Biochemistry,19,3270–3275 (1980)

[27] Modak, M.J.: Biochemistry,18,2679–2684 (1979)

[28] Nakamura, H., Tanabe, K., Yoshida, S., Morita, T.: J. Biol. Chem.,256,8745–8751 (1981)

[29] Okamura, S., Crane, F., Messner, H.A., Mak, T.W.: J. Biol. Chem.,253,3765–3767 (1978)

[30] Deibel, M.R., Coleman, M.S.: J. Biol. Chem.,255,4206–4212 (1980)

[31] Deibel, M.R., Coleman, M.S.: J. Biol. Chem.,254,8634–8649 (1979)

1 NOMENCLATURE

EC number
2.7.7.32

Systematic name
dTTP:alpha-D-galactose-1-phosphate thymidylyltransferase

Recommended name
Galactose-1-phosphate thymidylyltransferase

Synonyms
Thymidylyltransferase, galactose 1-phosphate
dTDPgalactose pyrophosphorylase
Galactose 1-phosphate thymidylyl transferase
Thymidine diphosphogalactose pyrophosphorylase
Thymidine triphosphate:alpha-D-galactose 1-phosphate thymidylyltrans-
ferase [1]

CAS Reg. No.
9023-25-0

---

2 REACTION AND SPECIFICITY

Catalyzed reaction
dTTP + alpha-D-galactose 1-phosphate →
→ diphosphate + dTDPgalactose

Reaction type
Nucleotidyl group transfer

Natural substrates
More (involved in biosynthesis of cell wall constituents in Streptococcus faecalis) [1]

Substrate spectrum
1 dTTP + alpha-D-galactose 1-phosphate [1]

Product spectrum
1 dTDPgalactose + ? [1]

Inhibitor(s)

Cofactor(s)/prosthetic group(s)/activating agents

Metal compounds/salts

Turnover number (min^{-1})

Specific activity (U/mg)

K$_m$-value (mM)

pH-optimum
  7 [1]

pH-range
  6.5–7.5 (70% of maximal activity at pH 6.5, 60% of maximal activity at pH
  7.5) [1]

Temperature optimum (°C)
  20 [1]

Temperature range (°C)

---

3 ENZYME STRUCTURE

Molecular weight
  80000–100000 (Streptococcus faecalis, sucrose density gradient centrifugation) [1]

Subunits

Glycoprotein/Lipoprotein
  –

---

4 ISOLATION/PREPARATION

Source organism
  Streptococcus faecalis [1]

Source tissue
  Cell [1]

Localization in source

Purification
  Streptococcus faecalis [1]

Crystallization
  –

Cloned
  –

Renatured
  –

5 STABILITY

pH
6–8 (complete inactivation outside this range) [1]

Temperature (°C)

Oxidation

Organic solvent

General stability information

Storage
3°C, 2 days, 50% remaining activity [1]; 3°C, 7 days, 20% remaining activity [1]; 3°C, 16 days, 5% remaining activity [1]

6 CROSSREFERENCES TO STRUCTURE DATABANKS

PIR/MIPS code

Brookhaven code

7 LITERATURE REFERENCES

[1] Pazur, J.H., Anderson, J.S.: J. Biol. Chem.,238,3155–3160 (1963)

1 NOMENCLATURE

EC number

2.7.7.33

Systematic name

CTP:D-glucose-1-phosphate cytidylyltransferase

Recommended name

Glucose-1-phosphate cytidylyltransferase

Synonyms

Cytidylyltransferase, glucose 1-phosphate

CDPglucose pyrophosphorylase

CDP-glucose pyrophosphorylase

Cytidine diphosphoglucose pyrophosphorylase

Cytidine diphosphate glucose pyrophosphorylase [2]

Cytidine diphosphate-D-glucose pyrophosphorylase [3]

CAS Reg. No.

9027-10-5

---

2 REACTION AND SPECIFICITY

Catalyzed reaction

CTP + D-glucose 1-phosphate →

→ diphosphate + CDPglucose

Reaction type

Nucleotidyl group transfer

Natural substrates

CTP + D-glucose 1-phosphate (branch-point in glucose 1-phosphate ana-
bolism [1], first of five enzymes committed to CDP-D-abequose biosynthesis
[4]) [1, 4]

Substrate spectrum

1 CTP + D-glucose 1-phosphate (r (equilibrium constants [1, 2]) [1–4], catalyses bimolecular group transfer reaction [4], specific for base and sugar [1], highly specific [2], best substrates [4]. No substrates of the reverse reaction: ADPglucose, dTDPglucose, UDPglucose [1, 2, 4], GDPglucose [1, 2], UDPgalactose, UDP-D-mannose, UDP-N-acetyl-D-glucosamine, dTDP-L-rhamnose [4], or CDPparatose [1], no substrates of the forward reaction: ATP, GTP [2–4], dATP, dGTP, dCTP, D-ribitol 5-phosphate, L-glycerol 3-phosphate [2], TTP [3], dTTP, alpha-D-galactose 1-phosphate [2, 4], alpha-D-mannose 1-phosphate, N-acetyl-alpha-D-glucosamine 1-phosphate [1, 2, 4], D-ribose 1-phosphate [4], or GTP and D-mannose 1-phosphate [1]) [1–4]

2 CTP + glucosamine 1-phosphate (reaction at about 15% [4] or 25% [2] the rate with glucose 1-phosphate, not [1]) [2, 4]

3 CTP + alpha-D-xylose 1-phosphate (poor substrate) [4]

4 UTP + alpha-D-glucose 1-phosphate (poor substrate [4], not [3]) [4]

Product spectrum

1 Diphosphate + CDPglucose [1, 2]

2 Diphosphate + CDPglucosamine [4]

3 Diphosphate + CDPxylose [4]

4 Diphosphate + UDPglucose [4]

Inhibitor(s)

CDPparatose (i.e. 3,6-dideoxy-D-glucose, feed-back inhibition, CDPglucose partially reverses) [1]; CDPascarylose (i.e. 3,6-dideoxy-L-mannose, strong, kinetics) [3]; CDPabequose (i.e. 3,6-dideoxy-D-galactose, strong) [3]; CDP-D-fucose/CDP-6-deoxy-D-glucose mixture [3]; CDP-4-keto-6-deoxy-D-glucose [3]; CDPglucose (product inhibition, kinetics) [2, 4]; Diphosphate (product inhibition, kinetics) [2, 4]; dTTP (strong, kinetics) [2]; Cytidine diphosphate 2-O-methyldeoxyaldose and its carboxylic acid ester [2]; CDP-D-galactose (weak) [2]; Phosphate (weak) [2]; dATP [2]; dGTP [2]; ATP [2]; GTP [2]; UTP [2]; dTDP (weak) [2]; dADP (weak) [2]; ADPglucose (weak [2], not [1]) [2]; CDP (weak [2], not [1]) [2]; More (no inhibition by NaF, pyrophosphatase, mercaptoethanol, paratose, CTP [1], abequose [3], little or no inhibition by CMP, GDPglucose, UDPglucose, dTDPglucose [1, 2], CDPethanolamine, CDPcholine, dTTPglucose, UDP-D-glucosamine, dCTP, dCDP, dTMP, dAMP, dGMP, dCMP, dTDP-L-rhamnose, dTDP-4-amino-4,6-dideoxy-D-glucose and its N-acetyl-derivative, UDP-D-galactose, UDP-D-glucuronic acid, UDP-N-acetyl-D-glucosamine, GDP-D-mannose, D-glucose 6-phosphate, D-glucose 1,6-diphosphate, D-galactose 1-phosphate, D-glucosamine 1-phosphate, D-mannose 1-phosphate, N-acetyl-D-glucosamine 1-phosphate, D-fructose 6-phosphate, D-fructose 1,6-diphosphate, D-ribitol 5-phosphate, L-glycerol 3-phosphate, phosphoenolpyruvate, pyruvate, ADP, AMP, GMP, UMP, beta-NAD$^+$ or beta-NADP$^+$ [2]) [1–3]

Cofactor(s)/prosthetic group(s)/activating agents

More (no activation by dTDP-L-rhamnose, dTDP-4-amino-4,6-dideoxy-D-glu-
cose and its N-acetyl-derivative, UDP-D-galactose, UDP-D-glucuronic acid,
UDP-N-acetyl-D-glucosamine, GDP-D-mannose, D-glucose 6-phosphate,
D-glucose 1,6-diphosphate, D-galactose 1-phosphate, D-glucosamine
1-phosphate, D-mannose 1-phosphate, N-acetyl-D-glucosamine 1-phos-
phate, D-fructose 6-phosphate, D-fructose 1,6-diphosphate, D-ribitol 5-phos-
phate, L-glycerol 3-phosphate, phosphoenolpyruvate, pyruvate, ADP, AMP,
GMP, UMP, beta-NAD^+ or beta-$NADP^+$) [2]

Metal compounds/salts

Co^{2+} (requirement, CDPglucose synthesis and pyrophosphorolysis [3], with
decreasing order of efficiency in CDPglucose synthesis: Co^{2+}, Mn^{2+}, Mg^{2+} [3],
can replace Mg^{2+} in pyrophosphorolysis [3], not [2]) [3]; Mg^{2+} (requirement
[1–3], 3 mM $MgCl_2$ [1], maximal reaction rates at molar ratios of $MgCl_2$:diphos-
phate or $MgCl_2$:CTP of 2:1 [2], can replace Co^{2+} in pyrophosphorolysis [3],
with decreasing order of efficiency in CDPglucose synthesis: Co^{2+}, Mn^{2+},
Mg^{2+} [3]) [1–3]; Mn^{2+} (requirement [2, 3], can replace Mg^{2+} with 87% efficiency [2], with decreasing order of efficiency in CDPglucose synthesis: Co^{2+},
Mn^{2+}, Mg^{2+} [3], less effective in pyrophosphorolysis [3]) [2, 3]; More (no activation by Ni^{2+} [2, 3] or Cu^{2+} [3]) [2, 3]

Turnover number (min^{-1})

Specific activity (U/mg)

0.0375 [2]; 0.143 [3]; 0.408 [1]; 10.3 [4]

K_m-value (mM)

0.015 (CDPglucose) [1]; 0.034 (CTP) [1]; 0.09 (glucose 1-phosphate) [1];
0.1 (diphosphate) [1]; 0.11 (CTP, glucose 1-phosphate, Mg^{2+} as cation [3],
CTPglucose [4]) [3, 4]; 0.28 (CTP) [4]; 0.41 (CDP-D-glucose, Mg^{2+} as cation) [3]; 0.43 (magnesium diphosphate) [3]; 0.56 (magnesium diphosphate)
[2]; 0.64 (alpha-D-glucose 1-phosphate) [4]; 0.7 (Mg2-CTP) [2]; 1.89
(diphosphate) [4]; 5 (glucosamine 1-phosphate) [2]

pH-optimum

7.8–8 [1]; 8–10 (broad, Mg^{2+} as activating cation) [3]; 8.3 (pyrophosphorolysis) [2]; 8.5 (CDPglucose synthesis) [2]

pH-range

6.2–9.5 (about half-maximal activity at pH 6.2 and about 70% of maximal activity at pH 9.5) [1]

Temperature optimum (°C)

30 (assay at) [3]; 37 (assay at) [1, 2, 4]

Temperature range (°C)

3 ENZYME STRUCTURE

Molecular weight
120000 (Pasteurella pseudotuberculosis type V, gel filtration) [3]

Subunits
? (x × 29035, Salmonella enterica LT2, calculated from amino acid composition deduced from nucleotide sequence, x × 31000, Salmonella enterica LT2, SDS-PAGE) [4]
Monomer (1 × 110000, Pasteurella pseudotuberculosis type V, SDS-PAGE) [3]

Glycoprotein/Lipoprotein
–

4 ISOLATION/PREPARATION

Source organism
Azotobacter vinelandii (strain O) [2]; Pasteurella pseudotuberculosis type V (rough mutant, strain 25VO) [3]; Salmonella enterica LT2 (recombinant, overproducing strain P9254) [4]; Salmonella paratyphi A [1]

Source tissue
Cell [1–4]

Localization in source

Purification
Azotobacter vinelandii (partial) [2]; Pasteurella pseudotuberculosis type V (partial) [3]; Salmonella enterica LT2 [4]; Salmonella paratyphi A [1]

Crystallization
–

Cloned
(Salmonella enterica LT2) [4]

Renatured
–

5 STABILITY

pH
6.5 (rapid and irreversible inactivation below) [3]

Temperature (°C)

Oxidation

Organic solvent

General stability information

During the final stages of purification the enzyme preparation is quite unstable to cold room conditions over a period of 2 to 3 days [3]; Stable to repeated freeze-thawing [3]

Storage

-20°C, about 30% loss of activity within 1 month [1]; -18°C, in EDTA-containing Tris-HCl buffer, at least 1 year [2]; Frozen, 2–3 months stable [3]

---

6 CROSSREFERENCES TO STRUCTURE DATABANKS

PIR/MIPS code

Brookhaven code

---

7 LITERATURE REFERENCES

[1] Mayer, R.M., Ginsburg, V.: J. Biol. Chem.,240,1900–1904 (1965)
[2] Kimata, K., Suzuki, S.: J. Biol. Chem.,241,1099–1113 (1966)
[3] Rubenstein, P.A., Strominger, J.L.: J. Biol. Chem.,249,3789–3796 (1974)
[4] Lindqvist, L., Kaiser, R., Reeves, P.R., Lindberg, A.A.: J. Biol. Chem.,269,122–126 (1994)

1 NOMENCLATURE

EC number
2.7.7.34

Systematic name
GTP:alpha-D-glucose-1-phosphate guanylyltransferase

Recommended name
Glucose-1-phosphate guanylyltransferase

Synonyms
GDPglucose pyrophosphorylase
Guanylyltransferase, glucose 1-phosphate
Glucose 1-phosphate guanylyltransferase
Guanosine diphosphoglucose pyrophosphorylase

CAS Reg. No.
9033-13-0

2 REACTION AND SPECIFICITY

Catalyzed reaction
GTP + alpha-D-glucose 1-phosphate →
→ diphosphate + GDPglucose

Reaction type
Nucleotidyl group transfer

Natural substrates
More (may be involved in biosynthesis of heparin in mast cell tissue [1], involved in microbial metabolism of guanosine sugar nucleotides in Streptomyces sp. [2]) [1, 2]

Substrate spectrum
1 GTP + alpha-D-glucose 1-phosphate (r [1, 2]) [1, 2]
2 GTP + D-mannose 1-phosphate (r, less effectiv than alpha-D-glucose 1-phosphate) [1]
3 More (ATP, CTP, TTP, UTP are ineffective) [1]

Product spectrum
1 Diphosphate + GDPglucose [1, 2]
2 Diphosphate + GDPmannose [1]
3 ?

Inhibitor(s)

D-Mannose 1-phosphate (0.07 mM: 18% inhibition in presence of 3.5 mM GDPglucose, 40% inhibition in presence of 0.7 mM GDPglucose, no inhibition in presence of more than 7 mM GDPglucose) [1]

Cofactor(s)/prosthetic group(s)/activating agents

Metal compounds/salts

Mg^{2+} (absolute requirement, highest activity at 8–10 mM) [1]

Turnover number (min^{-1})

Specific activity (U/mg)

0.046 [1]

K_m-value (mM)

0.1 (GDPglucose) [1]; 0.8 (diphosphate) [1]

pH-optimum

7.0 (assay at) [2]; 7.4 (broad) [1]

pH-range

6–8.7 (65% of maximal activity at pH 6, 60% of maximal activity at pH 8.7) [1]

Temperature optimum (°C)

30 (assay at [1, 2], reverse reaction [1]) [1, 2]; 38 (assay at, forward reaction) [1]

Temperature range (°C)

3 ENZYME STRUCTURE

Molecular weight

Subunits

Glycoprotein/Lipoprotein

–

4 ISOLATION/PREPARATION

Source organism

Mouse (mast cell tumor, grown in LAF mice) [1]; Streptomyces rimosus (IFO 3441) [2]; Streptomyces lavendulae (IFO 3145) [2]; Streptomyces griseus (IFO 3122, 3430, 3356) [2]; Streptomyces ruber (IFO 3310) [2]; Streptomyces olivaceus (IFO 3409) [2]; Streptomyces fradiae (IFO 3439, IFO 3123) [2]; Streptomyces purpurascens (IFO 3389) [2]; Streptomyces scabies (IFO 3111) [2]; Streptomyces sp. (strain AKU 2801) [2]; More (enzyme is also present in Eremothecium ashbyii [1, 2], Gleditishia maracantha seed [2], fresh peas [2], rat mammary gland [1, 2], calf liver [2]) [1, 2]

Source tissue
Mastocytoma tissue [1]; Cell [2]

Localization in source

Purification
Mouse (mast cell tumor, grown in LAF mice, partial) [1]; Streptomyces rimosus (IFO 3441, partial) [2]; Streptomyces lavendulae (IFO 3145, partial) [2]; Streptomyces griseus (IFO 3122, 3430, 3356, partial) [2]; Streptomyces ruber (IFO 3310, partial) [2]; Streptomyces olivaceus (IFO 3409, partial) [2]; Streptomyces fradiae (IFO 3439, IFO 3123, partial) [2]; Streptomyces purpurascens (IFO 3389, partial) [2]; Streptomyces scabies (IFO 3111, partial) [2]; Streptomyces sp. (strain AKU 2801, partial) [2]

Crystallization
–

Cloned
–

Renatured
–

5 STABILITY

pH

Temperature (°C)

Oxidation

Organic solvent

General stability information

Storage
–20°C, dried acetone powder, stable for at least two weeks [1]

6 CROSSREFERENCES TO STRUCTURE DATABANKS

PIR/MIPS code

Brookhaven code

7 LITERATURE REFERENCES

[1] Danishefsky, I., Heritier-Watkins, O.: Biochim. Biophys. Acta,139,349–357 (1967)
[2] Kawaguchi, K., Tanida, S., Matsuda, K., Tani, Y., Ogata, K. : Agric. Biol. Chem.,37,75–81 (1973)

1 NOMENCLATURE

EC number
2.7.7.35

Systematic name
ADP:D-ribose-5-phosphate adenylyltransferase

Recommended name
Ribose-5-phosphate adenylyltransferase

Synonyms
ADPribose phosphorylase
Adenylyltransferase, ribose 5-phosphate
Adenosine diphosphoribose phosphorylase

CAS Reg. No.
9054-55-1

2 REACTION AND SPECIFICITY

Catalyzed reaction
Phosphate + ADPribose →
→ ADP + D-ribose 5-phosphate

Reaction type
Nucleotidyl group transfer

Natural substrates

Substrate spectrum
1 Phosphate + ADPribose (ir, highly specific. No substrates are ADPglucose, UDPglucose [1, 2], polyadenylic acid, RNA (crude, from Euglena),
   NAD(H), NADP(H), coenzyme A, FAD [1]) [1–3]
2 Phosphate + deamino-NAD$^+$ (inosine diphosphoribose, not adenosine
   triphosphoribose) [1]
3 Phosphate + ADP (ADP/phosphate-exchange reaction, inorganic phosphate is incorporated into terminal position of ADP [2], IDP can replace
   ADP with 16% efficiency. No substrates are CDP, GDP, UDP [1, 2], AMP
   or ATP [2]) [1, 2]

Product spectrum
1 ADP + D-ribose 5-phosphate [1–3]
2 ?
3 ADP + phosphate [1, 2]

Inhibitor(s)

PCMB (ADP/phosphate-exchange, reversible by cysteine [1], ADP/phosphate-exchange and ADPribose phosphorolysis [2, 3]) [1–3]; Ag^{2+} [1]; Arsenate (kinetics) [2]; Selenate [1]; ADPribose (competitive to phosphate) [1]; AMP (kinetics) [2]; ATP (kinetics) [2]; ADP (weak, ADPribose phosphorolysis) [2]; IDP (weak, ADPribose phosphorolysis) [2]; Ribose 5-phosphate (kinetics [2]) [1, 2]; Deoxyribose 5-phosphate (ADP/phosphate-exchange) [1]; More (no inhibition by cyanide, 1,10-phenanthroline, 2,2'-dipyridyl, ribose 1-phosphate, ribose 1,5-diphosphate, glucose 1-phosphate, glucose 6-phosphate [1], $NADP^+$, EDTA, ribose [1, 2], adenosine, deoxyadenosine, 3'-AMP, CMP, CDP, UDP, NADH, NAD^+, ADPglucose, UDPglucose, IAA, NEM [2]) [1–3]

Cofactor(s)/prosthetic group(s)/activating agents

More (no activation by EDTA) [2]

Metal compounds/salts

More (no metal ion requirement [1, 3], no Mn^{2+} or Mg^{2+} requirement [2]) [1–3]

Turnover number (min^{-1})

Specific activity (U/mg)

0.51 [2]; 1.3 [1, 3]

K_m-value (mM)

More (kinetic study) [1]; 0.04 (ADPribose) [1]; 0.05 (ADPribose) [2]; 0.4 (phosphate) [1]; 0.5 (phosphate) [2]; 0.6 (ADP, ADP/phosphate exchange) [1–3]

pH-optimum

7.5 (ADP/phosphate-exchange) [2]; 7.5–7.8 (ADP/phosphate-exchange) [1]; 7.8–8 [3]; 8 (ADPribose-phosphorolysis) [1, 2]

pH-range

6.4–8.1 (ADP/phosphate-exchange, about half-maximal activity at pH 6.4 and 8.1) [1]; 6.7–8.5 (ADP/phosphate-exchange, about 60% of maximal activity at pH 6.7 and about half-maximal activity at pH 8.5) [2]; 6.8–9 (ADPribose phosphorolysis, about half-maximal activity at pH 6.8 and about 90% of maximal activity at pH 9) [1]; 6.8–9.4 (about half-maximal activity at pH 6.8 and 9.4) [2]

Temperature optimum (°C)

50 [2]

Temperature range (°C)

29–56 (about half-maximal activity at 29°C and 56°C) [2]

3 ENZYME STRUCTURE

Molecular weight

Subunits

Glycoprotein/Lipoprotein
–

---

4 ISOLATION/PREPARATION

Source organism
Euglena gracilis (green alga, strain Z [1, 3], var. bacillaris Pringsheim [2])
[1–3]; Ochromonas danica (slight activity) [1, 3]; More (not in Chlorella sp.,
Gloeocapsa sp., Spinacia oleracea, Pisum sativum [1], swiss chard, Chlorella pyrenoidosa [2]) [1, 2]

Source tissue
Cell (heterotrophically or autotrophically grown cells [1–3] or UV-mutant cells
incapable of forming chloroplasts [2]) [1–3]

Localization in source
Soluble [1–3]

Purification
Euglena gracilis (partial) [1–3]

Crystallization
–

Cloned
–

Renatured
–

---

5 STABILITY

pH

Temperature (°C)
60 (3 min, inactivation) [2]

Oxidation

Organic solvent

General stability information
Solution of lyophilized enzyme, 2 mg/ml, stable to repeated freeze-thaw cycles during 2 weeks [2]

Storage

Frozen, fairly stable in the presence of EDTA [1]; –15°C, undialyzed purified enzyme, lyophilized and placed under vacuum, 35% loss of activity within 2 months [2]

6 CROSSREFERENCES TO STRUCTURE DATABANKS

PIR/MIPS code

Brookhaven code

7 LITERATURE REFERENCES

[1] Evans, W.R., Pietro, A.S.: Arch. Biochem. Biophys.,113,236–244 (1966)
[2] Stern, A.I., Avron, M.: Biochim. Biophys. Acta,118,577–591 (1966)
[3] Evans, W.R.: Methods Enzymol.,23A,566–570 (1971) (Review)

1 NOMENCLATURE

EC number
2.7.7.36

Systematic name
ADP:aldose-1-phosphate adenylytransferase

Recommended name
Aldose-1-phosphate adenylyltransferase

Synonyms
Sugar-1-phosphate adenylyltransferase
ADPaldose phosphorylase
Adenylyltransferase, sugar 1-phosphate
Adenosine diphosphosugar phosphorylase
ADP sugar phosphorylase
Adenosine diphosphate glucose:orthophosphate adenylyltransferase [1]

CAS Reg. No.
37278-27-6

---

2 REACTION AND SPECIFICITY

Catalyzed reaction
ADP + aldose 1-phosphate →
→ phosphate + ADPaldose

Reaction type
Nucleotidyl group transfer

Natural substrates

Substrate spectrum
1 ADPglucose + phosphate [1]
2 dADPglucose + phosphate (72% of activity compared to ADPglucose) [1]
3 ADPxylose + phosphate (15% of activity compared to ADPglucose) [1]
4 ADP-beta-glucose + phosphate (10% of activity compared to ADPglucose) [1]
5 ADPmaltose + phosphate (2% of activity compared to ADPglucose) [1]
6 More (arsenate can partially substitute for phosphate, UDPglucose,
  UDPxylose, UDPacetylglucosamine, GDPmannose, dTDPglucose, NAD,
  NADP are no substrates) [1]

Product spectrum
1 ADP + glucose 1-phosphate [1]
2 dADP + glucose 1-phosphate
3 ADP + xylose 1-phosphate
4 ?
5 ADP + maltose 1-phosphate
6 ?

Inhibitor(s)

Cofactor(s)/prosthetic group(s)/activating agents

Metal compounds/salts

Turnover number (min^{-1})

Specific activity (U/mg)

K_m-value (mM)

pH-optimum
8.5 [1]

pH-range
7–10 (70% of maximal activity at pH 7, 50% of maximal activity at pH 10) [1]

Temperature optimum (°C)
37 (assay at) [1]

Temperature range (°C)

3 ENZYME STRUCTURE

Molecular weight

Subunits

Glycoprotein/Lipoprotein
–

4 ISOLATION/PREPARATION

Source organism
Wheat [1]

Source tissue
Germ [1]

Localization in source

Purification
    Wheat (partial) [1]

Crystallization
    −

Cloned
    −

Renatured
    −

---

5 STABILITY

pH

Temperature (°C)

Oxidation

Organic solvent

General stability information

Storage
    −15°C, several weeks, stable [1]

---

6 CROSSREFERENCES TO STRUCTURE DATABANKS

PIR/MIPS code

Brookhaven code

---

7 LITERATURE REFERENCES

[1] Dankert, M., Goncalves, I.R.J., Recondo, E.: Biochim. Biophys. Acta,81,78–85 (1964)

1 NOMENCLATURE

EC number
2.7.7.37

Systematic name
NDP:aldose-1-phosphate nucleotidyltransferase

Recommended name
Aldose-1-phosphate nucleotidyltransferase

Synonyms
Sugar-1-phosphate nucleotidyltransferase
NDPaldose phosphorylase
Nucleotidyltransferase, sugar 1-phosphate
Inosityltransferase, glucose 1-phosphate
NDP sugar phosphorylase
Nucleoside diphosphosugar phosphorylase
Nucleotidyltransferase, sugar phosphate
Sugar nucleotide phosphorylase
Nucleoside diphosphate sugar:orthophosphate nucleotidyltransferase [1]

CAS Reg. No.
9033-61-8

2 REACTION AND SPECIFICITY

Catalyzed reaction
NDP + aldose 1-phosphate →
→ NDPaldose + phosphate

Reaction type
Nucleotidyl group transfer

Natural substrates

Substrate spectrum

1 GDPmannose + phosphate [1]
2 UDPmannose + phosphate (290% of activity compared to GDPmannose) [1]
3 UDPglucose + phosphate (22% of activity compared to GDPmannose) [1]
4 UDPgalactose + phosphate (18% of activity compared to GDPmannose) [1]
5 UDPxylose + phosphate (24.5% of activity compared to GDPmannose) [1]
6 UDPacetylglucosamine + phosphate (5.1% of activity compared to GDPmannose) [1]
7 ADPmannose + phosphate (30% of activity compared to GDPmannose) [1]
8 dTDPmannose + phosphate (157% of activity compared to GDPmannose) [1]
9 dTDPglucose + phosphate (11% of activity compared to GDPmannose) [1]
10 GDPgalactose + phosphate (3% of activity compared to GDPmannose) [1]
11 More (GDPglucose, ADPglucose, ADPgalactose, ADPxylose, dADPglucose, IDPglucose are no substrates) [1]

Product spectrum

1 GDP + mannose 1-phosphate [1]
2 UDP + mannose 1-phospahte
3 UDP + glucose 1-phosphate
4 UDP + galactose 1-phosphate
5 UDP + xylose 1-phosphate
6 ?
7 ADP + mannose 1-phosphate
8 dTDP + mannose 1-phosphate
9 dTDP + glucose 1-phosphate
10 GDP + galactose 1-phosphate
11 ?

Inhibitor(s)

Inorganic phosphate (4.5 mM: 55% inhibition) [1]; Mannose 1-phosphate (1.25 mM: 22% inhibition, 4.2 mM: 40% inhibition) [1]; Mg^{2+} (17 mM: 32% inhibition) [1]; Zn^{2+} (5 mM: complete inhibition) [1]

Cofactor(s)/prosthetic group(s)/activating agents

Metal compounds/salts

More (0.1–1 mM of Mg^{2+} or 1 mM of Zn^{2+} has no effect) [1]

Turnover number (min^{-1})

Specific activity (U/mg)

0.178 [1]

K_m-value (mM)
   0.12 (phosphate) [1]; 0.25 (UDPmannose) [1]; 1.5 (UDPglucose) [1]; 6.2 (GDPmannose) [1]

pH-optimum
   8.5 [1]

pH-range
   6.5–9.5 (14% of maximal activity at pH 6.5, 43% of maximal activity at pH 9.5) [1]

Temperature optimum (°C)
   30 (assay at) [1]

Temperature range (°C)

---

3 ENZYME STRUCTURE

Molecular weight

Subunits

Glycoprotein/Lipoprotein
   –

---

4 ISOLATION/PREPARATION

Source organism
   Saccharomyces cerevisiae [1]

Source tissue
   Protoplasts [1]

Localization in source

Purification
   Saccharomyces cerevisiae (partial) [1]

Crystallization
   –

Cloned
   –

Renatured
   –

5 STABILITY

pH

Temperature (°C)
  55 (15 min: 40% remaining activity, 30 min: 27% remaining activity) [1]

Oxidation

Organic solvent

General stability information
  Ultrafiltration or ammonium sulfate precipitation results in heavy loss of activity [1]

Storage
  −20°C, concentrated enzyme fraction, stable for several months [1]

---

6 CROSSREFERENCES TO STRUCTURE DATABANKS

PIR/MIPS code

Brookhaven code

---

7 LITERATURE REFERENCES

[1] Cabib, E., Carminatti, H., Woyskovsky, N.M.: J. Biol. Chem.,240,2114–2121 (1965)

1 NOMENCLATURE

EC number
2.7.7.38

Systematic name
CTP:3-deoxy-D-manno-octulosonate cytidylyltransferase

Recommended name
3-Deoxy-manno-octulosonate cytidylyltransferase

Synonyms
CMP-3-deoxy-D-manno-octulosonate pyrophosphorylase
Cytidylyltransferase, 2-keto-3-deoxyoctonate
2-Keto-3-deoxyoctonate cytidylyltransferase
3-Deoxy-D-manno-octulosonate cytidylyltransferase
CMP-3-deoxy-D-manno-octulosonate synthetase
CMP-KDO synthetase
CTP:CMP-3-deoxy-D-manno-octulosonate cytidylyltransferase
Cytidine monophospho-3-deoxy-D-manno-octulosonate pyrophosphorylase

CAS Reg. No.
37278-28-7

---

2 REACTION AND SPECIFICITY

Catalyzed reaction
CTP + 3-deoxy-D-manno-octulosonate →
→ diphosphate + CMP-3-deoxy-D-manno-octulosonate

Reaction type
Nucleotidyl group transfer

Natural substrates
CTP + 3-deoxy-manno-octulosonate (involved in biosynthesis of cell wall li-
popolysaccharide constituent of gram-negative bacteria [2], supposed rate-
limiting step of this biosynthesis [4]) [2, 4]

Substrate spectrum
1 CTP + 3-deoxy-manno-octulosonate (i.e. KDO, r, strict specificity, only
  beta-KDO-pyranose, not alpha-form [4], slight activity with ITP or TTP [1].
  No substrates are dUTP [2], ATP, GTP, UTP, CDP, 3-deoxy-manno-octulo-
  sonate 8-phosphate, 3-deoxy-D-arabinoheptulosonate, 3-deoxy-D-erythro-
  hexulosonate, N-acetylneuraminate [1]) [1–4]
2 dCTP + 3-deoxy-manno-octulosonate (at pH 8 and 9.5) [2, 3]
3 UTP + 3-deoxy-manno-octulosonate (at pH 9.5, not [1]) [2, 3]

Product spectrum

1 Diphosphate + CMP-3-deoxy-manno-octulosonate (beta-pyranose) [1–4]
2 Diphosphate + dCMP-3-deoxy-manno-octulosonate
3 Diphosphate + UMP-3-deoxy-manno-octulosonate

Inhibitor(s)

2,6-Anhydro-3-deoxy-D-glycero-D-talo-octanoate (substrate analog, mechanism) [4]; Hg^{2+} (strong) [3]; Diphosphate (weak) [3]; More (no inhibition by CDP, CMP, 3-deoxy-manno-octulosonate 8-phosphate, N-acetylneuraminate [3] or 2,6-anhydro-3-deoxy-D-glycero-D-galacto-octanoate (substrate analog) [4]) [3, 4]

Cofactor(s)/prosthetic group(s)/activating agents

GSH (activation [1], no reducing agent required [3]) [1]

Metal compounds/salts

Mg^{2+} (requirement, 10 mM [3]) [1–3]; Mn^{2+} (activation, 38% as effective as Mg^{2+} [3], much less effective than Mg^{2+} [2]) [2, 3]; Cd^{2+} (activation, 42% as effective as Mg^{2+}) [3]; Zn^{2+} (activation, 32% as effective as Mg^{2+}) [3]; Ba^{2+} (activation, 20% as effective as Mg^{2+}) [3]; Ca^{2+} (activation, 14% as effective as Mg^{2+}) [3]; Co^{2+} (activation, 11% as effective as Mg^{2+} [3], much less effective than Mg^{2+} [2]) [2, 3]; More (no activation by Na^+, K^+, Hg^{2+}, Fe^{2+} or Ni^{2+}) [3]

Turnover number (min^{-1})

Specific activity (U/mg)

0.05 [1]; 9.3–9.6 [2, 3]

K_m-value (mM)

0.2 (CTP, pH 9.5 [3]) [2, 3]; 0.22 (CTP) [1]; 0.29 (3-deoxy-manno-octulosonate (+ CTP), pH 9.5 [3]) [3]; 0.34 (dCTP, pH 9.5) [3]; 0.39 (3-deoxy-manno-octulosonate (+ CTP)) [2]; 0.8 (3-deoxy-manno-octulosonate) [1]; 0.88 (UTP, pH 9.5) [3]

pH-optimum

More (pl: 4.15–4.4) [3]; 7.8 [1]; 9.3 (glycine-NaOH buffer) [3]; 9.5 (Tris-acetate or glycine-NaOH buffer) [2]; 9.6 (Tris-acetate buffer) [3]

pH-range

7–8.4 (about half-maximal activity at pH 7 and 8.4) [1]; 7.5–10.2 (about half-maximal activity at pH 7.5 and about 80% of maximal activity at pH 10.2, Tris-acetate buffer) [3]; 7.5–10.5 (about half-maximal activity at pH 7.5 and about 60% of maximal activity at pH 10.5, glycine-NaOH buffer) [3]

Temperature optimum (°C)

37 (assay at) [1–3]

Temperature range (°C)

3 ENZYME STRUCTURE

Molecular weight
    35000–40000 (E. coli, PAGE) [3]
    35000–44000 (E. coli, PAGE, sucrose density gradient centrifugation) [2]
    35000–45000 (E. coli, sucrose density gradient centrifugation) [3]
    40000–46000 (E. coli, gel filtration) [3]

Subunits
    Monomer (1×36000, E. coli, SDS-PAGE) [2, 3]

Glycoprotein/Lipoprotein
    –

4 ISOLATION/PREPARATION

Source organism
    E. coli (strain B [2, 3], strain 0111-B4 [1], strain D21 [4]) [1–4]; Salmonella
    typhimurium (mutant strain SL 1102) [4]

Source tissue
    Cell [1–4]

Localization in source
    Cytosol [3]

Purification
    E. coli (partial [1]) [1–3]

Crystallization
    –

Cloned
    –

Renatured
    –

5 STABILITY

pH
    7.4–8 (more stable in phosphate buffer than in Tris-acetate or Tris-HCl buffer)
    [3]

Temperature (°C)

Oxidation

Organic solvent

General stability information
Repeated freeze-thaw cycles result in substantial loss of activity [1]

Storage
–90°C, partially purified preparation in 0.05 M phosphate buffer, 0.5 mM DTT, up to 3 months [3]; –90°C, in 0.05 M phosphate buffer, pH 7.2, 0.5 mM DTT, up to 4 months [2]; –20°C, 20% loss of activity within 1 year [1]; –1°C, several months [1]; 4°C, crude and partially purified preparation, up to 1 month [2]; 4°C, partially purified preparation in 0.05 M phosphate buffer, 0.5 mM DTT, up to 3 months [3]

---

6 CROSSREFERENCES TO STRUCTURE DATABANKS

PIR/MIPS code
PIR2:A26322 (Escherichia coli); PIR2:C48492 (Escherichia coli)

Brookhaven code

---

7 LITERATURE REFERENCES

[1] Ghalambor, M.A., Heath, E.H.: J. Biol. Chem.,241,3216–3221 (1966)
[2] Ray, P.H., Benedict, C.D.: Methods Enzymol.,83,535–540 (1982) (Review)
[3] Ray, P.H., Benedict, C.D., Grasmuk, H.: J. Bacteriol.,145,1273–1280 (1981)
[4] Claesson, A., Luthman, K., Gustafsson, K., Bondesson, G.: Biochem. Biophys. Res. Commun.,143,1063–1068 (1987)

1 NOMENCLATURE

EC number
2.7.7.39

Systematic name
CTP:sn-glycerol-3-phosphate cytidylyltransferase

Recommended name
Glycerol-3-phosphate cytidylyltransferase

Synonyms
CDPglycerol pyrophosphorylase
Cytidylyltransferase, glycerol 3-phosphate
CDP-glycerol pyrophosphorylase
Cytidine diphosphoglycerol pyrophosphorylase
Cytidine diphosphate glycerol pyrophosphorylase [1]
CTP:glycerol 3-phosphate cytidylyltransferase [3]
Gro-PCT [4]

CAS Reg. No.
9027-11-6

---

2 REACTION AND SPECIFICITY

Catalyzed reaction
CTP + sn-glycerol 3-phosphate →
→ diphosphate + CDPglycerol (rapid equilibrium random order mechanism [3])

Reaction type
Nucleotidyl group transfer

Natural substrates
CTP + sn-glycerol 3-phosphate (CDPglycerol may function in: synthesis of lipids [1], repression of synthesis occurs at the onset of phosphate starvation and is accompanied by inhibition or inactivation of CDPglycerol pyrophosphorylase [2], enzyme of teichonic acid synthesis [3, 4], enzyme is inactivated under phosphate-limited conditions [5], enzyme is possibly a control point in synthesis of the cell wall in Bacillus licheniformis [6]) [1–6]

Substrate spectrum
1 CTP + sn-glycerol 3-phosphate [1–6]
2 dCTP + glycerol 3-phosphate (about 95% of the activity with CTP) [3]
3 More (not: cytidine diphosphate ribitol [1], ribitol phosphate [1], phosphocholine [1], phosphoethanolamine [1], ATP [1], ITP [1], UTP [1], GTP [3]) [1, 3]

Product spectrum
1 Diphosphate + CDPglycerol [1]
2 Diphosphate + dCDPglycerol
3 ?

Inhibitor(s)
Cd^{2+} [3]; Hg^{2+} [3]; Sn^{2+} [3]; Cu^{2+} [3]; Zn^{2+} [3]; Diphosphate [3]; CDPglycerol [3]

Cofactor(s)/prosthetic group(s)/activating agents
UDP-N-acetylglucosamine (stimulates, up to 5 mM) [6]; UDP-N-acetylmuramyl-L-alanyl-D-glutamyl-meso-diaminopimelyl-D-alanyl-D-alanine (slight stimulation below 2.5 mM, inhibition above) [6]

Metal compounds/salts
Mg^{2+} (divalent cation: Co^{2+}, Mg^{2+}, Mn^{2+} or Fe^{2+} required [3], Mg^{2+}, Mn^{2+} or Co^{2+} required [1]) [1, 3]; Mn^{2+} (divalent cation: Co^{2+}, Mg^{2+}, Mn^{2+} or Fe^{2+} required [3], Mg^{2+}, Mn^{2+} or Co^{2+} required [1]) [1, 3]; Co^{2+} (divalent cation: Co^{2+}, Mg^{2+}, Mn^{2+} or Fe^{2+} required [3], Mg^{2+}, Mn^{2+} or Co^{2+} required [1]) [1, 3]; Fe^{2+} (divalent cation: Co^{2+}, Mg^{2+}, Mn^{2+} or Fe^{2+} required) [3]

Turnover number (min^{-1})

Specific activity (U/mg)
85.7 [3]

K_m-value (mM)
3.23 (glycerol 3-phosphate) [3]; 3.85 (CTP) [3]

pH-optimum
6.9–9.5 [3]; 7 (Tis-maleate buffer) [1]; 8 (Tris-HCl buffer) [1]

pH-range

Temperature optimum (°C)
37 (assay at) [3]; 50 [3]

Temperature range (°C)

3 ENZYME STRUCTURE

Molecular weight
30500 (Bacillus subtilis, conditional lethal mutant 168, gel filtration) [4]
30900 (Bacillus subtilis, gene expression in E. coli, gel filtration) [3]

Subunits
Dimer (2 × 14800, Bacillus subtilis, gene expression in E. coli, SDS-PAGE
[3], 2 × 15271, Bacillus subtilis, calculated from the nucleotide sequence
[3]) [3, 4]

Glycoprotein/Lipoprotein
–

4 ISOLATION/PREPARATION

Source organism
Bacillus subtilis (W23 [2], BR151, gene expressed in E. coli [3], conditional
lethal mutant 168 [4]) [1–4]; Bacillus licheniformis (ATCC 9945) [5, 6]; Lactobacillus arabinosus [1]; Propionibacterium shermanii [1]; Chlorella vulgaris
[1]; Saccharomyces cerevisiae [1]; Staphylococcus aureus [1]; E. coli [1]

Source tissue
Cell culture [2]

Localization in source
Soluble [6]

Purification
Lactobacillus arabinosus [1]; Bacillus subtilis (gene expression in E. coli) [3]

Crystallization
–

Cloned
(Bacillus subtilis gene expressed in E. coli) [3]

Renatured
–

5 STABILITY

pH
6.5 (37°C, 4 h, about 30% loss of activity) [3]; 7.5–9.5 (37°C, 4 h, stable) [3];
10 (37°C, 4 h, about 10% loss of activity) [3]

Enzyme Handbook © Springer-Verlag Berlin Heidelberg 1997
Duplication, reproduction and storage in data banks are only
allowed with the prior permission of the publishers

Temperature (°C)
40 (not stable for longer than 30 min) [3]; More (thermostability of the gro-PCT in extract of strains bearing mutations in tagA, tagB and tagF genes) [4]

Oxidation

Organic solvent

General stability information

Storage
–80°C, pure enzyme stable for at least 7 months [3]; –20°C, about 20% loss of activity after 1 month [3]

---

6 CROSSREFERENCES TO STRUCTURE DATABANKS

PIR/MIPS code
PIR2:A49757 (Bacillus subtilis (strain 168))

Brookhaven code

---

7 LITERATURE REFERENCES

[1] Shaw, D.R.D.: Biochem. J.,82,297–312 (1962)
[2] Cheah, S.-C., Hussey, H., Baddiley, J.: Eur. J. Biochem.,118,497–500 (1981)
[3] Park, Y.S., Sweitzer, T.D., Dixon, J.E., Kent, C.: J. Biol. Chem.,268,16648–16654 (1993)
[4] Pooley, H.M., Abellan, F.-X., Karamata, D.: J. Gen. Microbiol.,137,921–928 (1991)
[5] Hussey, H., Sueda, S., Cheah, S.-C., Baddiley, J.: Eur. J. Biochem.,82,169–174 (1978)
[6] Anderson, R.G., Douglas, L.J., Hussey, H., Baddiley, J.: Biochem. J.,136,871–876 (1973)

1 NOMENCLATURE

EC number
    2.7.7.40

Systematic name
    CTP:D-ribitol-5-phosphate cytidylyltransferase

Recommended name
    D-Ribitol-5-phosphate cytidylyltransferase

Synonyms
    Cytidine diphosphate ribitol pyrophosphorylase [1]
    Cytidylyltransferase, ribitol 5-phosphate
    CDP-ribitol pyrophosphorylase
    Cytidine diphosphoribitol pyrophosphorylase
    Ribitol 5-phosphate cytidylyltransferase
    CDPribitol pyrophosphorylase

CAS Reg. No.
    9027-07-0

---

2 REACTION AND SPECIFICITY

Catalyzed reaction
    CTP + D-ribitol 5-phosphate →
    → diphosphate + CDPribitol

Reaction type
    Nucleotidyl group transfer

Natural substrates
    More (enzyme functions possibly in synthesis of teichonic acids, repression
    of enzyme synthesis occurs at the onset of phosphate starvation) [2]

Substrate spectrum
    1 CTP + D-ribitol 5-phosphate [1]

Product spectrum
    1 Dihosphate + CDPribitol [1]

Inhibitor(s)

Cofactor(s)/prosthetic group(s)/activating agents

Metal compounds/salts

Turnover number (min^{-1})

Specific activity (U/mg)

K_m-value (mM)

pH-optimum

pH-range

Temperature optimum (°C)

Temperature range (°C)

3 ENZYME STRUCTURE

Molecular weight

Subunits

Glycoprotein/Lipoprotein
–

4 ISOLATION/PREPARATION

Source organism
Streptococcus lactis [1]; Bacillus subtilis (W23 [2]) [1, 2]; Lactobacillus arabinosus [1]; Propionibacterium shermanii [1]; Chlorella vulgaris [1]; Saccharomyces cerevisiae [1]; Staphylococcus aureus [1]

Source tissue

Localization in source
Soluble [1]

Purification
Staphylococcus aureus [1]

Crystallization
–

Cloned
–

Renatured
–

5 STABILITY

pH

Temperature (°C)

Oxidation

Organic solvent

General stability information

Storage

---

6 CROSSREFERENCES TO STRUCTURE DATABANKS

PIR/MIPS code

Brookhaven code

---

7 LITERATURE REFERENCES

[1] Shaw, D.R.D.: Biochem. J.,82,297–312 (1962)
[2] Cheah, S.-C., Hussey, H., Baddiley, J.: Eur. J. Biochem.,118,497–500 (1981)

1 NOMENCLATURE

EC number
   2.7.7.41

Systematic name
   CTP:phosphatidate cytidylyltransferase

Recommended name
   Phosphatidate cytidylyltransferase

Synonyms
   Cytidylyltransferase, phosphatidate
   CDP-diacylglycerol synthase
   CDP-diacylglyceride synthetase
   Cytidine diphosphoglyceride pyrophosphorylase
   Phosphatidate cytidyltransferase
   Phosphatidic acid cytidylyltransferase
   CTP:1,2-diacylglycerophosphate-cytidyl transferase [3]
   CTP-diacylglycerol synthetase
   DAG synthetase [2]
   CDP-DG [11]
   CDPdiglyceride pyrophosphorylase

CAS Reg. No.
   9067-83-8

2 REACTION AND SPECIFICITY

Catalyzed reaction
   CTP + phosphatidate →
   → diphosphate + CDPdiacylglycerol (sequential mechanism [4, 12, 18], sequential bi-bi reaction [5, 13], ping-pong mechanism [10])

Reaction type
   Nucleotidyl group transfer

Natural substrates
   CTP + phosphatidate (CDPdiglyceride formation [3], enzyme is essential for phospholipid biosynthesis in all organisms [12], influence of exogenous and membrane-bound phosphatidate concentration on activity [19]) [3, 12, 19]

Enzyme Handbook © Springer-Verlag Berlin Heidelberg 1997
Duplication, reproduction and storage in data banks are only
allowed with the prior permission of the publishers

Substrate spectrum

1 CTP + phosphatidate (r [14, 18]) [1–19]
2 dCTP + phosphatidate (50% of the activity with CTP [1]) [1, 4, 10, 18]
3 CTP + 1-stearoyl-2-arachidonoylphosphatidic acid [2, 16]
4 CTP + lysophosphatidic acid [3]
5 CTP + 1,2-dipalmitoylphosphatidic acid [10, 16]
6 CTP + 1-stearoyl-2-oleoylphosphatidic acid [16]
7 CTP + 1-oleoyl-2-stearoylphosphatidic acid [16]
8 CTP + 1,2-dioleoyl phosphatidic acid [16]
9 CTP + 1-palmitoyl-2-oleoyl phosphatidic acid [16]
10 CTP + 1-arachidonoyl-2-stearoylphosphatidic acid [16]
11 CTP + 1,2-diarachidonoylphosphatidic acid [16]
12 CTP + 1,2-dicaproylphosphatidic acid [16]
13 CTP + 1,2-distearoylphosphatidic acid [16]
14 More (no substrate: lysophosphatidic acid [16], bis-phosphatidic acid [16], GTP [10], ATP [10], UTP [10], 1-acyl-sn-glycero-3-phosphate [4], phosphatidic acids with acyl chains shorter than 16 carbons are poor substrates [4], relative activities of phosphatidic acids with varying fatty acid composition [16]) [4, 10, 16]

Product spectrum

1 CDPdiacylglycerol + diphosphate [5, 9, 10, 16]
2 dCDPdiacylglycerol + diphosphate [10, 18]
3 CDP(1-stearoyl-2-arachidonoyl)glycerol + diphosphate
4 ? + diphosphate
5 CDPdipalmitoylglycerol + diphosphate
6 CDP(1-stearoyl-2-oleoyl)glycerol + diphosphate
7 CDP(1-oleoyl-2-stearoyl)glycerol + diphosphate
8 CDPdioleoylglycerol + diphosphate
9 CDP(1-palmitoyl-2-oleoyl)glycerol + diphosphate
10 CDP(1-arachidonoyl-2-stearoyl)glycerol + diphosphate
11 CDPdiarachidonoylglycerol + diphosphate
12 CDPdicaproylglycerol + diphosphate
13 CDPdistearoylglycerol + diphosphate
14 ?

Inhibitor(s)

Phosphatidylinositol (slight) [15]; Phosphatidylserine (slight) [15]; Cardiolipin (slight) [15]; Amphiphilic cationic drugs (inhibition at relatively high concentrations, noncompetitive to phosphatidate) [17]; CDPdipalmitin (inhibits pyrophosphorolysis) [18]; Zwitterionic quarternary ammonium sulfobetaine detergent (ZWT-12, ZWT-14) [15]; NEM (not [18]) [2]; 5,5'-Dithiobis(2-nitro-benzoic acid) [2]; PCMB [2, 7, 10, 13]; Palmitoyl-CoA [3]; dCTP [5, 14]; Thiophosphatidate [5]; Hg^{2+} [7, 10, 13]; Zn^{2+} [7]; Cd^{2+} [7]; EDTA (in excess of Mg^{2+} [18]) [17, 18]; Triton X-100 (above 1% w/v [18], above 20 mM [13]) [13,

18]; ATP (5 mM, no effect: adenosine 5'-[alpha,beta-methylene]triphosphate, adenosine 5'-[beta,gamma-imido]triphosphate) [9]; Tween-20 [15]; Sodium deoxycholate [15]; SDS (in excess of Mg^{2+} [18]) [15, 18]; Ca^{2+} [15]; Co^{2+} [15]; F^- [9, 15]; CDP [11]; Diphosphate [11, 14]; $FeSO_4$ [15]; Lysolecithin [15]; CMP (synthesis of CDPdiacylglycerol and dCDPdiacylglycerol) [10]; Hydrazine [1]; Phosphatidic acid (above 8 mM) [1]; Mg^{2+} (maximal stimulation at 10mM, 70% inhibition above 30 mM [1], divalent cation required, 30% inhibition at 25 mM [10]) [1, 10]; CHAPS (inhibition at 0.3%, maximal activity at 0.5%) [2]

Cofactor(s)/prosthetic group(s)/activating agents
Lecithin (stimulates, even in presence of optimal concentrations of cationic detergents) [15]; Sphingomyelin (stimulates, even in presence of optimal concentrations of cationic detergents) [15]; ATP (stimulates, only in presence of either lecithin or sphingomyelin) [5]; GTP (stimulates, only in presence of either lecithin or sphingomyelin) [5]; ITP (stimulates, only in presence of either lecithin or sphingomyelin) [5]; Norfenfluramine (stimulates) [17]; Chlorpromazine (stimulates) [17]; Non-ionic detergents (absolute requirement [10], marked stimulation [1]) [1, 10]; Cationic detergents (the enzyme requires phosphatidate emulsified in cationic detergent for optimum activity) [15]; CHAPS (inhibition at 0.3%, maximal activity at 0.5%) [2]; Phosphatidylcholine (activation of solubilized but not of microsomal enzyme) [2]; Triton X-100 (required [5, 13, 14], activity depends on Triton X-100 (5 mM) [7], maximal stimulation at 15 mM, inhibition above 20 mM [13]) [5, 7, 13, 14]; GTP (stimulates enzyme in rat liver microsomes, heat or proteolytic treatment or treatment with low levels of detergent of microsomes prevents stimulation [8], no effect: guanosine 5'-[beta,gamma-methylene]-triphosphate, guanosine 5'-[beta,gamma-imido]triphosphate, guanosine 3'-diphosphate 5'-diphosphate [9], no stimulation [15], F^- reverses GTP stimulation [9]) [8, 9]

Metal compounds/salts
K^+ (marked stimulation [1], stimulates [10]) [1, 10]; Mg^{2+} (divalent cation required [10, 11], only metal ion that stimulates [15], Mg^{2+} most effective in stimulation [10, 11], stimulates [1, 16, 17], $MgCl_2$ required [5], Mg^{2+} required [13, 14], Mg^{2+} (5 mM) or Mn^{2+} (1 mM) required [7], optimal concentration: 10 mM [1, 11], 3–6 mM [10], 20 mM [14–15], 60 mM [16], 70% inhibition at 30 mM [1], 30% inhibition at 25 mM [10]) [1, 5, 7, 10, 11, 13–17]; Mn^{2+} (stimulates [1], Mg^{2+} (5 mM) or Mn^{2+} (1 mM) required [7], 1 mM, 70% of the activity with Mg^{2+} [10]) [1, 7, 10]

Turnover number (min^{-1})

Specific activity (U/mg)
0.028 [18]; 133.3 [1]; 1.409 [5, 13]; 30.79 [12]; 0.45 [14]; 0.0001 [15]; More [4]

K_m-value (mM)

More [10, 11, 15]; 0.18 (CTP) [7]; 0.22 (phosphatidic acid) [7]; 0.26 (dCTP, CTP) [10]; 0.28 (phosphatidic acid) [4, 12]; 0.3 (phosphatidic acid) [1]; 0.5 (phosphatidate) [5, 13]; 0.58 (dCTP) [4, 12]; 1 (CTP) [5, 13]; 1.2 (CTP) [1]

pH-optimum

6.5 [13, 14]; 6.5–7.5 [7]; 6.8 (1,2-dioleoyl phosphatidic acid) [16]; 7.0 (Tris-HCl buffer [15]) [3, 15]; 7–8 [4]; 7.3 [12]; 7.4 (assay at) [4]; 7.5 (phosphate buffer [15]) [10, 15]

pH-range

5.5–8.5 (5.5: about 60% of activity maximum, 8.5: about 70% of activity maximum) [13]; 5.8–7.8 (5.8: 38% of activity maximum, 7.8: 35% of activity maximum) [1]; 6.2–8.7 (6.2: 54% of activity maximum, 8.7: 56% of activity maximum) [7]

Temperature optimum (°C)

25 (assay at) [10]; 30 (assay at) [4, 12]; 37 (assay at) [6, 15, 18]; 38 (assay at) [1]; 45 [1]; 50 [7]

Temperature range (°C)

20–60 (20°C: about 35% of activity maximum, 60°C: about 45% of activity maximum) [7]

---

3 ENZYME STRUCTURE

Molecular weight

114000 (Saccharomyces cerevisiae, radiation inactivation analysis) [5, 13]
More (MW of 400000, Saccharomyces cerevisiae, gel filtration, enzyme still associated with phospholipid, 2 major protein bands MW 19000 and 45000 in SDS-PAGE) [14]

Subunits

Dimer (2 × 56000, Saccharomyces cerevisiae, SDS-PAGE) [5]
? (x × 27000, E. coli, SDS-PAGE) [4, 12]
More (Saccharomyces cerevisiae: 2 subunits with MW of 56000 and 54000, SDS-PAGE, 54000 MW protein may possibly be a proteolysis product of the 56000 MW protein) [13]

Glycoprotein/Lipoprotein

–

4 ISOLATION/PREPARATION

Source organism
E. coli (overproducing strain [12]) [4, 12, 18]; Rat [6, 8, 9, 16, 17, 19]; Clostridium perfringens (type A, ATCC 3624) [7]; Micrococcus cerificans [1]; Bovine [2]; Chicken (embryo) [3]; Saccharomyces cerevisiae (S288C (alphagal2) [5]) [5, 13, 14]; Bacillus subtilis [10]; Catharanthus roseus [11]; Pig [15]

Source tissue
Kidney [16]; Heart [16]; Lung [16]; Small intestine [16]; Brain [2, 3, 16]; Liver [6, 8, 9, 16, 17, 19]; Cell suspension culture [11]; Mesenteric lymph node (lymphocytes) [15]

Localization in source
Microsomes (endoplasmic reticulum, mitochondrial and chloroplast envelope, plasma membrane [11]) [2, 8, 9, 11, 16, 17, 19]; Cell envelope [7]; Membranes [1, 10, 12, 13, 18]; Mitochondria (associated with inner membrane [6], membrane [14]) [6, 14]; More (mitochondrial and microsomal enzymes appear to be 2 distinct enzymes with different localization and regulatory characteristics) [6]

Purification
Clostridium perfringens [7]; Catharanthus roseus [11]; Micrococcus cerificans [1]; E. coli (overproducing strain [12]) [4, 12, 18]; Saccharomyces cerevisiae [5, 13, 14]

Crystallization
–

Cloned
–

Renatured
–

---

5 STABILITY

pH

Temperature (°C)
25 (30 min, no loss of activity) [10]; 30 (labile above) [13]; 40 (30 min (+ detergent), inactivation) [10]; 55 (30 min, irreversible inactivation) [1]; 57 (30 min, membrane-bound enzyme, no loss of activity, solubilized enzyme loses 70%, but only 19% if Mg^{2+} is added in excess of EDTA) [18]; 60 (30 min, membrane-associated enzyme stable up to) [10]

Oxidation

Organic solvent

General stability information
Phospholipid stabilizes [18]; Triton X-100 destabilizes [18]; Stable to at least 2 cycles of freezing and thawing [1, 7, 13]; Stable to freezing and thawing [18]

Storage
−70°C, solubilized enzyme stable for at least several months [18]; −80°C, 1 mM CTP, 90–100% stable for at least 3 months [5, 13]; −80°C, stable for at least 2 months [7]; 0–4°C, solubilized enzyme stable for at least several weeks [18]; −20°C, 24 h, enzyme in isolated microsomes, 75% loss of activity [15]; Stable for at least 9 months in intact lymphocytes [15]

---

6 CROSSREFERENCES TO STRUCTURE DATABANKS

PIR/MIPS code
PIR1:SYECDG (Escherichia coli); PIR2:C64248 (Mycoplasma genitalium (SGC3)); PIR3:JC4832 (Pseudomonas aeruginosa)

Brookhaven code

---

7 LITERATURE REFERENCES

[1] McCaman, R.E., Finnerty, W.R.: J. Biol. Chem.,243,5074–5080 (1968)
[2] Lin, C.H., Lin, J., Strickland, K.P.: Biochem. Int.,25,299–306 (1991)
[3] Petzold, G.L., Agranoff, B.W.: J. Biol. Chem.,242,1187–1191 (1967)
[4] Sparrow, C.P.: Methods Enzymol.,209,237–242 (1992) (Review)
[5] Carman, G.M., Kelley, M.J.: Methods Enzymol.,209,242–247 (1992) (Review)
[6] Mok, A.Y.P., McDougall, G.E., McMurray, W.C.: FEBS Lett.,312,236–240 (1992)
[7] Carman, G.M., Zaniewski, R.L., Cousminer, J.J.: Appl. Environ. Microbiol.,43,81–85 (1982)
[8] Liteplo, R.G., Sribney, M.: Biochim. Biophys. Acta,619,660–668 (1980)
[9] Sribney, M., Dove, J.L., Lyman, E.M.: Biochim. Biophys. Acta,79,749–755 (1977)
[10] Gaillard, J.-L., Lubochinsky, B., Rigomier, D.: Biochim. Biophys. Acta,753,372–380 (1983)
[11] Hanenberg, A., Heim, S., Wissing, J.B., Wagner, K. G.: Plant Sci.,88,13–18 (1993)
[12] Sparrow, C.P., Raetz, C.R.H.: J. Biol. Chem.,260,12084–12091 (1985)
[13] Kelley, M.J., Carman, G.M.: J. Biol. Chem.,262,14563–14570 (1987)
[14] Belendiuk, G., Mangnall, D., Tung, B., Westley, J., Getz, G.S.: J. Biol. Chem.,253, 4555–4565 (1978)
[15] Sribney, M., Hegadorn, C.A.: J. Biochem.,60,668–674 (1982)
[16] Bishop, H.H., Strickland, K.P.: Can. J. Biochem.,54,249–260 (1976)
[17] Sturton, R.G., Brindley, D.N.: Biochem. J.,162,25–32 (1977)
[18] Langley, K.E., Kennedy, E.P.: J. Bacteriol.,136,85–95 (1978)
[19] Van Heusden, G.P.H., Van den Bosch, H.: Eur. J. Biochem.,84,405–412 (1978)

1 NOMENCLATURE

EC number
2.7.7.42

Systematic name
ATP:[L-glutamate:ammonia ligase (ADP-forming)] adenylyltransferase

Recommended name
[Glutamate-ammonia-ligase] adenylyltransferase

Synonyms
Glutamine-synthetase adenylyltransferase
Adenylyltransferase, glutamine synthetase
Glutamine synthetase adenylyltransferase
ATP:glutamine synthetase adenylyltransferase [1, 2]
Adenosine triphosphate:glutamine synthetase adenylyltransferase [4]

CAS Reg. No.
9077-66-1

---

2 REACTION AND SPECIFICITY

Catalyzed reaction
ATP + [L-glutamate:ammonia ligase (ADP-forming)] →
→ diphosphate + adenylyl-[L-glutamate:ammonia ligase (ADP-forming)]
(mechanism [4])

Reaction type
Nucleotidyl group transfer

Natural substrates
More (inactivation of glutamine synthetase by attachment of the adenylyl
moiety of ATP) [4]

Substrate spectrum
1 ATP + glutamine synthetase (r [6], ADP, AMP, cAMP, UTP, CTP, ITP, NAD+
cannot replace ATP [1]) [1–6]

Product spectrum
1 Diphosphate + glutamine synthetase-(AMP) [1–6]

Enzyme Handbook © Springer-Verlag Berlin Heidelberg 1997
Duplication, reproduction and storage in data banks are only
allowed with the prior permission of the publishers

Inhibitor(s)

6-Diazo-5-oxonorleucine [1]; S-(2-Hydroxyethyl)-L-cysteine [1]; DL,2-Amino-butyric acid [1]; L-Methionine [1]; 4-Methyl-L-glutamate [1]; L-Tryptophan [1]; 2-Oxoglutarate (inhibition of adenylylation, activation of deadenylylation [6]) [3, 6]; ADP [1]; CTP [1]; UTP [1]; ITP [1]; Sulfate [1]; Phosphate [1]; Diphosphate [1]; D-Glutamine [1]; Glutamate (L- and D-isomer) [1]; 3-Phosphoglycerate [3]

Cofactor(s)/prosthetic group(s)/activating agents

E. coli PII regulatory protein (activity is modulated by a regulatory protein PII, which exists in two interconvertible forms, PIIA and PIID, the unmodified form PIIA stimulates the adenylylation of glutamine synthetase, the uridylated form PIID is required for deadenylylation [5]) [5, 6]; ATP (activator of adenylylation) [6]; Glutamine (stimulates [3], L-isomer, activator of adenylylation [6], in presence of saturating amounts of PIIA protein Mg^{2+}-supported activity is activated, Mn^{2+}-supported activity is almost unchanged [6]) [3, 6]

Metal compounds/salts

Mg^{2+} (Mg^{2+} or Mn^{2+} required [1], half-maximal activity at: 14 mM [1], stimulates [3, 6]) [1, 3, 6]; Mn^{2+} (Mn^{2+} or Mg^{2+} required [1], half-maximal activity at 1.4 mM [1], stimulates [6]) [1, 6]; More (Ca^{2+}, Zn^{2+} and Cu^{2+} at 10 mM are ineffective) [1]

Turnover number (min^{-1})

Specific activity (U/mg)

More (assay method [5]) [1, 5]

K_m-value (mM)

0.005 (glutamine synthetase) [1]; 0.150 (ATP) [1]

pH-optimum

7.6 [1]; 7.9 (assay at) [5]; 8.0–8.2 (adenylylation) [6]

pH-range

5.5–9.8 (5.5: about 50% of activity maximum, 9.8: about 35% of activity maximum) [1]

Temperature optimum (°C)

37 (assay at) [5]

Temperature range (°C)

3 ENZYME STRUCTURE

Molecular weight

64000 (E. coli, low and high speed sedimentation equilibrium, 115000 MW
enzyme form is slowly converted during storage at 4°C to a smaller protein
that is active only in adenylylation, not in deadenylylation) [6]
115000 (E. coli B, ultracentrifugation) [2]
145000 (E. coli B, gel filtration) [1]

Subunits

Monomer (1 × 114000, E. coli, high speed sedimentation study of the enzyme in 6 M guanidine–HCl) [6]

Glycoprotein/Lipoprotein

–

4 ISOLATION/PREPARATION

Source organism

E. coli (B [1, 2, 4], W [5]) [1–6]

Source tissue

Localization in source

Purification

E. coli (B [1]) [1, 6]

Crystallization

–

Cloned

–

Renatured

–

5 STABILITY

pH

4–9 (4°C, 12 h, no loss of activity) [1]

Temperature (°C)

Oxidation

Organic solvent

General stability information

Considerably less stable in Tris or imidazole buffer than in a magnesium phosphate buffer [6]; Bovine serum albumin, above 1 mg/ml, prevents inactivation at 4°C and 25°C and aggregation [1, 2]; No stabilization by ATP, CTP, Mn^{2+}, glutamine, cysteine or mercaptoethanol each at 20 mM, 2 mM DTT, 20% glycerol, sucrose, polyethyleneglycol or urea at 1 M [1]; Mg^{2+}, 20 mM protects to some extent against heat inactivation [1]

Storage

–80°C, stored after quick freezing with liquid N_2, potassium phosphate buffer, 10–100 mM, pH 7.6, 1 mM $MgCl_2$, stable for months at enzyme concentration above 0.1 mg/ml [6]; 0°C - 4°C, enzyme concentration above 1 mg/ml, stable for 12 days [6]

---

6 CROSSREFERENCES TO STRUCTURE DATABANKS

PIR/MIPS code

Brookhaven code

---

7 LITERATURE REFERENCES

[1] Ebner, E., Wolf, D., Gancedo, C., Elsässer, S., Holzer, H.: Eur. J. Biochem., 14,535–544 (1970)
[2] Wolf, D., Ebner, E., Hinze, H.: Eur. J. Biochem.,25,239–244 (1972)
[3] Wolf, D.H., Ebner, E.: J. Biol. Chem.,247,4208–4212 (1972)
[4] Wohlhueter, R.M., Ebner, E., Wolf, D.H.: J. Biol. Chem.,247,4213–4218 (1972)
[5] Rhee, S.G., Park, R., Wittenberger, M.: Anal. Biochem.,88,174–185 (1978)
[6] Caban, C.E., Ginsburg, A.: Biochemistry,15,1569–1580 (1976)

1 NOMENCLATURE

EC number
   2.7.7.43

Systematic name
   CTP:N-acylneuraminate cytidylyltransferase

Recommended name
   N-Acylneuraminate cytidylyltransferase

Synonyms
   CMPsialate pyrophosphorylase
   CMPsialate synthase
   Cytidine 5'-monophosphosialic acid synthetase [1]
   CMP-Neu5Ac synthetase [3]
   CMP-NeuAc synthetase [6]
   Cytidyltransferase, acylneuraminate
   Acylneuraminate cytidyltransferase
   CMP sialate pyrophosphorylase
   CMP-N-acetylneuraminate synthetase
   CMP-N-acetylneuraminate synthase
   CMP-N-acetylneuraminic acid synthase
   CMP-NANA synthetase
   CMP-sialate synthase
   CMP-sialate synthetase
   CMP-sialic synthetase
   Cytidine 5'-monophospho-N-acetylneuraminic acid synthetase
   Cytidine 5-monophosphate N-acetylneuraminic acid synthetase
   Cytidine 5'-monophosphosialic acid synthetase
   Cytidine monophosphate-N-acetylneuraminic acid synthetase
   Cytidine monophospho-sialic acid synthetase
   Cytidine monophosphoacetylneuraminic synthetase
   Cytidine monophosphosialate pyrophosphorylase
   Cytidine monophosphosialate synthetase
   Cytidylyltransferase, acetylneuraminate

CAS Reg. No.
   9067-82-7

---

2 REACTION AND SPECIFICITY

Catalyzed reaction
   CTP + N-acylneuraminate →
   → diphosphate + CMP-N-acylneuraminate

Reaction type
Nucleotidyl group transfer

Natural substrates
CTP + N-acetylneuraminate (enzyme activates N-acetylneuraminate to transfer to the nascent capsular polysaccharide in multiple group B Streptococcus serotypes [3]) [3, 7]

Substrate spectrum
1 CTP + N-acetylneuraminate (r [1]) [1–11]
2 CTP + N-glycoloylneuraminate (r [1]) [1–11]
3 N-Acetyl-7(8)-O-acetylneuraminic acid + CTP [8]
4 N-Acetyl-4-O-acetylneuraminic acid + CTP [8]
5 Fluoroacetylneuraminic acid + CTP [8]
6 N-Chloroacetylneuraminic acid + CTP [8]
7 4-O-Methyl-N-acetylneuraminic acid + CTP [10]
8 More (no activity with: N-acetylneuraminic acid 9-phosphate, N-glycolylneuraminic acid 9-phosphate, N-acetyl-4-O-acetylneuraminic acid, N-acetyl-7-O-acetylneuraminic acid, 2-keto-3-deoxygluconate, 2-keto-3-deoxyheptanoate, dCTP, ATP, GTP, ITP, UTP, TTP, ADP, GDP, IDP, UDP) [1]

Product spectrum
1 Diphosphate + CMP-N-acetylneuraminate [1–4, 9]
2 Diphosphate + CMP-N-glycoloylneuraminate [1]
3 Diphosphate + CMP-N-acetyl-7(8)-O-acetylneuraminate
4 Diphosphate + CMP-N-acetyl-4-O-acetylneuraminate
5 Diphosphate + CMP-N-fluoroacetylneuraminate
6 Diphosphate + CMP-N-chloroacetylneuraminate
7 CMP-4-O-methyl-N-acetylneuraminic acid + diphosphate [10]
8 ?

Inhibitor(s)
2-Deoxy-2,3-dehydro-N-acetylneuraminic acid [11]; EDTA (in absence of Mg^{2+} or Ca^{2+}) [9]; Diphosphate [11]; 5-Mercuri-CTP [6]; CTP-2',3'-dialdehyde [6]; CMP-N-acetylneuraminic acid [7]; CTP (high concentrations) [7]; CDP [7, 11]; CMP [7–9, 11]; ATP [7]; UTP [7]; GTP [7]; TTP [7]; Cu^{2+} [7, 9]; Zn^{2+} [7]; Hg^{2+} [7, 9]; Fe^{3+} [7]; PCMB [9]; Iodoacetate [9]

Cofactor(s)/prosthetic group(s)/activating agents
DTT (stimulates) [3]; 2-Mercaptoethanol (discrete activation maximum at 1 mM, higher and lower concentrations result in lower activity) [11]

Metal compounds/salts
More (inactive in stimulation: Cd^{2+} [1, 6], Cu^{2+} [1], Zn^{2+} [1, 6]) [1, 6]; Mg^{2+} (absolute requirement for a divalent cation, Mg^{2+} most effective [1], required [2, 3, 7], Mn^{2+} or Mg^{2+} required [6], Mg^{2+} or Ca^{2+} required [9], maximal activity: at 20 mM $MgCl_2$ [3], 50 mM [9], 20–40 mM [6]) [1–3, 6, 7, 9]; Ca^{2+} (Mg^{2+} or Ca^{2+} required, maximal activity at 50 mM [9], 20% of the activity with Mg^{2+} [1], no effect [6], can partially replace Mg^{2+} [7]) [1, 6, 7, 9]; Mn^{2+} (20% of the activity with Mg^{2+} [1], Mn^{2+} or Mg^{2+} required [6]) [1, 6]; Fe^{2+} (10% of the activity with Mg^{2+}) [1]; Co^{2+} (10% of the activity with Mg^{2+} [1], can partially replace Mg^{2+} [7]) [1, 7]

Turnover number (min^{-1})

Specific activity (U/mg)
2.1 [6]; 2.0 [7]; More [1–4, 9]

K_m-value (mM)
0.31 (CTP) [6]; 0.6 (CTP) [1]; 0.8 (N-acetylneuraminate) [1]; 1.4 (CTP) [3]; 1.7 (CTP) [4]; 2.1 (N-acetylneuraminic acid) [4]; 2.2 (N-acetylneuraminic acid) [2]; 2.3 (N-glycoloylneuraminic acid) [1, 9]; 2.9 (N-glycoloylneuraminic acid) [4]; 4 (CTP [2], N-acetylneuraminate [6]) [2, 6]; 7.6 (N-acetylneuraminate) [3]; More [7–9, 11]

pH-optimum
8.0 (in presence of 10 mM Mg^{2+} and 5 mM DTT) [7]; 8.3–9.4 [3]; 9.0 (enzyme immobilized in Sepharose 4B [11], assay at [1]) [1, 2, 9, 11]; 9.0–10 [6]; 9.3 [4]; 9.5 [8]

pH-range
More [8]

Temperature optimum (°C)
25–37 [3]; 28 [4]; 37 (assay at [1]) [1, 7, 9]; 40 (maximum of 40°C maintained to 60°C, enzyme immobilized on Sepharose 4B, activity of soluble enzyme decreased sharply above 40°C) [11]; 45 [7]

Temperature range (°C)
25–42 (4°C: no activity detectable, 25°C-27°C: activity maximum, 42°C: activity decreased by 88%) [3]; 33–42 (about 50% of activity maximum at 33°C and 42°C) [9]

3 ENZYME STRUCTURE

Molecular weight
116000 (rat, gel filtration) [7]
160000 (rainbow trout, gel filtration) [4]
163000 (Rana esculenta, gel filtration) [9]

Subunits
Monomer (1 × 160000, rainbow trout, SDS-PAGE [4], 1 × 163000, Rana esculenta, PAGE with or without SDS or urea [9]) [4, 9]
Dimer (2 × 58000, rat, SDS-PAGE) [7]
? (x × 50000, E. coli, SDS-PAGE) [6]

Glycoprotein/Lipoprotein
–

4 ISOLATION/PREPARATION

Source organism
Neisseria meningitidis (group B) [5]; Pig (hog [1]) [1, 8]; Bovine (cow) [8]; E. coli (O18:K1) [6]; Rat [7]; Human [2]; Streptococcus sp. (group B, high-producing type Ib strain) [3]; Rainbow trout [4]; Horse [8, 10]; Rana esculenta [9–11]

Source tissue
Submaxillary glands [1, 8, 10]; Placenta [2]; Liver [4, 7, 9–11]

Localization in source
Cytoplasm [6]; Nucleus (more than 85% of the enzyme activity is associated with) [7]; Soluble [9]

Purification
E. coli (O18:K1) [6]; Rat [7]; Pig [1]; Human (partial) [2]; Streptococcus sp. (partial) [3]; Rainbow trout [4]; Rana esculenta [9]

Crystallization
–

Cloned
[5]

Renatured
–

5 STABILITY

pH

Temperature (°C)
37 (at high enzyme concentrations the reaction rate decreases 10–15% between 45 and 60 min) [1]

Oxidation

Organic solvent

General stability information

Tolerates flash freezing and lyophilization [3]; Particularly sensitive to repeated freezing [7]; Glycerol, 20% v/v improves thermal stability [7]; If CTP or Mg^{2+} is present, enzyme is much more sensitive to thermal deactivation [7]; 2-Mercaptoethanol, 1 mM, improves stability [11]

Storage

4°C, pH 7.6, 17% loss of activity after 4 weeks [3]; –80°C, stable for 8 months [3]; 4°C, pH 7.0, stable for several weeks [6]; –20°C, 2 weeks, more than 50% loss of activity [7]; 2°C, enzyme immobilized on Sepharose 4B, 20% loss of activity after 4 months [11]

6 CROSSREFERENCES TO STRUCTURE DATABANKS

PIR/MIPS code

PIR2:A36509 (Escherichia coli); PIR2:A40198 (rat (fragment))

Brookhaven code

7 LITERATURE REFERENCES

[1] Kean, E.L., Roseman, S.: J. Biol. Chem.,241,5643–5650 (1966)
[2] Kolisis, F.N.: Arch. Int. Physiol. Biochim.,92,179–184 (1984)
[3] Haft, R.F., Wessels, M.R.: J. Bacteriol.,176,7372–7374 (1994)
[4] Schmelter, T., Ivanov, S., Wember, M., Stangier, P., Thiem, J., Schauer, R.: Biol. Chem. Hoppe-Seyler,374,337–342 (1993)
[5] Edwards, U., Frosch, M.: FEMS Microbiol. Lett.,96,161–166 (1992)
[6] Vann, W.F., Silver, R.P., Abeijon, C., Chang, K., Aaronson, W., Sutton, A., Finn, C.W., Lindner, W., Kotsatos, M.: J. Biol. Chem.,262,17556–17562 (1987)
[7] Rodriguez-Aparicio, L.B., Luengo, J.M., Gonzalez-Clemente, C., Reglero, A.: J. Biol. Chem.,267,9257–9263 (1992)
[8] Schauer, R., Wember, M.: Hoppe-Seyler's Z. Physiol. Chem.,354,1405–1414 (1973)
[9] Schauer, R., Haverkamp, J., Ehrlich, K.: Hoppe-Seyler's Z. Physiol. Chem.,361, 641–648 (1980)
[10] Haverkamp, J., Beau, J.-M., Schauer, R.: Hoppe-Seyler's Z. Physiol. Chem.,360, 159–166 (1979)
[11] Corfield, A.P., Schauer, R., Wember, M.: Biochem. J.,177,1–7 (1979)

1 NOMENCLATURE

EC number
2.7.7.44

Systematic name
UTP:1-phospho-alpha-D-glucuronate uridylyltransferase

Recommended name
Glucuronate-1-phosphate uridylyltransferase

Synonyms
Uridylyltransferase, glucuronate 1-phosphate
UDP-glucuronate pyrophosphorylase
UDP-D-glucuronic acid pyrophosphorylase
UDP-glucuronic acid pyrophosphorylase
Uridine diphosphoglucuronic pyrophosphorylase

CAS Reg. No.
52228-05-4

2 REACTION AND SPECIFICITY

Catalyzed reaction
UTP + 1-phospho-alpha-D-glucuronate →
→ diphosphate + UDPglucuronate (Theorell-chance mechanism [2])

Reaction type
Nucleotidyl group transfer

Natural substrates
UTP + D-glucuronic acid 1-phosphate (predominant route whereby UDPglu-curonic acid is termed in young barley seedlings) [1]

Substrate spectrum
1 UTP + D-glucuronic acid 1-phosphate (r) [1, 2]

Product spectrum
1 Diphosphate + UDP-D-glucuronic acid [1]

Inhibitor(s)
Mg^{2+} (best fulfills the requirement for a divalent cation [1], maximal activity at a Mg^{2+}/UTP ratio of 2:1 in the forward direction and a Mg^{2+} to diphosphate ratio of about 1:1 in the reverse direction, excess Mg^{2+} inhibits) [1]

Cofactor(s)/prosthetic group(s)/activating agents

Metal compounds/salts

Mg^{2+} (best fulfills the requirement for a divalent cation [1], maximal activity
at a Mg^{2+}/UTP ratio of 2:1 [1], 1:1 [2] in the forward direction and a Mg^{2+} to
diphosphate ratio of about 1:1 [1, 2] in the reverse direction [1, 2], excess
Mg^{2+} inhibits [1]) [1, 2]; Mn^{2+} (53% of the activation with Mg^{2+}) [1]; Co^{2+}
(40% of the activation with Mg^{2+}) [1]; Zn^{2+} (37% of the activation with Mg^{2+})
[1]; Ca^{2+} (36% of the activation with Mg^{2+}) [1]

Turnover number (min^{-1})

Specific activity (U/mg)

3.8 [1]

K_m-value (mM)

0.33 (D-glucuronic acid 1-phosphate) [1]; 0.5 (UDP-D-glucuronic acid) [1]

pH-optimum

8–9 [1]

pH-range

Temperature optimum (°C)

Temperature range (°C)

3 ENZYME STRUCTURE

Molecular weight

Subunits

Glycoprotein/Lipoprotein

–

4 ISOLATION/PREPARATION

Source organism

Hordeum vulgare (var. Larker) [1]; Typha latifolia [2]

Source tissue

Seedlings [1]; Pollen [2]

Localization in source

Purification

Hordeum vulgare (partial) [1]; Typha latifolia [2]

Crystallization

–

Cloned

–

Renatured

–

---

5 STABILITY

pH

Temperature (°C)

Oxidation

Organic solvent

General stability information

Storage

---

6 CROSSREFERENCES TO STRUCTURE DATABANKS

PIR/MIPS code

Brookhaven code

---

7 LITERATURE REFERENCES

[1] Roberts, R.M.: J. Biol. Chem.,246,4995–5002 (1971)
[2] Toshinobu, H., Akira, H., Tooru, F.: Plant Cell Physiol.,24,1535–1543 (1983)

1 NOMENCLATURE

EC number
    2.7.7.45

Systematic name
    GTP:GTP guanylyltransferase

Recommended name
    Guanosine-triphosphate guanylyltransferase

Synonyms
    Guanylyltransferase, guanosine triphosphate
    Diguanosine tetraphosphate synthetase
    GTP-GTP guanylyltransferase
    Gp_4G synthetase [1]
    Guanosine triphosphate-guanose triphosphate guanylyltransferase
    Synthetase, diguanosine tetraphosphate

CAS Reg. No.
    54576-89-5

---

2 REACTION AND SPECIFICITY

Catalyzed reaction
    2 GTP →
    → diphosphate + P^1,P^4-bis(5'-guanosyl)tetraphosphate (ping-pong kinetics
    with a covalent enzyme-guanylate intermediate containing a phosphorami-
    date linkage, probably phospholysine [3])

Reaction type
    Nucleotidyl group transfer

Natural substrates
    GTP + GTP (enzyme catalyzes synthesis of P^1,P^4-bis(5'-guanosyl)tetraphos-
    phate during oogenesis in Artemia [1], the structure and mechanism of this
    enzyme suggest an evolutionary relationship to mRNA capping enzymes
    [3]) [1, 3]

Substrate spectrum
1 GTP + GTP (r [2–4], in the reverse reaction certain phosphate analogs can substitute for diphosphate [3]) [1–4]
2 GDP + GDP (rate of the reaction is low compared with synthesis of P^1,P^4-bis(5'-guanosyl)tetraphosphate and dependent on other small molecular weight components of yolk platelets) [1]
3 GDP + P^1,P^4-bis(5'-guanosyl)tetraphosphate [2]
4 Guanosine 5'-tetraphosphate + guanosine 5'-tetraphosphate [3]
5 dGTP + dGTP [3]
6 5'-Guanylylimidodiphosphate + diphosphate [3]
7 GTP + XTP [3]
8 GTP + ITP [3]
9 GTP + ADP [3]
10 More (either 2 enzymes are present in guanosine-triphosphate guanylyltransferase preparations or 2 catalytic sites exist on one protein, one for the synthesis of P^1,P^3-bis(5'-guanosyl)triphosphate and one for the synthesis of P^1,P^4-bis(5'-guanosyl)tetraphosphate) [2]

Product spectrum
1 Diphosphate + P^1,P^4-bis(5'-guanosyl)tetraphosphate (i.e. Gp_4G) [1–4]
2 Diphosphate + P^1,P^3-bis(5'-guanosyl)triphosphate (i.e. Gp_3G) [1]
3 P^1,P^3-bis(5'-guanosyl)triphosphate + GTP [2]
4 Diguanosine 5',5'''-P^1,P^5-pentaphosphate + diphosphate (i.e. Gp_5G) [3]
5 Di(2'-deoxyguanosine)5',5'''-P^1,P^4-tetraphosphate + diphosphate [3]
6 GppNHppG + diphosphate [3]
7 Guanosine 5'-xanthosine + diphosphate [3]
8 Guanosine 5'-inosine + diphosphate [3]
9 Guanosine 5'-adenosine + diphosphate [3]
10 ?

Inhibitor(s)
Diphosphate [1]; XTP (uncompetitive inhibition of P^1,P^4-bis(5'-guanosyl)tetraphosphate synthesis) [3]; ITP (partially uncompetitive inhibition of P^1,P^3-bis(5'-guanosyl)triphosphate synthesis) [3]

Cofactor(s)/prosthetic group(s)/activating agents
DTT (required) [1]

Metal compounds/salts
Mg^{2+} (required [1], 10–15 mM required for maximal activity with GTP as substrate [1], 20 mM with GDP as substrate) [1]; Mn^{2+} (86% of the activity with Mg^{2+} in P^1,P^4-bis(5'-guanosyl)tetraphosphate synthesis) [1]; Cu^{2+} (28% of the activity with Mg^{2+} in P^1,P^4-bis(5'-guanosyl)tetraphosphate synthesis) [1]

Turnover number (min^{-1})
96 (GTP) [3]; 72 (GTP) [3, 4]

Specific activity (U/mg)
0.113 [3]; More [1]

K_m-value (mM)
0.84 (diphosphate) [2]; 1.06 (Gp$_4$G) [2]; 2.2 (GTP) [2]; 6.7 (GTP) [3, 4]

pH-optimum
5.9–6.0 [1]

pH-range
4.5–8.0 (no activity below and above) [1]

Temperature optimum (°C)
37 (P^1,P^3-bis(5'-guanosyl)triphosphate synthesis, partially purified enzyme, Sephadex G-25 fraction) [1]; 40–42 (P^1,P^4-bis(5'-guanosyl)tetraphosphate synthesis) [1]

Temperature range (°C)

---

3 ENZYME STRUCTURE

Molecular weight
480000 (Artemia sp., gel filtration) [3, 4]
490000 (Artemia salina, gel filtration) [1]

Subunits
Tetramer (2 × 142000 (alpha) + 2 × 80000 (beta), Artemia sp., SDS-PAGE) [3, 4]

Glycoprotein/Lipoprotein
–

---

4 ISOLATION/PREPARATION

Source organism
Artemia salina (brine shrimp) [1, 2]; Artemia sp. [3, 4]

Source tissue
Encysted embryos (yolk platelets of) [1, 2]; Cysts [3]

Localization in source
Mitochondria (small amount of activity) [2]; Yolk platelets (almost 80% of the enzyme activity [2]) [1–4]

Purification
Artemia salina [1]; Artemia sp. [3, 4]

Crystallization

–

Cloned

–

Renatured

–

---

5 STABILITY

pH

Temperature (°C)

Oxidation

Organic solvent

General stability information
Albumin, 10 mg/ml, glycerol, 30% or P^1,P^4-bis(5'-guanosyl)tetraphosphate, 0.6 mM, prevents loss of activity at 0°C, Gp_4G most effective [1]

Storage
0°C or –15°C, 50% loss of activity after 1 week in buffered 500 mM NaCl [1]

---

6 CROSSREFERENCES TO STRUCTURE DATABANKS

PIR/MIPS code

Brookhaven code

---

7 LITERATURE REFERENCES

[1] Warner, A.H., Beers, P.C., Huang, F.L.: Can. J. Biochem.,52,231–240 (1974)
[2] Warner, A.H., Huang, F.L.: Can. J. Biochem.,52,241–251 (1974)
[3] Liu, J.J., McLennan, A.G.: J. Biol. Chem.,269,11787–11794 (1994)
[4] Liu, J.J., McLennan, A.G.: Biochem. Soc. Trans.,22,219S (1994)

1 NOMENCLATURE

EC number
2.7.7.46

Systematic name
NTP:gentamicin 2"-nucleotidyltransferase

Recommended name
Gentamicin 2"-nucleotidyltransferase

Synonyms
Adenylyltransferase, gentamicin 2"-
Aminoglycoside adenylyltransferase [1]

CAS Reg. No.
62213-33-6

2 REACTION AND SPECIFICITY

Catalyzed reaction
NTP + gentamicin →
→ diphosphate + 2"-nucleotidylgentamicin

Reaction type
Nucleotidyl group transfer

Natural substrates

Substrate spectrum
1 ATP + gentamicin (substrates are gentamicin C_1, C_{1a}, C_2, C or A, dATP, CTP, ITP or GTP can act as donors) [1]
2 ATP + sisomicin [1]
3 ATP + dibekacin [1]
4 ATP + kanamycin A (poor substrate) [1]
5 ATP + tobramycin (poor substrate [1]) [1, 2]
6 ATP + dideoxykanamycin B (adenylates the 2-hydroxyl group of the 3-amino-3-deoxy-D-glucose moiety) [3, 4]
7 CTP + tobramycin [2]
8 GTP + tobramycin [2]
9 dATP + tobramycin [2]
10 dGTP + tobramycin [2]
11 dCTP + tobramycin [2]
12 dTTP + tobramycin [2]
13 More (no substrates are amikacin, neomycin, lividomycin, butirosin A or B) [1]

Product spectrum
  1 Diphosphate + 2''-adenylylgentamicin [1]
  2 ?
  3 ?
  4 ?
  5 ?
  6 Diphosphate + 3',4'-dideoxykanamycin B-2''-adenylate [3, 4]
  7 ?
  8 ?
  9 ?
  10 ?
  11 ?
  12 ?
  13 ?

Inhibitor(s)
  EDTA (above 0.5 mM) [1]; Nucleotides (free form) [2]

Cofactor(s)/prosthetic group(s)/activating agents

Metal compounds/salts
  Mg^{2+} (requirement, 2 mM [1], actual substrate: MgNTP [2]) [1–3]; Ca^{2+} (activation, 1 mM, can replace Mg^{2+}) [1]; K^+ (slight activation) [1]; Na^+ (slight activation) [1]; NH_4^+ (slight activation) [1]

Turnover number (min^{-1})

Specific activity (U/mg)
  0.134–0.274 [2]

K_m-value (mM)
  0.0028 (tobramycin) [2]; 0.03 (MgdATP, MgdGTP (+ tobramycin)) [2]; 0.09 (MgdTTP (+ tobramycin)) [2]; 0.404 (MgATP (+ tobramycin)) [2]; 2.01 (MgCTP (+ tobramycin)) [2]

pH-optimum
  More (pI: 5.7–6) [2]; 7.5–8 [3]; 7.9 [1]; 9.5 [2]

pH-range

Temperature optimum (°C)
  37 (assay at) [1, 3]

Temperature range (°C)

3 ENZYME STRUCTURE

Molecular weight

Subunits
? (x × 31500–32500, E. coli, SDS-PAGE) [2]

Glycoprotein/Lipoprotein
–

4 ISOLATION/PREPARATION

Source organism
Pseudomonas aeruginosa [1]; E. coli (strain pMY10/W677 [2], strain JR66/W677) [2–4]

Source tissue
Cell [1–4]

Localization in source
Soluble [3]

Purification
Pseudomonas aeruginosa (partially purified from recombinant E. coli K12 host strain bearing plasmid pK237, affinity chromatography on gentamicin-affinity gel) [1]; E. coli (partial, affinity chromatography on gentamicin-affinity gel) [2]

Crystallization
–

Cloned
–

Renatured
–

5 STABILITY

pH

Temperature (°C)

Oxidation

Organic solvent

General stability information
Highly unstable after purification [1]

Storage

6 CROSSREFERENCES TO STRUCTURE DATABANKS

PIR/MIPS code

PIR1:XNKBLS (Klebsiella pneumoniae plasmid pLST1000); PIR1:XNKBGP (Klebsiella pneumoniae transposon Tn4000 and plasmid pBWH1); PIR2:S35980 (Salmonella oranienburg)

Brookhaven code

7 LITERATURE REFERENCES

[1] Angelatou, F., Litsas, S.B., Kontomichalou, P.: J. Antibiot.,35,235–244 (1982)
[2] Van Pelt, J.E., Northrop, D.B.: Arch. Biochem. Biophys.,230,250–263 (1984)
[3] Yagisawa, M., Naganawa, H., Kondo, S., Hamada, M., Takeuchi, T., Umezawa, H.: J. Antibiot.,26,911–912 (1971)
[4] Naganawa, H., Yagisawa, M., Kondo, S., Takeuchi, T., Umezawa, H.: J. Antibiot.,26, 913–914 (1971)

1 NOMENCLATURE

EC number
2.7.7.47

Systematic name
ATP:streptomycin 3"-adenylyltransferase

Recommended name
Streptomycin 3"-adenylyltransferase

Synonyms
Adenylyltransferase, streptomycin 3"-
Streptomycin adenylate synthetase
Streptomycin adenyltransferase
Streptomycin adenylylase
Streptomycin adenylyltransferase
Streptomycin-spectinomycin adenylyltransferase
Synthetase, streptomycin adenylate
AAD (3") [1]
Aminoglycoside 3"-adenylyltransferase [1]

CAS Reg. No.
52660-23-8

2 REACTION AND SPECIFICITY

Catalyzed reaction
ATP + streptomycin →
→ diphosphate + 3"-adenylylstreptomycin

Reaction type
Nucleotidyl group transfer

Natural substrates

Substrate spectrum
1 ATP + streptomycin [1, 2]
2 ATP + spectinomycin [1, 2]
3 ATP + bluensomycin [2]
4 dATP + streptomycin [2]
5 More (ADP, AMP, UTP, GTP, CTP, TTP, dAMP, adenine, adenosine, S-adenosyl methionine, dADP, ADPglucose are ineffective) [2]

Product spectrum
1 ? + adenylylstreptomycin (adenylylstreptomycin demonstrated in crude extract [1], adenylation at streptobiosamine moiety [2]) [1, 2]
2 ? + adenylylspectinomycin (adenylation proposed [2]) [1, 2]
3 ? + adenylated bluensomycin (adenylation proposed) [2]
4 ?
5 ?

Inhibitor(s)

Cofactor(s)/prosthetic group(s)/activating agents
beta-Mercaptoethanol (activating) [2]; Tris-HCl (activating) [2]

Metal compounds/salts
Mg^{2+} (required, 8–10 mM optimal) [2]; More (monovalent cations and salts of Ni^{2+}, Co^{2+}, Ca^{2+}, Zn^{2+}, Mn^{2+} have no influence) [2]

Turnover number (min^{-1})

Specific activity (U/mg)
0.000046 (streptomycin) [1]

K_m-value (mM)
0.00032 (streptomycin) [1]

pH-optimum
8.3 [2]; 9 [1]

pH-range

Temperature optimum (°C)
37 (assay at) [2]; 50 [1]

Temperature range (°C)

---

3 ENZYME STRUCTURE

Molecular weight

Subunits

Glycoprotein/Lipoprotein
–

---

4 ISOLATION/PREPARATION

Source organism
E. coli [1, 2]

Source tissue
Cell [1, 2]

Localization in source
Periplasm [2]

Purification
E. coli (partial) [1, 2]

Crystallization
–

Cloned
–

Renatured
–

---

5 STABILITY

pH

Temperature (°C)

Oxidation

Organic solvent

General stability information

Storage

---

6 CROSSREFERENCES TO STRUCTURE DATABANKS

PIR/MIPS code
PIR2:I64916 (Escherichia coli); PIR2:I51989 (Escherichia coli (fragment));
PIR2:S05476 (Escherichia coli transposon Tn21); PIR1:XUECSA (Escheri-
chia coli transposon Tn7); PIR2:C37392 (Klebsiella pneumoniae transposon
Tn1331); PIR2:JQ1756 (plasmid R46); PIR2:S25252 (Salmonella
choleraesuis)

Brookhaven code

---

7 LITERATURE REFERENCES

[1] Kono, M., Ohmiya, K., Kanda, T., Noguchi, N., O'hara, K: FEMS Microbiol. Lett.,40,
223–228 (1987)
[2] Harwood, J.H., Smith, D.H.: J. Bacteriol.,3,1262–1271 (1969)

1 NOMENCLATURE

EC number
2.7.7.48

Systematic name
Nucleoside-triphosphate:RNA nucleotidyltransferase (RNA-directed)

Recommended name
RNA-directed RNA polymerase

Synonyms
RNA nucleotidyltransferase (RNA-directed)
Nucleotidyltransferase, ribonucleate, RNA-dependent
3D polymerase
PB1 proteins
PB2 proteins
Phage f2 replicase
Polymerase L
Proteins, PB 2
Proteins PB1
Proteins, specific or class, lambda3, of reovirus
Proteins, specific or class, PB 1
Proteins, specific or class, PB 2
Q-Beta replicase
Qbeta-replicase
Replicase, phage f2
Replicase, Qbeta
Ribonucleic acid replicase
Ribonucleic acid-dependent ribonucleate nucleotidyltransferase
Ribonucleic acid-dependent ribonucleic acid polymerase
Ribonucleic replicase
Ribonucleic synthetase
RNA replicase
RNA synthetase
RNA transcriptase
RNA-dependent ribonucleate nucleotidyltransferase
RDRP [11]
RNA-dependent RNA polymerase
RNA-dependent RNA replicase
Transcriptase
More (see also EC 2.7.7.6)

CAS Reg. No.
9026-28-2

2 REACTION AND SPECIFICITY

Catalyzed reaction
Nucleoside triphosphate + RNA$_n$ →
→ diphosphate + RNA$_{n+1}$

Reaction type
Nucleotidyl group transfer

Natural substrates
More (enzyme is responsible for replication of viral RNA [1, 2]) [1, 2]

Substrate spectrum
1 Nucleoside triphosphate + RNA$_n$ (NTP: GTP [1, 2], with all templates the initiating nucleoside triphosphate is GTP [2], ITP even at very high concentration cannot substitute for GTP [2], initiates new chains with purine ribonucleoside triphosphates [9], binds tenfold more tightly to Qbeta RNA than to nonhomologous RNA molecules [2], RNAs of brome mosaic virus and the closely related cowpea Chlorotic mottle virus are the most effective, some activity is also shown by certain other viral nucleic acids and polyribonucleotides [4], low activity with either single- or double-stranded DNA as template [6], transcription of RNA and DNA-oligonucleotide templates equally effective, differences in efficiency depend on nucleotide sequence rather than on RNA or DNA nature of the single-stranded nucleic acid [18], various species of RNA are effective as template [24], Encephalomyocarditis virus enzyme has no strict specificity towards EMC RNA template, can also use Qbeta RNA, rRNA of BHK cell, or poly(C) [8], a single-stranded molecule of RNA or polyribonucleotide is required as template [9], utilizes a variety of viral RNAs and CMV satellite RNA as template for minus-strand synthesis [14], dependent on and specific for brome mosaic virus RNAs [15], all RNAs and heteropolynucleotides are effective as templates [35], double-stranded RNA is active as template [35], DNA is inactive as template [35], rhinovirion RNA is copied with the highest efficiency by the rhinovirion enzyme, polyvirion RNA and globin mRNA with an efficiency of 50 to 60% [21], a primer can substitute for GTP to allow initiation [2], catalyzes poly(A)-dependent oligo(U)-primed poly(U)-polymerase activity as well as RNA polymerase activity [13], in presence of oligo(U) primer: synthesis of a full-length copy of either poliovirus or globin RNA templates [13], in absence of added primer: RNA products up to twice the length of the template are synthesized [13], reaction can be primed [18], unprimed transcription starts preferentially at the 3'-terminal nucleotides of the template [18], catalyzes RNA-template-directed extension of the 3'-end of an RNA strand by one nucleotide at a time, can initiate a chain de novo [2], deoxyribonucleoside triphosphates not accepted as substrates [17], double-stranded nucleic acids, e.g. poly(A), poly(U) and double stranded DNA are not transcribed [18], en-

zyme can add a single noncomplementary nucleotide to the 3'-terminus of about 50% of the runoff transcripts (AMP is preferred over GMP, CMP and UMP are terminally added at very low frequency) [18]) [1–35]

Product spectrum

1 Diphosphate + RNA_{n+1} (the 3'end of all product strands is an A residue [2], product has a size distribution similar to that of the template [8], both single-stranded and double-stranded RNA are present [10], characterization of products [16], at 40 min reaction time a major product of the reaction is double-stranded RNA or RNA that has a double-stranded core [24], product of the Cucumber mosaic virus RNA as template is heterogenous in size with a peak length of about 150 residues [29], single-stranded RNA transcripts are identical in size to the denatured parental double-stranded RNA segments [30], full-length negative strand black beetle virus RNAs are synthesized [33]) [2, 8, 10, 16, 24, 29, 30, 33]

Inhibitor(s)

NEM [17]; p-Hydroxymercuribenzoate [17]; RNA (high template RNA concentration) [15]; rRNA [14]; $(NH_4)_2SO_4$ (above 60 mM) [10]; Detergents (ionic and nonionic) [33]; Polyethylene sulfonate (inhibits initiation by competing with the template for binding to the enzyme, no inhibition of elongation of preinitiated RNA chains) [2]; Aurintricarboxylic acid (inhibits initiation by competing with the template for binding to the enzyme, no inhibition of elongation of preinitiated RNA chains) [2]; Poly(U) (inhibits initiation by competing with the template for binding to the enzyme, no inhibition of elongation of preinitiated RNA chains) [2]; K^+ (above 15 mM [11], 50 mM, 90% inhibition [21]) [11, 21]; $HgCl_2$ (inhibition reversed by DTT and 2-mercaptoethanol) [5]; Zn^{2+} (0.5–5 mM: inhibition, 0.05 mM: slight stimulation) [23]; Ca^{2+} (1 mM) [23]; PCMB (inhibition reversed by DTT and 2-mercaptoethanol) [5]; Diphosphate (sodium diphosphate [17]) [6, 17, 35]; Phosphonoacetic acid [22]; C-Substituted methylene biphosphonic acids and related tetrazole [22]; Putrescine [7]; Spermine [7, 17]; Cadaverine [7]; Guanidine [20]; Guanidine nucleotides [32]; Polylysine (strong) [7]; Polyornithine (strong) [7]; RNase [8, 24, 31]; RNase T1 [31]; Polyarginine (strong) [7]; Salmine (strong) [7]; Glycerol (addition of 30% v/v glycerol to the standard incubation mixture containing 5–10% glycerol reduces the activity down to 40%) [17]; Ethylene glycol (addition of 10% v/v ethylene glycol to the standard incubation mixture containing 5–10% glycerol reduces the activity down to 40%) [17]; Heparin [17]; Sucrose (addition of 10% v/v sucrose to the standard incubation mixture containing 5–10% glycerol reduces the activity down to 40%) [17]; High ionic strength (0.1, either NaCl or $(NH_4)_2SO_4$) [31]; More (not: actinomycin D [6, 8, 10, 17, 35], rifampicin [6, 8, 24, 31], rifampin [35], alpha-amanitin [6, 8, 17, 24], DNase (I [31]) [8, 10, 24, 31], exotoxin from Bacillus thuringiensis [24], deoxyribonuclease [6], phosphate [6], polysarcosine [7], polyglutamic acid [7], heparin sulfate [7]) [6–8]

Enzyme Handbook © Springer-Verlag Berlin Heidelberg 1997
Duplication, reproduction and storage in data banks are only
allowed with the prior permission of the publishers

Cofactor(s)/prosthetic group(s)/activating agents

Metal compounds/salts

Monovalent cations (increase the GTP requirement for initiation with all templates [2], no absolute requirement for K^+, above 15 mM: inhibition [11], enzyme activity could be increased by a factor 1.2–1.5 when final salt concentration is raised to 60 mM by the addition of NaCl, KCl or NH_4Cl, presence of 100 mM NaCl or KCl results in slight decrease of activity [17]) [2, 11, 17]; NaCl (maximal activity at 200 mM) [23]; Potassium acetate (60 mM, slight stimulation) [33]; Ammonium acetate (maximum activity at 50 to 100 mM) [23]; Mg^{2+} (required [8, 10, 11, 16, 17, 23–25, 27, 33], maximum concentration: 32 mM [4], 13 mM [6], 10–20 mM (magnesium acetate) [33], 8–20 mM [10], 2.8 [24], 10 mM [17], 3–30 mM [23], 5 mM [16], 10–12 mM [11], activity of bound enzyme increases sharply with Mg^{2+} concentration up to 10 mM and than levels off up to 50 mM, soluble enzyme is very sensitive to $MgCl_2$ with incorporation diminishing with increasing concentration [25]) [4, 6, 8, 10, 11, 16, 17, 23–25, 27, 33]; Mn^{2+} (reduces the template specificity of Qbeta replicase [2], reduces GTP initiation requirement for all templates [2], cannot effectively replace Mg^{2+} in activation [8], poor substitute for Mg^{2+}, in presence of 10 mM Mg^{2+}, 40% inhibition with 2 mM Mn^{2+} [10], can partially replace Mg^{2+} in activation [17, 23, 33], at optimal concentration of 1 mM, 20% of the activation with Mg^{2+} [17], maximum concentration: 1.6 mM [33]) [2, 8, 10, 17, 23, 33]; Sulfate (required) [27]; Zn^{2+} (stimulates [21], slight stimulation at 0.05 mM [23], inhibition at 0.5–5.0 mM [23], cannot substitute for Mg^{2+} [21, 23]) [21, 23]

Turnover number (min^{-1})

Specific activity (U/mg)

More [8, 15, 17, 28, 31, 33, 35]

K_m-value (mM)

0.001 (UTP (in presence of ATP, CTP and GTP, TMV RNA as template)) [17]; 0.035 (equimolar mixture of ATP, CTP and GTP (in presence of UTP, TMV RNA as template)) [17]; More [33]

pH-optimum

7.3–8.9 [25]; 7.6 [21]; 7.8 [17]; 8.0 [4, 8, 23]; 8.2 [10, 33]

pH-range

7.0–9.0 (7.0: about 70% of maximum activity, 9.0: about 40% of maximum activity) [17]; 7.4–8.0 (90% of maximum activity at pH 7.4 and 8.0) [21]; 7.9–8.5 (at least 90% of maximum activity at pH 7.9 and 8.5) [10]

Temperature optimum (°C)

29 (broad) [33]; 30 (assay at [1, 12, 31]) [1, 12, 25, 31]; 30–37 [23]; 37 (assay at [8, 11, 27, 29]) [8, 11, 17, 27, 29]

Temperature range (°C)
27–37 (27°C: 50% of maximum activity, 37°C: maximum activity) [17]

3 ENZYME STRUCTURE

Molecular weight
119000 (Lycopersicon esculentum, sucrose gradient centrifugation) [17]
120000 (Vigna unguiculata infected with cowpea mosaic virus, glycerol gradient centrifugation) [28]
130000 (E. coli infected with bacteriophage Qbeta, glycerol gradient sedimentation) [2]
160000 (maize infected with Maize dwarf mosaic virus, glycerol gradient centrifugation) [24]
More (molecular structure [12], nonstructural viral protein P1 is a component of the RNA-dependent RNA polymerase complex, P1 protein is associated with at least six proteins in the infected cell [15]) [12, 15]

Subunits
Monomer (1×130000, Vigna unguiculata infected with cowpea mosaic virus, analysis of the protein composition after glycerol gradient centrifugation) [28]
Tetramer (1×65000 (product of the phage genome) + 1×70000 (ribosomal protein S_1) + 1×45000 + 1×35000 (protein synthesis elongation factors EF-Tu and EF-Ts), in addition host-coded proteins called host factor (HF), which is a hexamer of 12500 MW subunits is also required in vitro Qbeta RNA replication, E. coli infected with bacteriophage Qbeta, SDS-PAGE [2], 1×74000 (ribosomal protein S_1, host-coded), + 1×47000 (elongation factor Tu, host-coded) + 1×36000 (elongation factor Ts, host-coded) + 60000 (product of the phage genome), E. coli Q13 infected with bacteriophage GA, SDS-PAGE [31]) [2, 31]
? ($x \times 128000$, Lycopersicon esculentum, SDS-PAGE [17], $x \times 100000$, Cucumber infected with Cucumber mosaic virus, SDS-PAGE, possibly other virus-induced or virus-coded polypeptides are lost during purification [29]) [17, 29]
More (molecular structure) [12]

Glycoprotein/Lipoprotein
–

4 ISOLATION/PREPARATION

Source organism
E. coli infected with bacteriophage Qbeta (QbetaamB86 [2]) [1–3, 5, 32]; Brome mosaic virus-infected barley [4, 15]; Cucumber mosaic virus-infected

cucumber (seedlings infected with Cucumber mosaic virus, no activity in healthy plant material [29]) [6, 27, 29]; Cucumber mosaic virus infected Nicotiana tabacum [14]; Foot- and mouth disease virus-infected baby hamster kidney cells [7]; Encephalomyocarditis virus-infected baby hamster kidney cells (EMC infected BHK-21 cells) [8]; Halobacterium cutirubrum [9]; Cowpea mosaic virus [10]; Kunjin virus-infected Vero cells [11]; Influenza virus (A/PR/8 [12]) [12, 22]; Poliovirus (expressed in E. coli) [13]; Poliovirus-infected HeLa cells (human) [20]; Foxtail mosaic potexvirus-infected Chenopodium quinoa [16]; Lycopersicon esculentum [17, 18]; Mouse hepatitis virus-infected mouse cells [19]; Rhinovirus-infected HeLa cells [21]; West nile virus-infected BHK-21/W12 cells [23]; Maize dwarf mosaic virus-infected maize [24]; Tobacco necrosis virus-infected Nicotiana tabacum [25]; Nicotiana tabacum (uninfected [25]) [25, 26]; Alfalfa mosaic virus-infected Nicotiana tabacum (enzyme activity in Nicotiana tabacum is mediated by a pre-existing host enzyme, possibly modified by virus-coded proteins) [26]; Vigna unguiculata infected with cowpea mosaic virus (host encoded enzyme) [28]; Nebraska calf diarrhea virus [30]; SA11 Rotavirus [30]; E. coli infected with bacteriophage GA [31]; Drosophila melanogaster infected with black beetle virus (no activity in uninfected cells) [33]; Measle virus [34]; Nicotiana tabacum (uninfected cells and cells infected with tobacco necrosis virus, tobacco contains the enzyme, the amount of which is increased by infection with RNA virus, without changes in its enzymatic properties) [35]

Source tissue

E. coli cells infected with bacteriophage Qbeta [1–3, 5, 32]; Infected barley leaves [4]; Cucumber mosaic virus infected cucumber plants [6]; Foot- and mouth disease virus-infected baby hamster kidney cells [7]; Encephalomyocarditis virus-infected baby hamster kidney cells [8]; Cowpea mosaic virus-infected cowpea leaves [10]; Cytoplasm of Kunjin-infected Vero cells [11]; Virus particles (core [30]) [12, 30, 34]; Cucumber mosaic virus-infected tobacco leaves [14]; Foxtail mosaic virus-infected leaves of Chenopodium quinoa [16]; Brome mosaic virus-infected barley [15]; Tomato leaves (systemically infected with potato spindle tuber viroid, the activity is increased about 3-fold compared with isolated healthy leaf tissue [17]) [17, 18]; Poliovirus-infected HeLa cells [20]; Rhinovirus-infected HeLa cells [21]; West Nile virus-infected BHK-21/W12 cells [23]; Maize leaves infected with Maize dwarf mosaic virus [24]; Leaves of Nicotiana tabacum infected with Tobacco necrosis virus and uninfected [25]; Leaves of Nicotiana tabacum infected with alfalfa mosaic virus and uninfected [26]; Leaves of Vigna unguiculata infected with cowpea mosaic virus [28]; Cucumber plant infected with Cucumber mosaic virus [29]; E. coli cells infected with bacteriophage GA [31]

Localization in source
    Membrane (of infected cells, smooth membranes of infected BHK-21 cells
    [8], cellular membrane of brome mosaic virus-infected barley [15], mem-
    brane-bound [10, 11, 16, 28]) [8, 10, 11, 15, 16, 19, 28]; Particulate [27, 33]

Purification
    E. coli infected with bacteriophage Qbeta [1–3]; E. coli infected with bacte-
    riophage GA [31]; Brome mosaic virus-infected barley (partial [4]) [4, 15];
    Cucumber mosaic virus-infected cucumber (partial [27], no activity in healthy
    plant material [29]) [6, 27, 29]; Encephalomyocarditis virus-infected baby
    hamster kidney cells [8]; Influenza virus A/PR/8 [12]; Poliovirus (expressed
    in E. coli) [13]; Foxtail mosaic potexvirus-infected Chenopodium quinoa
    (partial) [16]; Lycopersicon esculentum [17]; Rhinovirus-infected HeLa cells
    [21]; Vigna unguiculata [28]; Drosophila melanogaster infected with black
    beetle virus (no activity in uninfected cells) [33]; Measle virus [34]; Nicotiana
    tabacum (uninfected cells and cells infected with tobacco necrosis virus, to-
    bacco contains the enzyme, the amount of which is increased by infection
    with RNA virus, without changes in its enzymatic properties) [35]

Crystallization
    –

Cloned
    (poliovirus RNA polymerase expressed in E. coli) [13]

Renatured
    (EF-Tu * Ts complex rather than individual polypeptides function in the re-
    naturation of the Qbeta replicase [3], guanine nucleotides inhibit renatura-
    tion of 8 M urea denatured enzyme [32]) [1, 3, 32]

---

5 STABILITY

pH

Temperature (°C)

Oxidation

Organic solvent

General stability information
    Purification is vitiated by the great instability of the enzyme [4]; Withstands
    several freeze-thaw cycles [16]; Very dilute enzyme preparation loses signif-
    icant amounts of activity upon freezing and thawing [28]; Stability is ob-
    tained by adding bovine serum albumin 0.1 mg/ml in 25 or 50% glycerol
    buffers at –20°C or –80°C [28]; Purified enzyme loses most of its activity

Enzyme Handbook © Springer-Verlag Berlin Heidelberg 1997
Duplication, reproduction and storage in data banks are only
allowed with the prior permission of the publishers

within 45 min, when dialyzed against 10 mM Tris-HCl, pH 8.1, 0.025 M
NH_4Cl, 0.01 KCl, containing only 5% glycerol and no mercaptoethanol, but
little activity is lost under these conditions if the glycerol concentration is raised to 30% [35]

Storage
Unstable at 0°C and –15°C, partially purified enzyme [6]; In liquid nitrogen
stable for 5 months, partially purified enzyme [6]; –70°C, stable for several
months [16]; –20°C, 50% glycerol-containing buffer, 15% loss of activity per
month [17]; –80°C or in liquid N_2, protein concentration 0.05 mg/ml, 25%
glycerol, stable [28]; –20°C, half-life: 2 months, at any stage of purification
[31]; 0°C, 25% glycerol, half-life: 10 days, at any stage of purification [31];
–70°C, stable for several months [33]

6 CROSSREFERENCES TO STRUCTURE DATABANKS

PIR/MIPS code
PIR2:JQ1898 (Andean potato mottle virus (fragments)); PIR1:RRVQFL (beet
western yellows virus (isolate FL1)); PIR2:S25012 (garlic latent virus);
PIR2:B38196 (hepatitis E virus (fragment)); PIR1:RRVQLL (potato leaf roll virus (strain 1)); PIR1:RRVQWA (potato leaf roll virus (strain Wageningen));
PIR2:S18676 (Rift Valley fever virus); PIR3:S30026 (Rift Valley fever virus);
PIR2:S26765 (Saccharomyces cerevisiae virus L-A); PIR1:RRBWSC (south-
ern bean mosaic virus (strain cowpea)); PIR2:S01865 (tobacco rattle virus
(strain SYM)); PIR1:A45353 (apple chlorotic leaf spot virus (strain P863));
PIR1:A44059 (apple stem grooving virus (strain P-209)); PIR1:VFIHB2 (avian
infectious bronchitis virus (strain Beaudette)); PIR2:A29249 (avian infectious
bronchitis virus (strain KB8523) (fragment)); PIR2:JQ2034 (beet cryptic virus
3); PIR1:RRXRBT (bluetongue virus (serotype 10 American isolate));
PIR1:RRVUBY (bunyamwera virus); PIR1:A45389 (canine distemper virus
(strain Onderstepoort)); PIR1:RRVECV (carnation mottle virus); PIR2:JQ1246
(chrysanthemum virus B (fragment)); PIR1:RRVGCN (cucumber necrosis virus); PIR1:RRVGCR (Cymbidium ringspot virus); PIR1:RRWPEM (eggplant
mosaic virus); PIR1:RRWVEV (equine arteritis virus); PIR2:S10158 (equine
arteritis virus (fragment)); PIR1:JQ1555 (Erysimum latent virus);
PIR2:A40481 (feline calicivirus (fragment)); PIR1:RRWWF9 (feline calicivirus
(strain F9)); PIR2:S02068 (foot-and-mouth disease virus A); PIR2:JN0431
(foot-and-mouth disease virus A (strain A22)); PIR2:JQ1258 (foxtail mosaic
virus); PIR2:B49529 (human astrovirus type 1 (fragment)); PIR2:C49529 (human astrovirus type 1 (fragment)); PIR1:RRNZA2 (human respiratory syncytial virus (strain A2)); PIR1:RRXSIB (infectious bursal disease virus);
PIR1:RRXSI5 (infectious bursal disease virus (strain 52/70) (fragment));
PIR1:RRXSJA (infectious pancreatic necrosis virus (strain Jasper));
PIR1:RRXSSP (infectious pancreatic necrosis virus (strain Sp)); PIR2:S07418

(influenza A virus (strain A/FPV/Rostock/34 [H7N1])); PIR1:JQ0533 (Kennedya yellow mosaic virus (strain Jervis Bay)); PIR2:C46171 (Leishmania RNA virus 1); PIR1:RRXPLC (lymphocytic choriomeningitis virus (strain Armstrong 53b)); PIR1:RRIWMV (Marburg virus (strain Musoke)); PIR2:S44054 (Marburg virus (strain Popp)); PIR1:G48556 (measles virus (strain AIK-C)); PIR1:ZLNZMV (measles virus (strain Udem)); PIR1:A42548 (mumps virus (strain Miyahara)); PIR2:S15760 (murine hepatitis virus (strain A59)); PIR2:A32440 (murine hepatitis virus (strain A59) (fragment)); PIR2:A36388 (murine hepatitis virus (strain defective JHM) (fragment)); PIR2:A31167 (murine hepatitis virus (strain JHM defective-interfering particle)); PIR2:A39927 (murine hepatitis virus (strain MHV-A59 defective interfering particle) (fragment)); PIR2:JC4762 (Mycovirus FusoV); PIR1:RRWGNV (narcissus mosaic virus); PIR1:RRNZNV (Newcastle disease virus (strain Beaudette C)); PIR1:RRWPYM (Ononis yellow mosaic virus); PIR1:ZLNZP3 (parainfluenza virus type 3); PIR2:B46451 (parainfluenza virus type 3 (strain 47885) (fragment)); PIR1:RRXBPM (pea enation mosaic virus); PIR1:RRBPBQ (phage Q-beta); PIR2:A45392 (porcine reproductive and respiratory syndrome virus (fragment)); PIR2:S24285 (porcine respiratory virus (strain 86/137004) (fragment)); PIR2:S47422 (porcine transmissible gastroenteritis virus (fragment)); PIR2:A43489 (porcine transmissible gastroenteritis virus (strain FS772/70) (fragment)); PIR1:PN0093 (potato virus M (strain Russian)); PIR1:WMWGPV (potato virus X (strain X3)); PIR1:RRVUNE (Puumala virus (strain Hallnas B1)); PIR1:RRWWRH (rabbit hemorrhagic disease virus); PIR1:ZLVNPV (rabies virus (strain PV)); PIR1:ZLVNSB (rabies virus (strain SAD B19)); PIR1:B43684 (red clover necrotic mosaic virus (strain Australia)); PIR1:A43377 (rice dwarf virus); PIR2:S47307 (rinderpest virus); PIR2:S14223 (Saccharomyces cerevisiae virus L-A); PIR2:S12851 (Saccharomyces cerevisiae virus L-A); PIR1:ZLNZSE (Sendai virus (strain Enders)); PIR1:ZLNZSV (Sendai virus (strain Z)); PIR1:S16449 (Seoul virus (strain 80–39)); PIR1:JQ1734 (shallot virus X); PIR1:JQ1750 (simian paramyxovirus SV41 (strain Toshiba/Chanock)); PIR1:JQ1532 (simian paramyxovirus SV5 (strain 21004-WR)); PIR1:ZLVNSY (Sonchus yellow net virus (ATCC PV-263)); PIR1:RRWGSM (strawberry mild yellow edge-associated virus); PIR1:RRXPTV (Tacaribe virus); PIR2:S02643 (tobacco mosaic virus (fragment)); PIR1:RRWQTN (tobacco necrosis virus (strain A)); PIR1:RRWQTD (tobacco necrosis virus (strain D)); PIR1:RRVGCT (tomato bushy stunt virus (strain cherry)); PIR1:RRVUTW (tomato spotted wilt virus (strain BR-01)); PIR2:S29529 (Toscana virus); PIR1:RRVETC (turnip crinkle virus); PIR1:RRWPTM (turnip yellow mosaic virus (strain Australia)); PIR1:JQ1621 (Uukuniemi virus (strain S23)); PIR1:ZLVN (vesicular stomatitis Indiana virus (strain Mudd-Summers)); PIR1:ZLVNNJ (vesicular stomatitis New Jersey virus (strain Hazelhurst)); PIR1:A46309 (vesicular stomatitis New Jersey virus (strain Ogden)); PIR1:A46350 (white clover mosaic virus (strain O)); PIR2:A40894 (yeast (Saccharomyces cerevisiae) RNA replicon 20S);

PIR2:A40895 (yeast (Saccharomyces cerevisiae) RNA replicon 20S (fragment)); PIR2:A38149 (yeast (Saccharomyces cerevisiae) RNA replicon 23S); PIR1:P1IVDV (1 Dhori virus (strain Dhori/India/1313/61)); PIR2:PQ0408 (1 influenza A virus (fragment)); PIR1:P1IV61 (1 influenza A virus (strain A/Ann Arbor/6/60 [H2N2])); PIR1:B60011 (1 influenza A virus (strain A/Mallard/NY/6750/78)); PIR1:P1IV68 (1 influenza A virus (strain A/NT/60/68)); PIR1:P1IV34 (1 influenza A virus (strain A/PR/8/34)); PIR1:A60008 (1 influenza A virus (strain A/Victoria/3/75 [H3N2])); PIR1:P1IV33 (1 influenza A virus (strain A/WSN/33)); PIR2:PQ0414 (1 influenza A virus (strain A/Yamagata/120/86 [H1N1]) (fragment)); PIR2:S06212 (1 influenza A virus (strain avian/Kiev/59/79[H1N1])); PIR1:P1IVBC (1 influenza B virus (strain B/Ann Arbor/1/66 [cold-adapted])); PIR1:P1IVBW (1 influenza B virus (strain B/Ann Arbor/1/66 [wild-type])); PIR1:P1IVBL (1 influenza B virus (strain B/Lee/40)); PIR1:P1IV50 (1 influenza C virus (strain C/JJ/50)); PIR2:A36861 (1 Lelystad virus); PIR1:RRIHM2 (1a murine hepatitis virus (strain JHM)); PIR1:VFIHJH (1b murine hepatitis virus (strain JHM)); PIR1:P3IV61 (2 influenza A virus (strain A/Ann Arbor/6/60 [H2N2])); PIR2:JN0435 (2 influenza A virus (strain A/FPV/Weybridge [H7N7])); PIR2:PQ0421 (2 influenza A virus (strain A/Guizhou/54/89 [H3N2]) (fragment)); PIR2:PQ0409 (2 influenza A virus (strain A/Hebei/24/89 [H1N2]) (fragment)); PIR1:A60011 (2 influenza A virus (strain A/Mallard/NY/6750/78)); PIR1:P3IV68 (2 influenza A virus (strain A/NT/60/68)); PIR1:P3IV34 (2 influenza A virus (strain A/PR/8/34)); PIR1:B60008 (2 influenza A virus (strain A/Victoria/3/75 [H3N2])); PIR1:P3IV33 (2 influenza A virus (strain A/WSN/33)); PIR1:P3IVAK (2 influenza A virus (strain avian/Kiev/59/79 [H1N1])); PIR1:P3IVBC (2 influenza B virus (strain B/Ann Arbor/1/66 [cold-adapted])); PIR1:P3IVBW (2 influenza B virus (strain B/Ann Arbor/1/66 [wild-type])); PIR1:P3IV50 (2 influenza C virus (strain C/JJ/50)); PIR2:B36861 (2 Lelystad virus); PIR1:P2IV61 (3 influenza A virus (strain A/Ann Arbor/6/60 [H2N2])); PIR2:PQ0422 (3 influenza A virus (strain A/Guizhou/54/89 [H3N2]) (fragment)); PIR2:PQ0410 (3 influenza A virus (strain A/Hebei/24/89 [H1N2]) (fragment)); PIR1:C60011 (3 influenza A virus (strain A/Mallard/NY/6750/78)); PIR1:P2IV68 (3 influenza A virus (strain A/NT/60/68)); PIR1:P2IV34 (3 influenza A virus (strain A/PR/8/34)); PIR1:C60008 (3 influenza A virus (strain A/Victoria/3/75 [H3N2])); PIR1:P2IVWS (3 influenza A virus (strain A/WSN/33 [H1N1])); PIR2:PQ0416 (3 influenza A virus (strain A/Yamagata/120/86 [H1N1]) (fragment)); PIR1:P2IVBC (3 influenza B virus (strain B/Ann Arbor/1/66 [cold-adapted])); PIR1:P2IVBW (3 influenza B virus (strain B/Ann Arbor/1/66 [wild-type])); PIR1:P2IVBS (3 influenza B virus (strain B/Singapore/222/79)); PIR1:P2IV50 (3 influenza C virus (strain C/JJ/50)); PIR1:P2IVTV (3 Thogoto virus); PIR1:RRVQBM (38.8K chain barley yellow dwarf virus (strain MAV-PS1)); PIR1:RRVQCM (38.8K chain barley yellow dwarf virus (strain P-PAV)); PIR1:RRVQC2 (60.5K chain barley yellow dwarf virus); PIR2:PN0108 (60K protein barley stripe mosaic virus); PIR2:PN0102 (74K protein barley stripe mosaic virus); PIR2:PN0107 (85K protein barley stripe mosaic virus (strain 12–2)); PIR2:PN0105 (85K protein barley stripe mosaic virus (strain 4–2)); PIR2:PN0106 (85K protein barley stripe mosaic virus (strain 7–2)); PIR2:S08020 (beta chain phage fr); PIR1:RRBPBG (beta chain phage GA); PIR1:RRBPBM (beta chain phage MS2); PIR2:S04930 (chain PB2 influenza A virus (strain Chile/1/83[H1N1])); PIR1:P3IV2A (chain PB2 influenza A virus (strain FPV/Rostock/34 [H7N1])); PIR1:S00946 (39K barley yellow dwarf virus); PIR2:A40473 (/coat fusion protein clover yellow mosaic virus); PIR2:A40473 (/coat fusion protein clover yellow mosaic virus)

Brookhaven code

---

7 LITERATURE REFERENCES

[1] Blumenthal, T.: Methods Enzymol.,60,628–638 (1979) (Review)
[2] Blumenthal, T. in "The Enzymes",3rd Ed. (Boyer, P.D., Ed.) 15,267–279 (1982) (Review)
[3] Brown, S., Blumenthal, T.: J. Biol. Chem.,251,2749–2753 (1976)
[4] Hadidi, A., Fraenkel-Conrat, H.: Virology,52,363–372 (1973)
[5] Ohki, K., Hori, K.: Biochim. Biophys. Acta,281,233–243 (1972)
[6] Clark, G.L., Peden, K.W.C., Symons, R.H.: Virology,62,434–444 (1974)
[7] Lazarus, L.H., Itin, A.: Arch. Biochem. Biophys.,156,154–160 (1973)
[8] Traub, A., Diskin, B., Rosenberg, H., Kalmar, E.: J. Virol.,18,375–382 (1976)
[9] Louis, B.G., Fitt, P.S.: Biochem. J.,128,755–762 (1972)
[10] Zabel, P., Weenen-Swaans, H., van Kammen, A.: J. Virol.,14,1049–1055 (1974)
[11] Chu, P.W.G., Westaway, E.G.: Virology,157,330–337 (1987)
[12] Honda, A., Mukaigawa, J., Yokoiyama, A., Kato, A., Ueda, S., Nagata, K., Krystal, M., Nayak, D.P., Ishihama, A.: J. Biochem.,107,624–628 (1990)
[13] Plotch, S.J., Palant, O., Gluzman, Y.: J. Virol.,63,216–225 (1989)
[14] Quadt, R., Jaspars, E.M.J.: FEBS Lett.,279,273–276 (1991)
[15] Quadt, R., Jaspars, E.M.J.: Virology,178,189–194 (1990)
[16] Rouleau, M., Bancroft, J.B., Mackie, G.A.: Virology,197,695–703 (1993)
[17] Schiebel, W., Haas, B., Marinkovic, S., Klanner, A., Sänger, H.L.: J. Biol. Chem.,268, 11851–11857 (1993)
[18] Schiebel, W., Haas, B., Marinkovic, S., Klanner, A., Sänger, H.L.: J. Biol. Chem.,268, 11858–11867 (1993)
[19] Brayton, P.R., Stohlman, S.A., Lai, M.M.C.: Virology,133,197–201 (1984)
[20] Tershak, D.R.: J. Virol.,41,313–318 (1982)
[21] Morrow, C.D., Lubrinski, J., Hocko, J., Gibbons, G.F., Dasgupta, A.: J. Virol.,53, 266–272 (1985)
[22] Hutchinson, D.W., Naylor, M., Semple, G., Cload, P.A.: Biochem. Soc. Trans.,13, 752–753 (1985)
[23] Grun, J.B., Brinton, M.A.: J. Virol.,60,1113–1124 (1986)
[24] Donofrio, J.C., Kuchta, J., Moore, R., Kaczmarczyk, W.: Can. J. Microbiol.,32, 637–644 (1986)
[25] Stussi-Garaud, C., Lemius, J., Fraenkel-Conrat, H.: Virology,81,224–236 (1977)
[26] Clerx, C.M., Bol, J.F.: Virology,91,453–463 (1978)
[27] Gill, D.S., Kumarasamy, R., Symons, R.H.: Virology,113,1–8 (1981)

[28] Dorssers, L., Zabel, P., van der Meer, J., van Kammen, A.: Virology,116,236–249 (1982)
[29] Kumarasamy, R., Symons, R.H.: Virology,96,622–632 (1979)
[30] Bernstein, J.M., Hruska, J.F.: J. Virol.,37,1071–1074 (1981)
[31] Yonesaki, T., Haruna, I.: J. Biochem.,89,741–750 (1981)
[32] Blumenthal, T.: Biochim. Biophys. Acta,478,201–208 (1977)
[33] Guarino, L.A., Kaesberg, P.: J. Virol.,40,379–386 (1981)
[34] Seifried, A.S., Albrecht, P., Milstien, J.B.: J. Virol.,25,781–787 (1978)
[35] Ikegami, M., Fraenkel-Conrat, H.: J. Biol. Chem.,254,149–154 (1979)

1 NOMENCLATURE

EC number
2.7.7.49

Systematic name
Deoxynucleoside-triphosphate:DNA deoxynucleotidyltransferase (RNA-directed)

Recommended name
RNA-directed DNA polymerase

Synonyms
DNA nucleotidyltransferase (RNA-directed)
Reverse transcriptase
Revertase
Nucleotidyltransferase, deoxyribonucleate, RNA-dependent
RNA revertase
RNA-dependent DNA polymerase
RNA-instructed DNA polymerase
RT [2]
More (see EC 2.7.7.7)

CAS Reg. No.
9068-38-6

---

2 REACTION AND SPECIFICITY

Catalyzed reaction
Deoxynucleoside triphosphate + DNA_n →
→ diphosphate + DNA_{n+1} (mode of action [19], mechanism [26])

Reaction type
Nucleotidyl group transfer

Natural substrates
Deoxynucleoside triphosphate + DNA_n (high mutation rate in retroviruses: several models how DNA polymerases make mistakes, fidelity of reverse transcriptases [3], role in the life cycle of RNA tumor viruses [5, 11, 26], enzyme is responsible for the viral infectivity and synthesis of the viral DNA provirus intermediate [6], chicken embryo enzyme may play a role in normal differentiation [7]) [3, 5–7, 11, 26]

Substrate spectrum

1 Deoxynucleoside triphosphate + DNA$_n$ (catalyzes RNA-template-directed extension [1–37] of the 3'-end of a DNA strand by one deoxynucleotide at a time [6], cannot initiate a chain de novo, requires an RNA or DNA primer [5, 6, 8, 11], primer: efficiency of various primers [8], for homopolymeric templates the primer used is often a complementary deoxyribo-oligomer, if poly(A) is the template, only an oligomer of T can act as primer [5], in the virion, the primer for the transcription of 60–70 S viral RNA is transfer RNA [5], can be as short as 4–8 nucleotides [5], provides the 3'-OH end to form a phosphodiester bond with the substrate [5], the direction of the synthesis is from 5' to 3' [5], template: DNA can also serve as template [5, 6, 11, 26, 37], RNA∗DNA hybrid as template [11], preference [6], size [6], efficiency of various templates [8], transcribes both homopolymers and heteropolymers [5], reaction is dependent on the presence of all four deoxynucleoside triphosphates for maximal activity [10, 15, 35], synthetic partial ovalbumin mRNA with a synthetic DNA oligonucleotide annealed to the 3'-end of the RNA as model substrate [19], enzyme prefers the template-primer poly(rA)∗oligo(dT) over poly(rC)∗oligo(dG), only marginal activity with poly(rCm)∗oligo(dG), no activity with poly(dA)∗oligo(dT) [20], high affinity for template primers $(rC)_n$∗$(dG)_{12}$ and $(rCm)_n$∗$(dG)_{12}$ compared to $(rA)_n$∗$(dT)_{12}$ [29], transcribes $(rAm)_n$∗$(dT)_{12}$ very efficiently [29], activated DNA as a heteropolymeric substrate is used more efficiently than the homopolymeric substrate poly(rA)∗oligo(dT) which in turn is used 2fold more effectively as the template primer than poly(dC)∗oligo(dG) [37], Mauriceville plasmid enzyme initiates minus-strand DNA synthesis directly at the 3' end of the plasmid transcript rather than at a tRNA primer complementary to an internal region of the template RNA [21], reverse transcriptase also shows degradative activity characterized by ribonuclease H [5, 26, 27, 37]) [1–37]

Product spectrum

1 Diphosphate + DNA$_{n+1}$ [1–37]

Inhibitor(s)

Mn^{2+} (optimal activation at 0.5–1.0 mM, strong inhibition above 2 mM) [27]; Inorganic phosphate [27]; Mg^{2+} (optimum concentration: 0.5 mM, inhibition at higher concentrations) [27]; Diphosphate [27]; SH inhibitors [6]; Rifamycin SV derivatives [5, 6]; Na^+ (above 40 mM, SR-RSV enzyme assayed with RSV RNA, poly(rA)∗poly(dT) and DNA [6], 15–20% stimulation by 20–30 mM, inhibition at higher concentrations [7]) [6, 7]; K^+ (above 40 mM, SR-RSV enzyme assayed with RSV RNA, poly(rA)∗poly(dT) and DNA, 100 mM KCl, 2-fold stimulation of AMV enzyme, 5 mM K^+, 30% stimulation of R-MLV enzyme assayed with poly(dA-dT) [6], 15–20% stimulation by 20–30 mM, inhibition at higher concentrations [7], 5 mM KCl: 20% stimulation, 80 mM: 50% inhibition [10]) [6, 7, 10]; NH_4^+ (above 40 mM, SR-RSV enzyme assayed with RSV RNA, poly(rA) * poly(dT) and DNA) [6]; Streptovaricins [5]; NF 346 (suramin analogue) [20]; NF 345 (suramin analogue) [20]; Alkaloid extract [5]; Pyran copolymer [5]; Actinomycin D [5]; Daunomycin [5]; Distamycin [5]; Ethidium bromide [5]; Chromamycin [5]; Parsomycin [5]; Adriamycin [5]; Cinerubin [5]; Proflavin [5]; Tilurone [5]; Acridine orange [5]; Congo red [5]; Histone [5]; Protamine [5]; Thymidylate derivatives [5]; Polyribonucleotides (potency of inhibition: poly(U) > poly(G) > poly(A), poly(C)) [5]; Modified polyribonucleotides [5]; Thiolated polycytidylate [5]; 2'-O-Alkylated polyadenylic acid [5]; Streptonigrin (and amide-type derivatives [34], no inhibition by methylstreptonigrin [34]) [5, 34]; Bleomycin [5]; Heparin [5, 20]; N-Methylisatin-beta-thiosemicarbazone [5, 6]; Ara-CTP [5]; Ara-ATP [22]; Dideoxythymidine triphosphate [5]; 1,10-Phenanthroline [5]; NEM [5, 6, 27]; p-Substituted mercuribenzoate [6]; 2-Mercaptoethanol [6]; Nalidixic acid (noncompetitive to polyriboadenylic acid and to TTP, higher inhibition with polyriboadenylic acid than with polyribocytidylic acid as synthetic template) [13]; Flavonoids (flavonols and flavanonols are very active, flavones and flavanones display very low activity) [14]; Fisetin [14]; Quercetin [14]; Myricetin [14]; Kaempferol [14]; Morin [14]; Taxifolin [14]; (+)-Catechin [14]; (-)Epicatechin [14]; HPA 23 [16]; 3'-Azido-2',3'-dideoxythymidine 4'-triphosphate (AZTTP) (most potent and selective inhibitor) [18]; 2',3'-Didehydro-2',3'-dideoxythymidine 5'-triphosphate [18]; 3'-Azido-2',3'-dideoxythymidine 5'-diphosphate [18, 31]; 3'-Azido-2',3'-dideoxythymidine 5'-triphosphate [18]; 2',3'-Didehydro-2',3'-dideoxycytidine 5'-triphosphate [18]; 2',3'-Dideoxythymidine 5'-triphosphate [31]; 2',3'-Dideoxy-2',3'-dehydrothymidine 5'-triphosphate [33]; 3'-Amino-2',3'-dideoxythymidine 5'-triphosphate [31]; 2'-Deoxyxylofuranosylthymine 5'-triphosphate [31]; 2',3'-Dextran sulfate [20]; Suramin [20, 22, 23]; Chloroquin [22]; Phosphonoformate [22]; Foscarnet [23]; alpha-Anomeric oligonucleotides (interference with the primer binding sites) [25]; 3'-Hydroxymethyl 2'-deoxynucleoside 5'-triphosphates (highly specific inhibitors for AMV reverse transcriptase) [28]; 5-(p-Chlorobenzyl)-6-aminouracil [30]; 2',5'-Oligoadenylate (potency of inhibition is more marked in the absence than in the presence of sulfhydryl agents) [32]; Polynucleotides ((U)$_n$ and a series of (U)$_n$ analogs) [36]; Dideoxyadenosine triphosphate (competitive to dATP, noncompetitive to dCTP, dGTP and dTTP) [37]

Cofactor(s)/prosthetic group(s)/activating agents

SH reagents (required [6, 7, 10, 11, 15], e.g. beta-mercaptoethanol or DTT [6, 7, 10], optimum concentration: 1 mM [11]) [6, 7, 10, 11, 15]; Stimulatory protein (increases the rate and yield of DNA synthesized, the viral enzyme in conjugation with this protein transcribes extended single-stranded regions of DNA and permits the enzyme to initiate synthesis from single-strand breaks in DNA) [10]; Detergent (activates, optimal concentration: 0.05%-1%) [35]

Metal compounds/salts

Mg^{2+} (required [35], divalent cation Mg^{2+} or Mn^{2+} required [7, 10, 11, 27], activates [5, 6, 22, 35], prefers Mg^{2+} as divalent ion to transcribe $(rC)_n*(dG)_{12}$ and $(rA)_n*(dT)_{12}$ [29], K_m: 2.5 mM (R-MLV enzyme assayed with poly(dA-dT)) [6], optimum concentration: 0.5 mM (inhibition at higher concentrations) [27], 2–10 mM (depending on source of enzyme and template-primer used) [5], 5 mM (poly(rA)*oligo(dT) as substrate [20]) [20, 35], 5–9 mM [11], 10 mM (AMV enzyme assayed with AMV RNA [6]) [6, 10, 22], 5–10 mM (RNA tumor virus) [7], 15 mM (chicken [7], poly(rC)*oligo(dG) as substrate [20]) [7, 20], no stimulatory effect by Mg^{2+} and Mn^{2+} together [6]) [5–7, 10, 11, 20, 22, 27, 29, 35]; Mn^{2+} (divalent cation Mg^{2+} or Mn^{2+} required [7, 10, 11, 27], Mn^{2+} is a better activator than Mg^{2+} [27], activates [5, 6, 22], optimal concentration 0.8 mM [11], 0.1–2.0 mM (depending on source of enzyme and template-primer used) [5], 0.5–1.0 mM (strong inhibition above 2 mM) [27], 1–2 mM (RNA tumor virus) [7], 1 mM (chicken, 30% of the maximal activation with Mg^{2+} [7]) [7, 10], 10 mM [22], no stimulatory effect of Mg^{2+} and Mn^{2+} together [6], inhibitory in presence of stimulatory protein [10], almost no activation [20]) [5–7, 10, 11, 22, 27]; Zinc (zinc metalloenzyme) [5]; K^+ (100 mM KCl: 2-fold stimulation of AMV enzyme, 5 mM K^+: 30% stimulation of R-MLV enzyme assayed with poly(dA-dT)) [6], 15–20% stimulation by 20–30 mM, inhibition at higher concentrations [7], 5 mM KCl: 20% stimulation, 80 mM: 50% inhibition [10], stimulates [20], optimal concentration: 50 mM (with poly(rA)*oligo(dT) as substrate) [20], 10–100 mM (with poly(rC)*oligo(dG) as substrate) [20]) [6, 7, 10, 20]; Na^+ (15–20% stimulation by 20–30 mM, inhibition at higher concentrations [7], optimum concentration: 60–80 mM [27]) [7, 27]; More (enzyme requires a monovalent and a divalent cation) [15]

Turnover number (min^{-1})

Specific activity (U/mg)

More [10, 11, 20, 37]

K_m-value (mM)

0.0028 (dTTP) [37]; 0.003 (dATP) [37]; 0.0047 (dCTP) [37]; 0.0055 (dGTP) [37]; 0.01–0.03 (deoxynucleoside triphosphate) [10]; 0.015 (dATP, R-MLV enzyme assayed with poly(dA-dT)) [6]; 0.02 (dTTP, SR-RSV enzyme assayed with RSV RNA) [6]; 0.026 (dTTP, R-MLV enzyme assayed with poly(dA-dT)) [6]; 0.03 (dTTP) [31]

pH-optimum

7.5 (RSV enzyme assayed with RSV RNA, poly(rA)*poly(dT) and DNA) [6]; 7.8 [35]; 7.8–8.5 [11]; 8 (broad [7], poly(rA)*oligo(dT) as substrate [20]) [7, 20, 22]; 8.0–8.2 [37]; 8.0–9.5 (poly(rC)*oligo(dG) as substrate) [20]; 8.2 (AMV enzyme assayed with AMV RNA [6]) [6, 10]

pH-range

7.5–8.5 (rapid fall of activity below and above) [11]; 7.2–9.8 (50% of activity maximum at pH 9.8 and 7.2) [10]

Temperature optimum (°C)

25 [15, 16]; 37–40 (mammalian viruses) [7]; 40–45 (avian viruses) [7]; 42 [22]

Temperature range (°C)

---

3 ENZYME STRUCTURE

Molecular weight

68000 (gibbon ape leukemia virus) [26]
70000 (Rauscher mouse leukemia virus, gel filtration [6], Simian-sarcoma virus [26]) [6, 26]
70000–84000 (reticuloendotheliosis virus) [5, 26]
71000 (Moloney murine leukemia virus, enzyme expressed in E. coli, glycerol density gradient centrifugation) [27]
74000 (mouse leukemia virus) [9]
80000–100000 (Mason-Pfizer monkey virus, squirrel monkey retrovirus, Po-1-LV (Langur monkey), type D retroviruses) [26]
90000 (Rauscher mouse leukemia virus, glycerol density gradient centrifugation) [6]
95000–98000 (AIDS virus HTLV-III, gel filtration) [29]
100000 (mouse mammary tumor virus) [26]
120000 (hamster leukemia virus) [5, 26]
120000–130000 (Simian immunodeficiency virus TYO-7, velocity sedimentation analysis) [20]
170000 (avian RNA tumor viruses, glycerol density gradient sedimentation) [5, 26]
More (murine leukemia virus, 3 polypeptides MW 82000, MW 68000, MW 60000 do not appear in equimolar ratio, the 60000 and the 68000 polypeptide probably are degradation products of the 82000 polypeptide [5], avian myeloma virus (65000–70000 and 95000–110000 [6], 69000 and 110000 [11]) and Rous sarcoma virus exist in two forms with different size and polypeptide composition [6, 11]) [5, 6, 11]

Subunits

Monomer (1 × 70000–84000, Reticuloendotheliosis virus [5], 1 × 100000, mouse mammary tumor virus [26], 1 × 71000, Moloney murine leukemia virus, enzyme expressed in E. coli, SDS-PAGE [27], 1 × 80000–100000, Mason-Pfizer monkey virus, squirrel monkey retrovirus, Po-1-LV (Langur monkey), type D retroviruses [26]) [5, 26, 27]

Dimer (1 × 65000 (alpha) + 1 × 95000 (beta), avian RNA tumor viruses,
SDS-PAGE [5], 1 × 53000 + 1 × 68000, hamster leukemia virus [5, 26],
1 × 64000 + 1 × 50000, Simian immunodeficiency virus TYO-7, SDS-PAGE
[20]) [5, 20, 26]
? (x × 48000 + x × 64000, Simian immunodeficiency virus, denaturing PAGE)
[37]

Glycoprotein/Lipoprotein

–

---

4 ISOLATION/PREPARATION

Source organism

RNA tumor viruses [1–3, 5–7, 26]; AIDS virus HTLV-III [29]; HIV-1 (human
immunodeficiency virus type 1) [3]; Mammalian C-type viruses [6]; Murine
leukemia virus (Molony (expression in E. coli [27]) [14, 18, 25, 27], MuLV [5,
19]) [5, 14, 18, 19, 25, 27]; Rauscher mouse leukemia virus (R-MLV) [1, 6,
31]; Rous sarcoma virus [1, 2, 6, 11, 33]; Lymphadenopathy associated virus (LAV) [35]; Rous-associated virus-1 [6]; Rous-associated virus-0 [6];
Rous-associated virus 2 [19]; Avian sarcoma-leukosis viruses [6, 17]; Avian
myeloma virus (AMV, high rate of misincorporations [5]) [3, 5, 6, 11, 28, 32];
Avian myeloblastosis virus [8, 10, 11, 13, 19, 26, 30, 33, 34, 36]; Avian myeloma virus [6]; Viper C-type ribodeoxyvirus [6]; Mouse mammary tumor viruses [6, 26]; Mouse leukemia virus [9]; Visna virus [6]; Syncytium-forming
viruses [6]; Simian immunodeficiency virus (TYO-7, isolated from African
green monkey [20]) [20, 37]; Simian-sarcoma virus [5, 26]; Reticuloendotheliosis virus (REV [5]) [5, 6, 26]; Spleen necrosis virus (SNV) [5]; Duck infectious anemia virus [5]; Chick syncytial virus [5]; Duck hepatitis B virus [22];
Human T-cell lymphotrophic virus/lymphadenopathy-associated virus (expressed in E. coli) [23]; Hamster leukemia virus (HaLV) [5, 26]; E. coli [4];
Stigmatella aurantiaca [4]; Myxococcus xanthus [4]; Chicken (embryo cells,
noninfected) [5, 7]; Mouse (bone marrow cells infected with MuLV, spleen
cells infected with MuLV) [5]; Human (milk particles, leukemic cells) [5];
Xenopus laevis [12]; Drosophila melanogaster (retrovirus-like particles [15,
16], retrotransposon 1713 nucleotide sequence cloned and expressed in E.
coli [24]) [15, 16, 24]; Neurospora crassa (containing Mauriceville plasmids,
retroid elements that propagate in the mitochondria) [21]; Mouse mammary
tumor virus [26]; Mason-Pfizer monkey virus [26]; Squirrel monkey retrovirus
[26]; Po-1-LV (Langur monkey) [26]; Type D retroviruses [26]; Lymphadenopathy associated virus (LAV) [35]; Gibbon ape leukemia virus [26];
More (not: noninfectious hamster C-type RNA tumor virus (D9), mouse sarcoma viruses, presence of the enzyme in amphibian oocytes is not unam-
biguously proved) [5]

Source tissue

Core of the virion [5, 26]; RSV-infected cells [5]; BALB/3T3 cells (producing MuLV [5], MSV-transformed [8]) [5, 8]; Mouse bone marrow cells infected with MuLV [5]; Mouse spleen cells infected with MuLV [5]; NC37 cells infected by and producing simian sarcoma virus-1 or GaLV [5]; Mammalian cells chronically producing RD-114 virus [5]; Milk particles (human) [5]; Human leukemic cells [5]; Plasma of infected animals [6]; Embryos (chicken, noninfected) [7]; Ovaries [12]; Retrovirus-like particles (in Drosophila melanogaster) [15, 16]; Liver (duck, infected with duck hepatitis B virus) [22]

Localization in source

Purification

Avian myeloma virus [6]; Rauscher mouse leukemia virus [6]; Rous sarcoma virus (Schmidt-Ruppin strain) [6]; RNA tumor virus [7]; Chicken [7]; Avian myeloblastosis virus [8, 10, 11]; Avian sarcoma-leukosis virus (3 forms of enzyme) [17]; Mouse leukemia virus (partial) [9]; Human T-cell lymphotrophic virus/lymphadenopathy-associated virus (expressed in E. coli) [23]; Xenopus laevis [12]; Drosophila melanogaster (retrovirus-like particles) [16]; Simian immunodeficiency virus (TYO-7, isolated from African green monkey, partial [20]) [20, 37]; Neurospora crassa (containing Mauriceville plasmids, retroid elements that propagate in the mitochondria) [21]; Moloney murine leukemia virus (expression in E. coli) [27]; AIDS virus HTLV-III (enzyme is biochemically heterogenous) [29]; Lymphadenopathy associated virus (LAV) [35]; More (review of purification methods of the virus enzyme) [5, 6]

Crystallization

–

Cloned

[23, 24, 27]

Renatured

[9]

---

5 STABILITY

pH

Temperature (°C)

30 (15 min stable, with poly(rA) ∗ oligo(dT) as substrate, 20 min with poly(rC) ∗ oligo(dG) as substrate) [20]; 37 (10 min stable with poly(rA) ∗ oligo(dT) as substrate) [20]; 42 (15 min, 50% loss of activity) [27]; 45 (25 min: complete loss of pol(rA) ∗ oligo(dT)-dependent activity, 80% loss of poly(rC) ∗ oligo(dG)-dependent activity) [20]; 46 ($t_{1/2}$: 6.5 min (MuLV, with and without template), 7.5 min (alpha form of AMV, with and without template), 7.0 min (alpha,beta form of AMV, without template), 15.5 min (alpha,beta form of AMV, with template)) [5]

Oxidation

Organic solvent

General stability information
   Ammonium sulfate, 0.2 M stabilizes [10]; Rapid inactivation by repeated
   freezing and thawing [11]; Stable against freezing and thawing [23]; Stable
   against overnight dialysis [23]; Stable against high dilution [23]

Storage
   -20°C, 5 weeks, 90% recovery of AMV polymerase activity [6]; -10°C, 1
   month, 30–50% recovery of R-MLV polymerase activity [6]; -20°C, 50% glycerol, 2 mM dithioerythritol, less than 20% loss of activity after 2 months [10];
   -20°C or -70°C, 50% glycerol, retains full activity for about 6 months [11];
   -20°C; 50% glycerol, 0.2 mg/ml bovine serum albumin [12]; -20°C, 50%
   glycerol, stable [23, 26]; -70°C, 15% glycerol, stable [26]

---

6 CROSSREFERENCES TO STRUCTURE DATABANKS

PIR/MIPS code
   PIR2:S01651 (Chlamydomonas reinhardtii mitochondrion); PIR2:S27771 (Af-
   rican malaria mosquito transposon RT1 (fragment)); PIR2:S23312 (Arabi-
   dopsis thaliana retrotransposon Ta1–1 (fragment)); PIR1:RREC (Escherichia
   coli); PIR2:S28006 (Escherichia coli); PIR2:S19248 (Escherichia coli);
   PIR2:S32139 (human immunodeficiency virus type 1 (fragment));
   PIR2:S32071 (human immunodeficiency virus type 1 (fragment));
   PIR2:S32079 (human immunodeficiency virus type 1 (fragment));
   PIR2:S32077 (human immunodeficiency virus type 1 (fragment));
   PIR2:S32075 (human immunodeficiency virus type 1 (fragment));
   PIR2:S32074 (human immunodeficiency virus type 1 (fragment));
   PIR2:S32073 (human immunodeficiency virus type 1 (fragment));
   PIR2:S32085 (human immunodeficiency virus type 1 (fragment));
   PIR2:S32070 (human immunodeficiency virus type 1 (fragment));
   PIR2:S32067 (human immunodeficiency virus type 1 (fragment));
   PIR2:S32088 (human immunodeficiency virus type 1 (fragment));
   PIR2:S32080 (human immunodeficiency virus type 1 (fragment));
   PIR2:S32082 (human immunodeficiency virus type 1 (fragment));
   PIR2:S32083 (human immunodeficiency virus type 1 (fragment));
   PIR2:S32072 (human immunodeficiency virus type 1 (fragment));
   PIR2:S32087 (human immunodeficiency virus type 1 (fragment));
   PIR2:S32134 (human immunodeficiency virus type 1 (fragment));
   PIR2:S32081 (human immunodeficiency virus type 1 (fragment));
   PIR2:S32086 (human immunodeficiency virus type 1 (fragment));
   PIR2:S32138 (human immunodeficiency virus type 1 (fragment));
   PIR2:S32084 (human immunodeficiency virus type 1 (fragment));

PIR2:S32122 (human immunodeficiency virus type 1 (fragment));
PIR2:S32117 (human immunodeficiency virus type 1 (fragment));
PIR2:S32120 (human immunodeficiency virus type 1 (fragment));
PIR2:S32151 (human immunodeficiency virus type 1 (fragment));
PIR2:S32096 (human immunodeficiency virus type 1 (fragment));
PIR2:S32066 (human immunodeficiency virus type 1 (fragment));
PIR2:S32140 (human immunodeficiency virus type 1 (fragment));
PIR2:S32119 (human immunodeficiency virus type 1 (fragment));
PIR2:S32063 (human immunodeficiency virus type 1 (fragment));
PIR2:S32137 (human immunodeficiency virus type 1 (fragment));
PIR2:S32059 (human immunodeficiency virus type 1 (fragment));
PIR2:S32053 (human immunodeficiency virus type 1 (fragment));
PIR2:S32095 (human immunodeficiency virus type 1 (fragment));
PIR2:S32056 (human immunodeficiency virus type 1 (fragment));
PIR2:S32055 (human immunodeficiency virus type 1 (fragment));
PIR2:S32064 (human immunodeficiency virus type 1 (fragment));
PIR2:S32054 (human immunodeficiency virus type 1 (fragment));
PIR2:S32052 (human immunodeficiency virus type 1 (fragment));
PIR2:S32062 (human immunodeficiency virus type 1 (fragment));
PIR2:S32061 (human immunodeficiency virus type 1 (fragment));
PIR2:S32058 (human immunodeficiency virus type 1 (fragment));
PIR2:S32131 (human immunodeficiency virus type 1 (fragment));
PIR2:S32060 (human immunodeficiency virus type 1 (fragment));
PIR2:S32057 (human immunodeficiency virus type 1 (fragment));
PIR2:S32051 (human immunodeficiency virus type 1 (fragment));
PIR2:S32089 (human immunodeficiency virus type 1 (fragment));
PIR2:S32049 (human immunodeficiency virus type 1 (fragment));
PIR2:S32090 (human immunodeficiency virus type 1 (fragment));
PIR2:S32048 (human immunodeficiency virus type 1 (fragment));
PIR2:S32098 (human immunodeficiency virus type 1 (fragment));
PIR2:S32091 (human immunodeficiency virus type 1 (fragment));
PIR2:S32050 (human immunodeficiency virus type 1 (fragment));
PIR2:S32047 (human immunodeficiency virus type 1 (fragment));
PIR2:S32092 (human immunodeficiency virus type 1 (fragment));
PIR2:S32093 (human immunodeficiency virus type 1 (fragment));
PIR2:S32094 (human immunodeficiency virus type 1 (fragment));
PIR2:S32159 (human immunodeficiency virus type 1 (fragment));
PIR2:S32157 (human immunodeficiency virus type 1 (fragment));
PIR2:S32160 (human immunodeficiency virus type 1 (fragment));
PIR2:S32078 (human immunodeficiency virus type 1 (fragment));
PIR2:S32135 (human immunodeficiency virus type 1 (fragment));
PIR2:S32136 (human immunodeficiency virus type 1 (fragment));
PIR2:S32069 (human immunodeficiency virus type 1 (fragment));
PIR2:S32068 (human immunodeficiency virus type 1 (fragment));

PIR2:S32126 (human immunodeficiency virus type 1 (fragment));
PIR2:S32076 (human immunodeficiency virus type 1 (fragment));
PIR2:S32127 (human immunodeficiency virus type 1 (fragment));
PIR2:S32132 (human immunodeficiency virus type 1 (fragment));
PIR2:S32118 (human immunodeficiency virus type 1 (fragment));
PIR2:S32129 (human immunodeficiency virus type 1 (fragment));
PIR2:S32133 (human immunodeficiency virus type 1 (fragment));
PIR2:S32128 (human immunodeficiency virus type 1 (fragment));
PIR2:S32065 (human immunodeficiency virus type 1 (fragment));
PIR2:S32152 (human immunodeficiency virus type 1 (fragment));
PIR2:S32097 (human immunodeficiency virus type 1 (fragment));
PIR2:A47330 (human immunodeficiency virus type 1 (strain ERS100) (fragment)); PIR2:B47330 (human immunodeficiency virus type 1 (strain ERS101) (fragment)); PIR2:C47330 (human immunodeficiency virus type 1 (strain ERS103) (fragment)); PIR2:D47330 (human immunodeficiency virus type 1 (strain ERS104) (fragment)); PIR2:E47330 (human immunodeficiency virus type 1 (strain ERS200) (fragment)); PIR2:F47330 (human immunodeficiency virus type 1 (strain ERS201) (fragment)); PIR2:S02391 (human T-cell lymphotropic virus type 1 (fragment)); PIR2:S27768 (maize transposon (fragment)); PIR2:A05073 (mouse mammary tumor virus (fragment)); PIR2:A05072 (squirrel monkey retrovirus (fragment)); PIR2:A42383 (Stigmatella aurantiaca); PIR2:B35890 (51K chain human immunodeficiency virus type 1 (fragment)); PIR2:A35890 (66K chain human immunodeficiency virus type 1 (fragment)); PIR2:B27672 (homolog (R1) silkworm); PIR2:E41830 (msDNA-Ec73 specific phage phi-R73); PIR1:RRYC62 (msDNA-Mx162 specific Myxococcus xanthus); PIR1:RRYC65 (msDNA-Mx65 specific Myxococcus xanthus)

Brookhaven code

1HAR (Human Immunodeficiency virus type 1 (Hiv-1) strain: bh10 expression system: (Escherichia coli) plasmid: prt21); 1HMI (Human Immunodeficiency virus type 1 (Bh10 isolate) expressed in (Escherichia coli)); 1HMV (Organism: Human immunodeficiency virus type 1; Strain: bh10 isolate; Expression system: escherichia coli; Strain: ar120; Plasmid: potskf33; Gene: hiv-1 pol); 1HNI (Hiv-1 (Bh10 Isolate) expressed in (Escherichia coli)); 1HNV (Hiv-1 (Bh10 Isolate) expressed in (Escherichia coli)); 3HVT (Hiv-1 (Bh10 Isolate)); 1RDH (Human Immunodeficiency virus type 1 recombinant form expressed in (Escherichia coli)); 1RVL (Hiv-I (Bh10 Isolate)); 1RVM (Hiv-I (Bh10 Isolate)); 1RVN (Hiv-I (Bh10 Isolate)); 1RVO (Hiv-I (Bh10 Isolate)); 1RVP (Hiv-I (Bh10 Isolate)); 1RVQ (Hiv-I (Bh10 Isolate)); 1RVR (Hiv-I (Bh10 Isolate))

7 LITERATURE REFERENCES

[1] Baltimore, D.: Nature,226,1209–1211 (1970)
[2] Temin, H.M., Mizutani, S.: Nature,226,1211–1213 (1970)
[3] Bebenek, K., Kunkel, T.A.: Cold Spring Harbor Monogr. Ser.,23,85–102 (1993) (Review)
[4] Inouye, S., Inouye, M.: Cold Spring Harbor Monogr. Ser.,23,391–410 (1993) (Review)
[5] Verma, I.M.: Biochim. Biophys. Acta,473,1–38 (1977) (Review)
[6] Temin, H.M., Mizutani, S. in "The Enzymes",3rd Ed. (Boyer, P.D., ed.) 10,211–235, Academic, New York (1974) (Review)
[7] Mizutani, S., Kang, C.-Y., Temin, H.M.: Methods Enzymol.,29E,119–124 (1974) (Review)
[8] Verma, I.M., Baltimore, D.: Methods Enzymol.,29E,125–130 (1974) (Review)
[9] Scolnick, E.M., Parks, W.P.: Methods Enzymol.,29E,130–143 (1974) (Review)
[10] Leis, J., Hurwitz, J.: Methods Enzymol.,29E,143–150 (1974) (Review)
[11] Kacian, D.L., Spiegelman, S.: Methods Enzymol.,29E,150–173 (1974) (Review)
[12] Brown, R.D., Tocchini-Valentini, G.P.: Methods Enzymol.,29E,173–177 (1974) (Review)
[13] Aoyama, H.: Mol. Cell. Biochem.,108,169–174 (1991)
[14] Chu, S.C., Hsieh, Y.S., Lin, J.Y.: J. Nat. Prod.,55,179–183 (1992)
[15] Lescault, A., Becker, J.L., Barre-Sinoussi, F., Chermann, J.C., Best-Belpomme, M., Ono, K.: Cell. Mol. Biol.,35,163–171 (1989)
[16] Becker, J.L., Barre-Sinoussi, F., Dormont, D., Best-Belpomme, M., Chermann, J.C.: Cell. Mol. Biol.,33,225–235 (1987)
[17] Kato, A., Ishihama, A., Noda, A., Ueda, S.: J. Virol. Methods,9,325–339 (1984)
[18] Ono, K., Nakane, H., Herdewijn, P., Balzarini, J., De Clerq, E.: Nucleic Acids Res., 20,5–6 (1988)
[19] Oyama, F., Kikuchi, R., Crouch, R.J., Uchida, T.: J. Biol. Chem.,264,18808–18817 (1989)
[20] Lüke, W., Hoefer, K., Moosmayer, D., Nickel, P., Hunsmann, G., Jentsch, K.-D.: Biochemistry,29,1764–1769 (1990)
[21] Wang, H., Kennell, J.C., Kuiper, M.T.R., Sabourin, J.R., Saldanha, R., Lambowitz, A.M.: Mol. Cell. Biol.,12,5131–5144 (1992)
[22] Offensperger, W.-B., Walter, E., Offensperger, S., Zeschnigk, C., Blum, H.E., Gerok, W.: Virology,164,48–54 (1988)
[23] Hansen, J., Schulze, T., Moelling, K.: J. Biol. Chem.,262,12393–12396 (1987)
[24] Champion, S., Maisonhaute, C., Kim, M.H., Best-Belpomme, M.: Eur. J. Biochem., 209,523–531 (1992)
[25] Lavignon, M., Bertrand, J.-R., Rayner, B., Imbach, J.-L., Malvy, C., Paoletti, C.: Biochem. Biophys. Res. Commun.,161,1184–1190 (1989)
[26] Verma, I.M. in "The Enzymes",3rd ed. (Boyer, P.D., Ed.) 14,87–103 (1981) (Review)
[27] Roth, M.J., Tanese, N., Goff, S.P.: J. Biol. Chem.,260,9326–9335 (1985)
[28] Kutateladze, T.V., Kritzyn, A.M., Florentjev, V.L., Kavsan, V.M., Chidgeavadze, Z.G., Beabealashvilli, R.Sh.: FEBS Lett.,207,205–212 (1986)
[29] Chandra, A., Gerber, T., Chandra, P.: FEBS Lett.,197,84–88 (1986)
[30] Wright, G.E., Brown, N.C.: Biochem. Biophys. Res. Commun.,126,109–116 (1985)
[31] Ono, K., Ogasawara, M., Iwata, Y., Nakane, H., Fujii, T., Sawai, K., Saneyoshi, M.: Biochem. Biophys. Res. Commun.,140,498–507 (1986)
[32] Liu, D.K., Owens, G.F.: Biochem. Biophys. Res. Commun.,145,291–297 (1987)

[33] Dyatkina, N., Minassian, S., Kukhanova, M., Krayevsky, A., von Janta-Lipinsky, M.,
     Chidgeavadze, Z., Beabealashvilli, R.: FEBS Lett.,219,151–155 (1987)
[34] Okada, H., Inouye, Y., Nakamura, S.: J. Antibiot.,40,230–232 (1987)
[35] Rey, M.A., Spire, B., Dormont, D., Barre-Sinoussi, F., Montagnier, L., Chermann,
     J.C.: Biochem. Biophys. Res. Commun.,121,126–133 (1984)
[36] Warwick-Koochaki, P.E., Bobst, A.M.: Arch. Biochem. Biophys.,228,425–430 (1984)
[37] Kraus, G., Behr, E., Baier, M., König, H., Kurth, R.: Eur. J. Biochem.,192,207–213
     (1990)

1 NOMENCLATURE

EC number
2.7.7.50

Systematic name
GTP:mRNA guanylyltransferase

Recommended name
mRNA guanylyltransferase

Synonyms
Messenger RNA guanylyltransferase [11]
Protein lambda2 [20]
mRNA capping enzyme (has mRNA guanylyltransferase and RNA 5'-tri-phosphatase activity [3, 6], activities associated with vaccinia capping enzyme complex: 1. GTP-RNA guanylyltransferase, 2. RNA (guanine-7-)-methyltransferase, 3. RNA triphosphatase, 4. GTP-diphosphate exchange, 5. nucleoside triphosphate phosphorylase [7, 9])

CAS Reg. No.
56941-23-2

2 REACTION AND SPECIFICITY

Catalyzed reaction
GTP + (5')ppPur-mRNA →
→ diphosphate + G(5')pppPur-mRNA (mRNA containing a guanosine residue linked 5' through three phosphates to the 5' position of the terminal residue, mechanism [7, 8, 18] of capping [18])

Reaction type
Nucleotidyl group transfer

Natural substrates
More (messenger RNA capping enzyme [3], specific post-transcriptional modification of the 5'-terminus of mRNA [4], RNA polymerase II primary transcripts are substrates for the cellular capping enzyme [5]) [3–5]

Substrate spectrum

1 GTP + ppRNA (r [1, 7, 19], specific for GTP [1, 7], only the alpha-phos-
   phate is transferred [4], acceptor: diphosphate terminated poly(A) (HeLa
   cells [7]) [7, 15, 18], unmethylated vaccinia virus mRNA [4], no apparent
   base specificity for the penultimate nucleotide, a variety of synthetic ho-
   moribopolymers and naturally occuring mRNAs are effective substrates
   (vaccinia virus) [7], enzyme can modify synthetic poly(A) to form the
   structure $m^7G(5')ppp(5')AmP-$ [4]) [1-20, 22]
2 GTP + pppRNA (r [21], specifically requires 5'-triphosphate-terminated
   RNA chains [21]) [12, 13, 21]
3 dGTP + ppRNA (not [18]) [7, 19]
4 ITAP + ppRNA [18]
5 GTP gamma S + ppRNA [7]
6 PppG + pppA(pA)$_n$ [21]
7 Dinucleotide + GTP (e.g.: pppGpC [7], pppApG [7], ppApG (vaccinia virus) [7, 12, 18], ppGpC [12]) [7, 12, 18]
8 Diphosphate + (7Me)GpppA(pA)$_n$ [21]
9 More (no donor: 7-methylGTP [7], ATP [7, 18], CTP [7, 18], UTP [7, 18],
   GDP [7, 18], GMP [7], ADP [7], m^7GTP [18], no acceptor: RNA with a single 5'-terminal phosphate [1, 7, 16, 18], RNA with 5'-hydroxyl terminus [7],
   ppGp [12], specificity overview [18], enzyme also catalyzes GTP-diphosphate exchange [7, 12] in absence of acceptor RNA, the enzyme forms a
   nucleotidyl intermediate by phosphoamidate linkage of GMP [12], enzyme forms a covalent enzyme-GTP intermediate of apparent MW 45000
   [14], little sequence specificity for RNA acceptor [16], lacks strict sequence specificity, homoribonucleotides containing purines are preferred,
   in presence of diphosphate the enzyme catalyzes the phosphorolysis of
   the dinucleoside triphosphate G(5')ppA but not of $m^7(5')ppA$ [19]) [1, 7,
   12, 14, 16, 18, 19]

Product spectrum

1 G(5')pppRNA + diphosphate [1-20]
2 G(5')ppppRNA + diphosphate [13]
3 ?
4 ?
5 ?
6 ?
7 GpppA(pA)$_n$ + phosphate + diphosphate [21]
8 (7Me)GTP + ? [21]
9 ?

Inhibitor(s)

Zn^{2+} (above 10 mM) [8]; Co^{2+} (above 0.1 mM) [8]; Diphosphate [1, 7, 18];
NaCl (in excess of 0.1 M) [7]; Phosphate (above 20 mM) [7]; N-Ethylmaleimide [7]; EDTA [8]; Mn^{2+} (can partially replace Mg^{2+} in activation, inhibits in
presence of Mg^{2+}) [12]

Cofactor(s)/prosthetic group(s)/activating agents

S-Adenosylmethionine (stimulates) [7]; Bovine serum albumin (stimulates) [21]

Metal compounds/salts

Mg^{2+} (required [1, 4, 7, 12, 14, 21], can partially replace Mn^{2+} in activation [17], maximal activity: 3 mM [12], 0.5 mM [14], 2–5 mM [17], 0.5–2 mM [21]) [1, 4, 7, 12, 14, 17, 21]; Mn^{2+} (required [17], can partially replace Mg^{2+} in activation [1, 4, 7, 12, 21], 11% [1], 10% [12] of the activity with Mg^{2+}, inhibits in presence of Mg^{2+} [12], maximal activation: 1 mM [12], 2 mM [17]) [1, 4, 7, 12, 17, 21]; Na^+ (stimulates, maximal activation at 50–75 mM NaCl) [21]; K^+ (stimulates, maximal activation at 6 mM KCl) [21]; Ca^{2+} (can partially replace Mg^{2+} in activation) [21]; More (no activation by Ca^{2+} or Zn^{2+}) [7]

Turnover number (min^{-1})

Specific activity (U/mg)

More [2, 9, 11, 13, 15, 16, 20]; 0.000066 [17]; 7.27 [3]

K_m-value (mM)

0.000014 (ppA(pA)$_n$) [16]; 0.000019 (diphosphate-ended poly(A)) [18]; 0.000140 (lambdac17RNA) [18]; 0.0002 (termini of 5'-triphosphate poly(A)) [21]; 0.000285 (ppApGp [7, 18], HeLa cells [7]) [7, 18]; 0.0005 (2 K_m values: 0.0005 and 0.004, ppGCC(A$_2$,U$_2$G)$_n$) [12]; 0.0011 (GTP) [18]; 0.0027 (GTP) [16]; 0.004 (2 K_m values: 0.0005 and 0.004, ppGCC(A$_2$,U$_2$G)$_n$) [12]; 0.015 (GTP, Vaccinia virus) [7]; 0.017 (GTP) [21]; 0.019 (diphosphate terminated poly(A) with an average chain length of 2000 nucleotides) [7]

pH-optimum

7.0 [12]; 7.5 (HeLa cells) [7, 17]; 7.8 (Tris-HCl buffer, Vaccinia virus [7]) [1, 7]

pH-range

6.4–7.9 (6.4: about 30% of activity maximum, 7.9: about 25% of activity maximum) [12]; 6.5–8.5 (6.5: about 45% of activity maximum, 8.5: about 15% of activity maximum) [17]

Temperature optimum (°C)

37 (assay at) [2, 17]

Temperature range (°C)

3 ENZYME STRUCTURE

Molecular weight

48500 (human, HeLa cells, sucrose density gradient sedimentation) [7, 17]
65000 (rat, gel filtration) [7]
120000 (Vaccinia virus, copurifies with S-adenosylmethionine mRNA (guanine-7)-methyltransferase, sucrose density gradient centrifugation) [22]

127000 (Vaccinia virus, gel filtration, sucrose density gradient sedimentation) [2]
130000 (Saccharomyces cerevisiae, gel filtration) [11]
140000 (Saccharomyces cerevisiae, glycerol gradient sedimentation) [11]
180000 (Saccharomyces cerevisiae, glycerol gradient sedimentation) [3]
More (the capping enzyme has 2 subunits: MW 95000 and 31000, the 95000 MW subunit of the Vaccinia virus capping enzyme has guanylyltransferase activity, glycerol gradient centrifugation, the isolated 95000 MW guanylyltransferase can be converted to an active 60000 MW form in vitro by limited proteolysis with trypsin, the guanylyltransferase domain is localized to the amino two-thirds of the 95000 MW polypeptide) [10]

Subunits

More (Vaccinia virus, SDS-PAGE, 95000 MW and 31400 MW polypeptides are polypeptide components of the 127000 MW enzyme system [2], activities associated with vaccinia capping enzyme complex: 1. GTP-RNA guanylyltransferase, 2. RNA (guanine-7)-methyltransferase, 3. RNA triphosphatase, 4. GTP-diphosphate exchange, 5. nucleoside triphosphate phosphorylase [7, 9]) [2, 7, 9]
? (x × 59000, Vaccinia virus, guanylyltransferase lacking 7-methyltransferase activity, SDS-PAGE [22], x × 80000 (beta, RNA 5'-triphosphatase) + x × 52000 (alpha, mRNA guanylyltransferase activity), Saccharomyces cerevisiae, SDS-PAGE [3], x × 65000, bovine, SDS-PAGE [15]) [3, 15, 22]
Oligomer (x × 45000 (alpha) + x × 39000 (beta), Saccharomyces cerevisiae, SDS-PAGE, probably alpha$_2$beta$_2$) [11]

Glycoprotein/Lipoprotein

–

---

4 ISOLATION/PREPARATION

Source organism

Vaccinia virus (strain WR [2], expression in E. coli [9, 10]) [1, 2, 4, 7, 9, 10, 19, 21, 22]; Saccharomyces cerevisiae (expressed in E. coli [6], pep4, protease-deficient mutant [14]) [3, 6, 11–14]; Human (HeLa cells) [5, 7, 17, 18]; Rat [7]; Tobacco mosaic virus [8]; Bovine (calf) [15]; Wheat [16]; Reovirus serotype 3 (from vaccinia virus strain WR into whose TK gene of the reovirus lamda2 genome segment under the control of the CPV AT, protein gene promoter has been inserted, possesses neither nucleoside nor RNA triphosphatase activity nor methyltransferase activity) [20]

Source tissue

Liver [7]; Cores [1, 4, 19, 22]; HeLa cells [5, 7, 17, 18]; Thymus [15]; Germ [16]

Localization in source
Nucleus [3, 5, 7, 17]

Purification
Vaccinia virus (partial [9, 10], copurifies with S-adenosylmethionine mRNA (guanine-7)-methyltransferase [2, 22]) [2, 9, 10, 22]; Saccharomyces cerevisiae (large scale, enzyme has mRNA guanylyltransferase activity [3], partial [14], physically associated with mRNA 5'-triphosphatase activity [11, 12, 14], contains little or no RNA 5'-triphosphatase or methyltransferase activity [13]) [3, 11–14]; Human (HeLa cells) [17]; Reovirus serotype 3 [20]; Bovine (calf) [15]; Wheat [16]

Crystallization
–

Cloned
(mRNA guanylyltransferase subunit) [6]

Renatured
–

---

5 STABILITY

pH

Temperature (°C)

Oxidation

Organic solvent

General stability information
Can be frozen and thawed several times without apparent loss of activity [15]; Stable after a few cycles of freezing and thawing [17]

Storage
–80°C, stable for at least 6 months [3]; 4°C, stable for more than 6 months [15]; –20°C, stable for more than 6 months, considerably longer at –70°C [16]; –70°C, stable for at least 9 months [17]; –20°C, 10 mM Tris-HCl buffer, pH 8, 10 mM 2-mercaptoethanol, 50% glycerol, 0.1 mg/ml gelatin, enzyme concentration: 0.1 mg/ml, stable [20]; 0°C, 48 h, 90% loss of activity, 59000 MW protein which lacks 7-methyltransferase activity [21]

6 CROSSREFERENCES TO STRUCTURE DATABANKS

PIR/MIPS code

PIR1:RMXRR3 (reovirus type 3); PIR2:S59731 (yeast (Saccharomyces cerevisiae)); PIR1:A45391 (large chain African swine fever virus (strain BA71V)); PIR1:QQVZRA (large chain rabbit fibroma virus); PIR2:S33105 (small chain variola virus)

Brookhaven code

7 LITERATURE REFERENCES

[1] Martin, S.A., Moss, B.: J. Biol. Chem.,250,9330–9335 (1975)
[2] Martin, S.A., Paoletti, E., Moss, B.: J. Biol. Chem.,250,9322–9329 (1975)
[3] Itho, N., Yamada, H., Kaziro, Y., Mizumoto, K.: J. Biol. Chem.,262,1989–1995 (1987)
[4] Ensinger, M.J., Martin, S.A., Paoletti, E., Moss, B.: Proc. Natl. Acad. Sci. USA,72, 2525–2529 (1975)
[5] Groner, Y., Aviv, H.: Biochemistry,17,977–982 (1978)
[6] Shibagaki, Y., Gilboa, E., Itho, N., Yamada, H., Nagata, S., Mizumoto, K.: J. Biol. Chem.,267,9521–9528 (1992)
[7] Shuman, S., Hurwitz, J. in "The Enzymes",3rd Ed. (Eds., Boyer, P.D.) 15,245–265 (1982) (Review)
[8] Dunigan, D.D., Zaitlin, M.: J. Biol. Chem.,265,7779–7786 (1990)
[9] Shuman, S.: J. Biol. Chem.,265,11960–11966 (1990)
[10] Shuman, S., Morham, S.G.: J. Biol. Chem.,265,11967–11972 (1990)
[11] Itoh, N., Mizumoto, K., Kaziro, Y.: J. Biol. Chem.,259,13923–13929 (1984)
[12] Itoh, N., Mizumoto, K., Kaziro, Y.: J. Biol. Chem.,259,13930–13936 (1984)
[13] Wang, D., Shatkin, A.J.: Nucleic Acids Res.,12,2303–2315 (1984)
[14] Itoh, N., Mizumoto, K., Kaziro, Y.: FEBS Lett.,155,161–166 (1983)
[15] Nishikawa, Y., Chambon, P.: EMBO J.,1,485–492 (1982)
[16] Keith, J.M., Venkatesan, S., Gershowitz, A., Moss, B.: Biochemistry,21,327–333 (1982)
[17] Venkatesan, S., Gershowitz, A., Moss, B.: J. Biol. Chem.,255,2829–2834 (1980)
[18] Vankatesan, S., Moss, B.: J. Biol. Chem.,255,2835–2842 (1990)
[19] Martin, S.A., Moss, B.: J. Biol. Chem.,251,7313–7321 (1976)
[20] Mao, Z., Joklik, W.K.: Virology,185,377–386 (1991)
[21] Monroy, G., Spencer, E., Hurwitz, J.: J. Biol. Chem.,253,4490–4498 (1978)
[22] Monroy, G., Spencer, E., Hurwitz, J.: J. Biol. Chem.,253,4481–4489 (1978)

1 NOMENCLATURE

EC number
2.7.7.51

Systematic name
Adenylylsulfate:ammonia adenylyltransferase

Recommended name
Adenylylsulfate-ammonia adenylyltransferase

Synonyms
Adenylyltransferase, adenylylsulfate-ammonia
Adenylyl sulfate:ammonia adenylyl transferase [1]
APSAT [1]

CAS Reg. No.
79121-94-1

2 REACTION AND SPECIFICITY

Catalyzed reaction
Adenylylsulfate + NH_3 →
→ adenosine 5'-phosphoramidate + sulfate

Reaction type
Nucleotidyl group transfer

Natural substrates
Adenylylsulfate + NH_3 (the product adenosine 5'-phosphoramidate replaces AMP as an activator of phosphorylase b, threonine dehydratase and adenylate cyclase) [3]

Substrate spectrum
1 Adenylylsulfate + NH_3 (a large variety of ammonia analogs such as amines, amides etc. will not replace ammonia [1], very high specificity for the adenylylsulfate and ammonia [3]) [1–4]

Product spectrum
1 Adenosine 5'-phosphoramidate + sulfate [1–4]

Inhibitor(s)

Cofactor(s)/prosthetic group(s)/activating agents

Metal compounds/salts

Turnover number (min⁻¹)

Specific activity (U/mg)
  More [3]

K_m-value (mM)
  0.82 (adenylylsulfate) [1, 3, 4]; 10 (ammonia) [1, 3, 4]

pH-optimum
  8.8 [1, 3, 4]

pH-range

Temperature optimum (°C)
  30 (assay at) [3, 4]

Temperature range (°C)

---

3 ENZYME STRUCTURE

Molecular weight
  60000–65000 (Chlorella pyrenoidosa [3, 4], Chlorella sp. [1], PAGE [1, 3, 4])
  [1, 3, 4]

Subunits
  Trimer (1 × 26000 + 1 × 21000 + 1 × 17000, Chlorella sp. [1], Chlorella pyrenoidosa [3, 4], SDS-PAGE [1, 3, 4]) [1, 3, 4]

Glycoprotein/Lipoprotein
  –

---

4 ISOLATION/PREPARATION

Source organism
  Chlorella sp. [1]; Plants [2]; Microorganisms [2]; Chlorella pyrenoidosa [3, 4]; Euglena gracilis var. bacillaris [4]; Euglena sp. [3]; Spinacia oleracea [3, 4]; Hordeum vulgare [3, 4]; Dictyostelium discoideum [3, 4]; E. coli [3, 4]

Source tissue

Localization in source

Purification
  Chlorella sp. [1]; Chlorella pyrenoidosa [3, 4]

Crystallization
  –

Cloned
  –

2

Renatured

–

---

5 STABILITY

pH

Temperature (°C)

Oxidation

Organic solvent

General stability information

Storage

---

6 CROSSREFERENCES TO STRUCTURE DATABANKS

PIR/MIPS code

Brookhaven code

---

7 LITERATURE REFERENCES

[1] Frankhauser, H., Schiff, J.A.: Plant Physiol.,65S,17 (1979)
[2] Frankhauser, H., Garber, L., Schiff, J.A.: Plant Physiol.,63S,162 (1979)
[3] Frankhauser, H., Schiff, J.A., Garber, L.J., Saidha, T.: Methods Enzymol.,143,354–361 (1987) (Review)
[4] Frankhauser, H., Schiff, J.A., Garber, L.J.: Biochem. J.,195,545–560 (1981)

1 NOMENCLATURE

EC number
2.7.7.52

Systematic name
UTP:RNA uridylyltransferase

Recommended name
RNA uridylyltransferase

Synonyms
Terminal uridylyltransferase
Uridylyltransferase, terminal
TUT [1]

CAS Reg. No.
78519-53-6

2 REACTION AND SPECIFICITY

Catalyzed reaction
$UTP + RNA_n \rightarrow$
$\rightarrow diphosphate + RNA_{n+1}$

Reaction type
Nucleotidyl group transfer

Natural substrates
More (RNA uridylyltransferase might function in uridylating specific proteins,
RNA is not a natural substrate [1], acts as a host factor to initiate RNA syn-
thesis by poliovirus RNA polymerase in vitro [3]) [1, 3]

Substrate spectrum
1 $UTP + RNA_n$ (the enzyme requires a single-stranded oligoribonucleotide
or polyribonucleotide with a free terminal 3'-OH as primer, e.g. $oligoA_{20}$,
$tRNA^{Asp}$, E. coli RNA, alfalafa mosaic virus RNA 4 [1], 3'-poly(A) of virion
RNA [3], marked specificity for UTP [1]) [1, 3]

Product spectrum
1 $Diphosphate + RNA_{n+1}$ [1]

Inhibitor(s)
Heparin [1, 2]; Cibacron blue F3GA [1]; Aurintricarboxylic acid [1]; Diphosphate [1]; Ionic strength (quite sensitive to ionic strength, activity decreases by 50% at about 40 mM $(NH_4)_2SO_4$ or 125 mM potassium acetate) [1]; More (not inhibitory: actinomycin C, rifamycin, alpha-amanitin, phosphate) [1]

Cofactor(s)/prosthetic group(s)/activating agents

Metal compounds/salts
Mg^{2+} (divalent cation required, Mg^{2+} or Mn^{2+}, 4–12 mM stimulates) [1]; Mn^{2+} (divalent cation required, Mg^{2+} or Mn^{2+}, 1.1–1.5 mM stimulates) [1]

Turnover number (min^{-1})

Specific activity (U/mg)
More [1]

K_m-value (mM)

pH-optimum

pH-range

Temperature optimum (°C)
30 [1]

Temperature range (°C)

3 ENZYME STRUCTURE

Molecular weight
50000 (Vigna unguiculata, velocity sedimentation) [1]
68000 (human, gel electrophoresis) [3]

Subunits

Glycoprotein/Lipoprotein
–

4 ISOLATION/PREPARATION

Source organism
Vigna unguiculata [1]; Leishmania tarentola [2]; Human (HeLa cells) [3]

Source tissue
Leaf [1]; HeLa cells [3]

Localization in source
Membrane [1]; Mitochondria [2]

Purification
  Vigna unguiculata [1]; Human [3]

Crystallization
  –

Cloned
  –

Renatured
  –

---

5 STABILITY

pH

Temperature (°C)

Oxidation

Organic solvent

General stability information

Storage
  4°C, does not tolerate more than several h at [3]; –70°C, 25% glycerol, stable [3]

---

6 CROSSREFERENCES TO STRUCTURE DATABANKS

PIR/MIPS code

Brookhaven code

---

7 LITERATURE REFERENCES

[1] Zabel, P., Dorssers, L., Wernars, K., van Kammen, A.: Nucleic Acids Res.,9,
    2433–2453 (1981)
[2] Bakalara, N., Simpson, A.M., Simpson, L.: J. Biol. Chem.,264,18679–18686 (1989)
[3] Andrews, N.C., Baltimore, D.: Proc. Natl. Acad. Sci. USA,83,221–225 (1986)

1 NOMENCLATURE

EC number
2.7.7.53

Systematic name
ADP:ATP adenylyltransferase

Recommended name
ATP adenylyltransferase

Synonyms
Adenylyltransferase, adenine triphosphate
Bis(5'-nucleosyl)-tetraphosphate phosphorylase (NDP-forming)
Diadenosinetetraphosphate alphabeta-phosphorylase
Diadenosine 5',5'''-P^1,P^4-tetraphosphate alpha,beta-phosphorylase
(ADP-forming)
Dinucleoside oligophosphate alpha,beta-phosphorylase [7]

CAS Reg. No.
96697-71-1

---

2 REACTION AND SPECIFICITY

Catalyzed reaction
ADP + ATP →
→ phosphate + P^1,P^4-bis(5'-adenosyl) tetraphosphate

Reaction type
Nucleotidyl group transfer

Natural substrates
P^1,P^4-bis(5'-adenosyl) tetraphosphate + phosphate (involved in catabolism
of dinucleoside polyphosphates) [2]

Substrate spectrum
1 P^1,P^4-bis(5'-adenosyl) tetraphosphate + phosphate (i.e. Ap_4A or Appp-
pA, r, substrate specificity [5, 7], arsenate [1, 6], vanadate, molybdate,
chromate [6, 7] or tungstate, not sulfate, can substitute for phosphate
[6], $ApppCH_2pA$ can substitute for Ap_4A [5, 7]. No substrates are
phosphonate analogs of Ap_4A [5], ADP, ATP, NAD^+ [1, 6], p_4A, poly(A),
3',5'-cAMP, $NADP^+$, Ap_2A, thymidine 5'-monophosphate-p-nitrophenyles-
ter [1], AMP, p-nitrophenylthymidine 5'-triphosphate, bis-p-nitrophenyl
phosphate [6]) [1–7]

2 P^1,P^4-bis(5'-adenosyl) triphosphate + phosphate (i.e. Ap_3A, not [6]) [7]

3 P^1,P^4-bis(5'-adenosyl) pentaphosphate + phosphate (i.e. Ap_5A, arsenate can replace phosphate) [1, 7]

4 P^1,P^4-bis(5'-guanosyl) tetraphosphate + phosphate (i.e. Gp_4G, arsenate can replace phosphate [1]) [1, 7]

5 P^1,P^4-bis(5'-adenosyl) tetraphosphate + arsenate (arsenolysis, at about 80% the rate of phosphorolysis [1]) [1, 6]

6 Ap_4C + phosphate [2]

7 Ap_4G + phosphate [2]

8 ADP + ATP (r) [2, 6]

9 NTP + N'DP (N is A or G (not C, U or dA), N' is A, C, G, U or dA, asymmetric catalytic site: synthesis of Ap_4G or Ap_4C is much faster from ATP plus CDP (or GDP) than from CTP (or GTP) plus ADP) [2]

10 NDP + phosphate (exchange reaction between beta-phosphate of NDP and phosphate from the medium [4]) [4, 7]

11 NDP + arsenate [4, 7]

12 ADP + arsenate [6]

Product spectrum

1 ADP + ATP [1–7]

2 ?

3 ADP + p_4A [1]

4 GDP + GTP [1]

5 AMP + ATP [1]

6 CDP + ATP (not CTP + ADP) [2]

7 GDP + ATP [2]

8 P^1,P^4-bis(5'-adenosyl) tetraphosphate + phosphate [2]

9 Np_4N'+ phosphate (N is A or G, N' is A, C, G, U or dA) [2]

10 Phosphate + NDP [4, 7]

11 NMP + phosphate + ? [4, 7]

12 AMP + phosphate + ? [6]

Inhibitor(s)

Iodoacetate [1]; p-Hydroxymercuribenzoate [1]; EDTA [2]; N-Ethylmaleimide [1]; Dithioerythritol [1]; Ap_4A (ADP/phosphate-exchange) [4]; AMP (ADP/phosphate-exchange) [4]; Ap_3A (ADP/phosphate-exchange) [4]; Mg^{2+} (only NDP-arsenolysis or NDP/phosphate-exchange reaction) [4]; Cd^{2+} (only NDP-arsenolysis or NDP/phosphate-exchange reaction) [4]; Mn^{2+} (weak, NDP-arsenolysis or NDP/phosphate-exchange reaction) [4]; Ca^{2+} (weak, NDP-arsenolysis or NDP/phosphate-exchange reaction) [4]; Co^{2+} (weak, NDP-arsenolysis or NDP/phosphate-exchange reaction) [4]; More (no inhibition by F^- or adenosine 5'-tetraphosphate) [6]

Cofactor(s)/prosthetic group(s)/activating agents

Metal compounds/salts

Mn^{2+} (requirement, Ap_4A-synthesis [2], 10 mM [6], not NDP-arsenolysis or NDP/phosphate-exchange reaction [4], less effective than Mg^{2+} or Co^{2+} [6]) [1, 2, 4, 6]; Mg^{2+} (activation, Ap_4A-synthesis [2], less effective than Mn^{2+} [1, 2] or Ca^{2+} [2]) [1, 2, 6]; Ca^{2+} (requirement, less effective than Mn^{2+} [1], Co^{2+} [6] or Mg^{2+} [1, 6], as good as Mn^{2+} (Ap_4A-synthesis [2]) [2, 6]) [1, 2, 6]; Co^{2+} (activation, less effective than Mn^{2+}, Ca^{2+} [1] or Mg^{2+} [1, 6], slight activation of Ap_4A-synthesis [2]) [1, 2, 6]; Zn^{2+} (slight activation of Ap_4A-synthesis [2], not [1]) [2, 6]; Cd^{2+} (slight activation) [1, 6]; Divalent metal cations (requirement) [7]; More (no activation by Ni^{2+} [1, 2], Cu^{2+} [2], no divalent metal cation requirement for NDP-arsenolysis or NDP/phosphate-exchange reaction [4, 7]) [1, 2, 4, 7]

Turnover number (min^{-1})

Specific activity (U/mg)

0.002 (Scenedesmus obliquus) [6]; 18.8 [3]; 55 [1]

K_m-value (mM)

0.0053 (Ap_4A) [6]; 0.025 (Ap_3A) [7]; 0.027 (Ap_4A) [7]; 0.06 (Ap_4A) [1]; 0.07 (vanadate) [6]; 0.13 (arsenate) [6]; 0.16 (phosphate) [6]; 0.31 (ADP (+ ATP)) [2]; 0.45 (molybdate) [6]; 0.5 (phosphate) [7]; 0.52 (chromate) [6]; 0.7 (ADP, ADP/phosphate-exchange) [4]; 1 (phosphate) [1]; 1.75 (tungstate) [6]; 2 (phosphate, ADP/phosphate-exchange) [4]; 3 (arsenate) [1]; 5.7 (ATP (+ ADP)) [2]

pH-optimum

More (pH-dependence of NDP/phosphate-exchange reaction) [4]; 5.9 (Ap_4A-synthesis) [2]; 6.5 (UDP (+ phosphate), NDP/phosphate-exchange, ADP, GDP or CDP (+ arsenate), NDP-arsenolysis) [4]; 7 (ADP or CDP (+ phosphate), NDP/phosphate-exchange) [4]; 7.5–9.5 (Scenedesmus obliquus) [6]; 8 (Ap_4A-phosphorolysis [1], Euglena gracilis [7], GDP (+ phosphate), NDP/phosphate-exchange [4]) [1, 4, 7]

pH-range

6.5–9 (about half-maximal activity at pH 6.5 and 9, Ap_4A-phosphorolysis) [1]; 6.5–9.8 (about 20% of maximal activity at pH 6.5 and about 80% of maximal activity at pH 9.8, Scenedesmus obliquus) [6]

Temperature optimum (°C)

25 (assay at, Ap_4A-phosphorolysis) [6]; 30 (assay at, Ap_4A-synthesis) [6]; 37 (assay at) [1–4]

Temperature range (°C)

3 ENZYME STRUCTURE

Molecular weight
30000 (Euglena gracilis) [7]
40000 (Saccharomyces cerevisiae, gel filtration) [1]
46000–48000 (Scenedesmus obliquus, gel filtration) [6]

Subunits
Monomer (1 × 40000, Saccharomyces cerevisiae, SDS-PAGE [1], 1 × 46000, Scenedesmus obliquus, SDS-PAGE [6]) [1, 6]

Glycoprotein/Lipoprotein
–

4 ISOLATION/PREPARATION

Source organism
Saccharomyces cerevisiae (strain CGY 339 [3]) [1–5]; Scenedesmus obliquus (green alga) [6]; Scenedesmus basiliensis (green alga) [6]; Scenedesmus quadricauda (green alga) [6]; Chlorella vulgaris (green alga) [6]; Euglena gracilis (green alga) [7]; Acanthamoeba castellanii [7]

Source tissue
Cell [1–7]

Localization in source

Purification
Saccharomyces cerevisiae (immunoaffinity chromatography) [1, 3]; Scenedesmus obliquus [6]; Euglena gracilis [7]

Crystallization
–

Cloned
–

Renatured
–

5 STABILITY

pH

Temperature (°C)

Oxidation

Organic solvent

General stability information

Glycerol, 10%, stabilizes [1, 4]; Most stable in Tris or phosphate buffered saline, 20 mM potassium phosphate, pH 6.8, NaCl solution or 0.2 M glycine plus 0.1 mM EGTA, pH 2.7 [3]; Dithioerythritol stabilizes [4]

Storage

−20°C, 30 mM HEPES/KOH buffer, pH 7.8, 50% glycerol, stable [6]

---

6 CROSSREFERENCES TO STRUCTURE DATABANKS

PIR/MIPS code

PIR1:XXBYP1 (I yeast (Saccharomyces cerevisiae)); PIR2:A37836 (II yeast (Saccharomyces cerevisiae)); PIR1:XXBYP1 (I yeast (Saccharomyces cerevisiae))

Brookhaven code

---

7 LITERATURE REFERENCES

[1] Guranowski, A., Blanquet, S.: J. Biol. Chem.,260,3542–3547 (1985)
[2] Brevet, A., Coste, H., Fromant, M., Plateau, P., Blanquet, S.: Biochemistry,26, 4763–4768 (1987)
[3] Avila, D.M., Kaushal, V., Barnes, L.D.: Biotechnol. Appl. Biochem.,12,276–283 (1990)
[4] Guranowski, A., Blanquet, S.: J. Biol. Chem.,261,5943–5946 (1986)
[5] Guranowski, A., Biryukov, A., Tarussova, N.B., Khomutov, R.M., Jakubowski, H.: Biochemistry,26,3425–3429 (1987)
[6] McLennan, A.G., Mayers, E., Hankin, S., Thorne, N.M.H., Prescott, M., Powls, R.: Biochem. J.,300,183–189 (1994)
[7] Guranowski, A., Starzynska, E., Wasternack, C.: Int. J. Biochem.,20,449–455 (1988)

1 NOMENCLATURE

EC number
2.7.7.54

Systematic name
ATP:L-phenylalanine adenylyltransferase

Recommended name
Phenylalanine adenylyltransferase

Synonyms
Adenylyltransferase, phenylalanine
L-Phenylalanine adenylyltransferase [1, 2]

CAS Reg. No.
98285-55-3

---

2 REACTION AND SPECIFICITY

Catalyzed reaction
ATP + L-phenylalanine →
→ diphosphate + N-adenylyl-L-phenylalanine

Reaction type
Nucleotidyl group transfer

Natural substrates
More (part of the system for biosynthesis of the alkaloid cyclopeptine in Penicillium cyclopium) [1, 2]

Substrate spectrum
1 ATP + L-phenylalanine [1, 2]

Product spectrum
1 N-Adenylyl-L-phenylalanine + diphosphate (i.e. L-phenylalanylAMP) [2]

Inhibitor(s)

Cofactor(s)/prosthetic group(s)/activating agents

Metal compounds/salts

Turnover number (min^{-1})

Specific activity (U/mg)

K_m-value (mM)

pH-optimum
    7.5 (assay at) [2]

pH-range

Temperature optimum (°C)
    35 (assay at) [2]

Temperature range (°C)

---

3 ENZYME STRUCTURE

Molecular weight

Subunits

Glycoprotein/Lipoprotein
    –

---

4 ISOLATION/PREPARATION

Source organism
    Penicillium cyclopium (strain SM72) [1, 2]

Source tissue
    Hyphae (surface culture, activity is measurable at the beginning of the idio-
    phase and reaches a maximum 6 days after inoculation) [1]

Localization in source

Purification

Crystallization
    –

Cloned
    –

Renatured
    –

5 STABILITY

pH

Temperature (°C)

Oxidation

Organic solvent

General stability information

Storage

6 CROSSREFERENCES TO STRUCTURE DATABANKS

PIR/MIPS code

Brookhaven code

7 LITERATURE REFERENCES

[1] Lerbs, W., Luckner, M.: J. Basic Microbiol.,25,387–391 (1985)
[2] Gerlach, M., Schwelle, N., Lerbs, W., Luckner, M.: Phytochemistry,24,1935–1939
(1985)

1 NOMENCLATURE

EC number
2.7.7.55

Systematic name
ATP:anthranilate N-adenylyltransferase

Recommended name
Anthranilate adenylyltransferase

Synonyms
Adenylyltransferase, anthranilate
Anthranilic acid adenylyltransferase [1, 2]

CAS Reg. No.
70248-64-5

2 REACTION AND SPECIFICITY

Catalyzed reaction
ATP + anthranilate →
→ diphosphate + N-adenylylanthranilate

Reaction type
Nucleotidyl group transfer

Natural substrates
More (part of the system for biosynthesis of the alkaloid cyclopeptine in Penicillium cyclopium) [1, 2]

Substrate spectrum
1 ATP + anthranilate [1, 2]

Product spectrum
1 N-Adenylylanthranilate + diphosphate (i.e. anthranilylAMP) [2]

Inhibitor(s)

Cofactor(s)/prosthetic group(s)/activating agents

Metal compounds/salts

Turnover number (min⁻¹)

Specific activity (U/mg)

K_m-value (mM)

pH-optimum
    7.5 (assay at) [2]

pH-range

Temperature optimum (°C)
    35 (assay at) [2]

Temperature range (°C)

---

3 ENZYME STRUCTURE

Molecular weight

Subunits

Glycoprotein/Lipoprotein
    −

---

4 ISOLATION/PREPARATION

Source organism
    Penicillium cyclopium (strain SM72 [1, 2]) [1–3]

Source tissue
    Hyphae (surface culture, activity is measurable at the beginning of the idio-
    phase and reaches a maximum 6 days after inoculation) [1]; Conidiospores
    (constitutive) [3]

Localization in source

Purification

Crystallization
    −

Cloned
    −

Renatured
    −

5 STABILITY

pH

Temperature (°C)

Oxidation

Organic solvent

General stability information

Storage

6 CROSSREFERENCES TO STRUCTURE DATABANKS

PIR/MIPS code

Brookhaven code

7 LITERATURE REFERENCES

[1] Lerbs, W., Luckner, M.: J. Basic Microbiol.,25,387–391 (1985)
[2] Gerlach, M., Schwelle, N., Lerbs, W., Luckner, M.: Phytochemistry,24,1935–1939 (1985)
[3] Voigt, S., El Kousy, S., Schwelle, N., Nover, L., Luckner, M.: Phytochemistry,17, 1705–1709 (1978)

1 NOMENCLATURE

EC number
2.7.7.56

Systematic name
tRNA:orthophosphate nucleotidyltransferase

Recommended name
tRNA nucleotidyltransferase

Synonyms
Phosphate-dependent exonuclease
RNase PH
Nuclease, ribo-, PH
Ribonuclease PH
More (not identical with EC 2.7.7.8)

CAS Reg. No.
116412-36-3

---

2 REACTION AND SPECIFICITY

Catalyzed reaction
$tRNA_{n+1}$ + phosphate $\rightarrow$
$\rightarrow tRNA_n$ + a nucleoside diphosphate

Reaction type
Nucleotidyl group transfer

Natural substrates
More (possible role in tRNA processing and RNA degradation [4], brings
about the final exonucleolytic trimming of the 3'-terminus of tRNA precursors
in E. coli by phosphorolysis, producing a mature 3'-terminus on tRNA and
nucleoside diphosphate [1–3], implicated in the 3'-processing of tRNA precursors [5]) [1–5]

Substrate spectrum
1 $tRNA_{n+1}$ + phosphate (r [2, 4, 5], $tRNA\text{-}C\text{-}C\text{-}A\text{-}C_n$ [2, 3], phosphorolysis of
poly(A) 15times more rapidly than of $tRNA\text{-}C\text{-}C\text{-}A\text{-}C_n$ [2]) [1–5]

Product spectrum
1 tRNA + a nucleoside diphosphate

Inhibitor(s)
KCl (enzyme works optimally at 50 mM KCl, inhibition at 200 mM) [1];
N-Ethylmaleimide [4]; p-Hydroxymercuribenzoate [4]; Phosphate [5]; More
(not: diphosphate) [5]

Cofactor(s)/prosthetic group(s)/activating agents

Metal compounds/salts
Mg^{2+} (required [1, 2], optimal activity at: 3 mM [1], 3–10 mM [2], 5–10 mM
[4]) [1, 2, 4]; KCl (enzyme works optimally at 50 mM KCl, inhibition at 200
mM [1], 50–75 mM stimulates 2fold [4]) [1, 4]; Mn^{2+} (40% of the activity with
Mg^{2+}, 2 mM somewhat more effective than 5 mM [4], no effect [1]) [4]; Co^{2+}
(30% of the activity with Mg^{2+}, 2 mM somewhat more effective than 5 mM
[4], no effect [1]) [4]; More (no effect: Cd^{2+} [1], Zn^{2+} [4]) [1, 4]

Turnover number (min^{-1})

Specific activity (U/mg)
0.14 (formation of CDP from tRNA-C-C-A-C_n) [3]; 1.6 (phosphorolytic cleavage of poly(A)) [3]

K_m-value (mM)
0.001 (tRNA-C-C-A-C_{2-3}) [4]; 2 (phosphate (+ tRNA-C-C-A-C_{2-3})) [4]

pH-optimum
8–9 [1, 4]

pH-range

Temperature optimum (°C)
37 (assay at) [1–4]

Temperature range (°C)

3 ENZYME STRUCTURE

Molecular weight
45000–50000 (E. coli, gel filtration) [1, 4]
120000 (E. coli, sucrose density gradient centrifugation) [3]
200000 (E. coli, gel filtration) [3]

Subunits
More (native protein is composed of 2 or more subunits) [3]

Glycoprotein/Lipoprotein
–

4 ISOLATION/PREPARATION

Source organism
E. coli (strain deficient in 5 ribonucleases [1], overproducing strain [3]) [1–5]

Source tissue

Localization in source

Purification
E. coli (overproducing strain) [3]

Crystallization
–

Cloned
–

Renatured
–

5 STABILITY

pH

Temperature (°C)
45 (10 min, fairly stable up to) [4]; 55 (10 min, about 50% loss of activity)
[4]; 65 (10 min, complete inactivation) [4]

Oxidation

Organic solvent

General stability information

Storage
–20°C, 50% glycerol [3]

6 CROSSREFERENCES TO STRUCTURE DATABANKS

PIR/MIPS code

Brookhaven code

7 LITERATURE REFERENCES

[1] Cudny, H., Deutscher, M.P.: J. Biol. Chem.,263,1518–1523 (1988)
[2] Deutscher, M.P., Marshall, G.T., Cudney, H.: Proc. Natl. Acad. Sci. USA,85,
4710–4714 (1988)
[3] Jensen, K.F., Andersen, J.T., Poulsen, P.: J. Biol. Chem.,267,17147–17152 (1992)
[4] Kelly, K.O., Deutscher, M.P.: J. Biol. Chem.,267,17153–17158 (1992)
[5] Ost, K.A., Deutscher, M.P.: Biochimie,72,813–818 (1990)

1 NOMENCLATURE

EC number
2.7.7.57

Systematic name
CTP:N-methylethanolamine-phosphate cytidylyltransferase

Recommended name
N-Methylphosphoethanolamine cytidylyltransferase

Synonyms
Cytidylyltransferase, monomethylethanolamine phosphate
CTP:P-MEA cytidylyltransferase [1]

CAS Reg. No.
119345-28-7

2 REACTION AND SPECIFICITY

Catalyzed reaction
CTP + N-methylethanolamine phosphate →
→ diphosphate + CDP-N-methylethanolamine

Reaction type
Nucleotidyl group transfer

Natural substrates
More (may be involved in biosynthesis of phosphatidylcholine in certain
plants, e.g. carrot, Lemna, soybean) [1]

Substrate spectrum
1 CTP + N-methylethanolamine phosphate [1]
2 CTP + N,N-dimethylethanolamine phosphate [1]
3 CTP + choline phosphate [1]

Product spectrum
1 CDP-N-methylethanolamine + diphosphate (CDP-N-ethanolamine demonstrated in crude extract) [1]
2 CDP-N,N-dimethylethanolamine + diphosphate (CDP-N,N-dimethyletha-
nolamine demonstrated in crude extract) [1]
3 CDP-choline + diphosphate (CDPcholine demonstrated in crude extract)
[1]

Inhibitor(s)

Cofactor(s)/prosthetic group(s)/activating agents

Metal compounds/salts

Turnover number (min^{-1})

Specific activity (U/mg)

K_m-value (mM)

pH-optimum
  6.4 (assay at) [1]

pH-range

Temperature optimum (°C)
  30 (assay at) [1]

Temperature range (°C)

---

3 ENZYME STRUCTURE

Molecular weight

Subunits

Glycoprotein/Lipoprotein
  –

---

4 ISOLATION/PREPARATION

Source organism
  Lemna paucicostata [1]; Glycine max [1]; Daucus carota [1]

Source tissue
  Plant (Lemna paucicostata, homogenized) [1]; Suspension culture (Glycine
  max and Daucus carota) [1]

Localization in source

Purification
  Lemna paucicostata (partial) [1]; Glycine max (partial) [1]; Daucus carota
  (partial) [1]

Crystallization

–

Cloned

–

Renatured

–

---

5 STABILITY

pH

Temperature (°C)

Oxidation

Organic solvent

General stability information

Storage

---

6 CROSSREFERENCES TO STRUCTURE DATABANKS

PIR/MIPS code

Brookhaven code

---

7 LITERATURE REFERENCES

[1] Datko, A.H., Mudd, S.H.: Plant Physiol.,88,1338–1348 (1988)

Enzyme Handbook © Springer-Verlag Berlin Heidelberg 1997
Duplication, reproduction and storage in data banks are only
allowed with the prior permission of the publishers

1 NOMENCLATURE

EC number
2.7.7.58

Systematic name
ATP:2,3-dihydroxybenzoate adenylyltransferase

Recommended name
(2,3-Dihydroxybenzoyl)adenylate synthase

Synonyms
2,3-Dihydroxybenzoate-AMP ligase
Synthetase, (2,3-dihydroxybenzoyl)adenylate

CAS Reg. No.
122332-73-4

---

2 REACTION AND SPECIFICITY

Catalyzed reaction
ATP + 2,3-dihydroxybenzoate →
→ diphosphate + (2,3-dihydroxybenzoyl)adenylate

Reaction type
Nucleotidyl group transfer

Natural substrates
ATP + 2,3-dihydroxybenzoate (activation of 2,3-dihydroxybenzoic acid in the biosynthesis of siderophore enterobactin) [1]

Substrate spectrum
1 ATP + 2,3-dihydroxybenzoate [1, 2]
2 ATP + 2-hydroxybenzoic acid (i.e. salicylic acid, not 3- or 4-derivative) [1]
3 ATP + 2,4-dihydroxybenzoic acid [1]
4 ATP + 2,5-dihydroxybenzoic acid (i.e. gentisic acid) [1]
5 ATP + 2,3,4-trihydroxybenzoic acid [1]
6 More (no substrates are benzoic acid, 2,6-dihydroxybenzoic acid, 2,4,6-trihydroxybenzoic acid, anthranilic acid, thiosalicylic acid) [1]

Product spectrum
1 Diphosphate + (2,3-dihydroxybenzoyl)adenylate (product remains enzyme-bound for further reaction in overall biosynthesis of enterobactin) [1]
2 Diphosphate + (2-hydroxybenzoyl)adenylate
3 Diphosphate + (2,4-dihydroxybenzoyl)adenylate
4 Diphosphate + (2,5-dihydroxybenzoyl)adenylate
5 Diphosphate + (2,3,4-trihydroxybenzoyl)adenylate
6 ?

Inhibitor(s)

Cofactor(s)/prosthetic group(s)/activating agents
More (no activation by DTT) [1]

Metal compounds/salts

Turnover number (min^{-1})

Specific activity (U/mg)
8.33 [1]

K_m-value (mM)
0.0027 (2,3-dihydroxybenzoic acid) [1]; 0.091 (2-hydroxybenzoic acid) [1]; 0.093 (2,3,4-trihydroxybenzoic acid) [1]; 0.242 (2,4-dihydroxybenzoic acid) [1]; 0.552 (2,5-dihydroxybenzoic acid) [1]; 1.12 (ATP) [1]

pH-optimum

pH-range

Temperature optimum (°C)
37 (assay at) [1]

Temperature range (°C)

3 ENZYME STRUCTURE

Molecular weight
115000 (E. coli, gel filtration) [1]

Subunits
Dimer (2 × 59000, E. coli, SDS-PAGE [1], 2 × 59299, E. coli, calculated from DNA-sequence [2]) [1, 2]

Glycoprotein/Lipoprotein
–

4 ISOLATION/PREPARATION

Source organism
E. coli (overproducing recombinant strain pSF105/JM105 [1]) [1, 2]

Source tissue
Cell [1]

Localization in source

Purification
E. coli [1]

Crystallization
–

Cloned
(E. coli, structural gene entE subcloned into multi-copy plasmid pKK223–3 under control of tac-promoter) [1]

Renatured
–

5 STABILITY

pH

Temperature (°C)

Oxidation

Organic solvent

General stability information
Glycerol, 50%, stabilizes dilute enzyme solutions [1]

Storage
–70°C – 2°C, in 25 mM Tris-HCl buffer, pH 8, 5 mM DTT, 10 mM $MgCl_2$, 50% glycerol, several months [1]

6 CROSSREFERENCES TO STRUCTURE DATABANKS

PIR/MIPS code

Brookhaven code

7 LITERATURE REFERENCES

[1] Rusnak, F., Faraci, W.S., Walsh, C.T.: Biochemistry,28,6827–6835 (1989)
[2] Staab, J.F., Elkins, M.F., Earhart, C.F.: FEMS Microbiol. Lett.,59,15–20 (1989)

1 NOMENCLATURE

EC number
2.7.8.1

Systematic name
CDPethanolamine:1,2-diacylglycerol ethanolaminephosphotransferase

Recommended name
Ethanolaminephosphotransferase

Synonyms
EPT [20]
Ethanolaminephosphotransferase, diacylglycerol
CDPethanolamine diglyceride phosphotransferase
Diacylglycerol ethanolaminephosphotransferase
Ethanolamine phosphotransferase
Phosphorylethanolamine-glyceride transferase

CAS Reg. No.
9026-19-1

---

2 REACTION AND SPECIFICITY

Catalyzed reaction
CDPethanolamine + 1,2-diacylglycerol →
→ CMP + a phosphatidylethanolamine

Reaction type
Substituted phospho group transfer

Natural substrates
More (significant selectivity which may be of considerable importance in maintaining the characteristic composition of fatty acyl chains in membrane phospholipids [24], metabolic pathway for the synthesis of 1-alk-1-enyl-2-acyl-sn-glycero-3-phosphorylethanolamine [5], reversibility of phospho-ethanolamine transferase and phosphocholine transferase permits the inter-conversion of the diacylglycerol moieties of choline and ethanolamine glycerophospholipids [8], nonessential enzyme [12], catalyzes the final step in synthesis of phosphatidylethanolamine [13]) [5, 8, 12, 13, 24]

1

Substrate spectrum

1 CDPethanolamine + 1,2-diacylglycerol (r [1, 8, 13, 14, 20, 21], in presence of 5'-CMP [20]) [1–32]
2 CDPethanolamine + 1-alkyl-2-acyl-sn-glycerol [5]
3 Phosphatidylcholine + CMP (88% of the activity with phosphatidylethanolamine) [13]
4 Lysophosphatidylcholine + CMP (75% of the activity with phosphatidylethanolamine) [13]
5 Lysophosphatidylethanolamine + CMP (74% of the activity with phosphatidylethanolamine) [13]
6 CDPethanolamine + 1,2-dioleoylglycerol (highest activity [23]) [6, 20, 23, 25, 27]
7 CDPethanolamine + 1-stearoyl-2-oleoylglycerol [23]
8 CDPethanolamine + dipalmitoylglycerol [27]
9 CDPethanolamine + 1-heptadecanoylglycerol [25]
10 CDPethanolamine + 1-stearoylglycerol [25]
11 CDPethanolamine + 2-oleoylglycerol [25]
12 CDPethanolamine + 1-oleoyl-2-lauroylglycerol [25]
13 CDPethanolamine + 1-oleoyl-2-palmitoylglycerol [25]
14 CDPethanolamine + 1-oleoyl-2-stearoyl-sn-glycerol [25]
15 More (1-acyl,2-oleoylglycerols ranging from 1-lauroyl to 1-heptadecanoyl species: 1-heptadecanoyl and 1-stearoyl species used most actively [1], fully saturated diacylglycerols like dipalmitoylglycerol are poorly utilized [1], several 1-oleoyl,2-saturated types of diacylglycerol are effectively used [1], no discrimination towards the molecular species of endogenous pool of diacylglycerols [7], enzyme selects 1-palmitoyl-2-linoleoylglycerol as a preferred substrate [10], overview: utilization of endogenous phospholipids in the reverse reaction [15], substrate specificity overview [16, 24, 25], 1,2-diacylglycerols containing fatty acids 6, 12 and 14 carbons in length are no substrates [23], distinctly higher reaction rates with the combined 1-saturated 2-docosahexaenoyl precursors as compared to the corresponding monoenoic, dienoic or tetraenoic substrates, a selectivity towards 1-stearoyl-2-arachidonoylglycerol and 1-stearoyl-2-oleoyl glycerols over their 1-palmitoyl homologues [24], lung enzyme is relatively nonselective, the concentration of diacylglycerol and the physical state in which it is presented to the enzyme can effect the apparent selectivity of the enzyme for diacylglycerols [27], 16-fold selectivity for 1-O-alk-1-enyl-2-acyl-sn-glycerol, the primary determinant of substrate selectivity is the covalent nature of the sn-1 aliphatic group of diradyl glycerol acceptors [28]) [1, 7, 10, 12, 15, 16, 18, 23–25, 27, 28]

Product spectrum

1 CMP + a phosphatidylethanolamine (1,2-diacyl-sn-glycero-3-phosphorylethanolamine [5]) [1–32]
2 CMP + 1-alkyl-2-acyl-sn-glycero-3-phosphorylethanolamine [5]
3 CDPcholine + diacylglycerol
4 ?
5 ?
6 CMP + dioleoylphosphatidylethanolamine
7 CMP + 1-stearoyl-2-oleoylphosphatidylethanolamine
8 CMP + dipalmitoylphosphatidylethanolamine
9 CMP + ?
10 CMP + ?
11 CMP + ?
12 CMP + 1-oleoyl-2-lauroylphosphatidylethanolamine
13 CMP + 1-oleoyl-2-palmitoylphosphatidylethanolamine
14 CMP + 1-oleoyl-2-stearoyl-sn-glycerol-3-phosphorylethanolamine
15 ?

Inhibitor(s)

Ca^{2+} ($CaCl_2$ [23], half-maximal inhibition: 0.015 mM (with Mg^{2+} as cofactor), 5 mM (with Mn^{2+} as cofactor) [9], Mg^{2+}- or Mn^{2+}-activated enzyme [31]) [2, 9, 23, 31, 32]; Cytidine nucleotides (exposure of glomerular particles to) [2]; 1-Alkyl-2-acyl-sn-glycerol (inhibits formation of 1,2-diacyl-sn-glycero-3-phosphorylethanolamine) [2]; CMP [6, 20, 23]; CDPethanolamine (inhibits degradation of lecithins and phosphatidylethanolamine) [15]; Norepinephrine (substrate: diacylglycerol or alkylacylglycerol) [19]; 5-Hydroxytryptamine (substrate: diacylglycerol) [19]; Acetylcholine (substrate: diacylglycerol) [19]; ATP [19]; cAMP [19]; PCMB [32]; p-Hydroxymercuribenzoate (reversal by monothioglycerol) [20]; Triton WR 1339 [23]; Tween 20 [23]; Acetone [23]; Methanol [23]; Dioxane [23]; DTT [23]; 1,2-Dilaurin (slight) [20]; 5'-CMP [20]; CDP (5'- [20]) [20, 23]; CTP (5'- [20]) [20, 23]; 5'-AMP (slight) [20]; 5'-UMP (slight) [20]; 5'-GMP (slight) [20]; Microsomal phospholipids [22]; CDPcholine [22, 23, 32]; Ethylene glycol-bis(beta-aminoethyl ether)-N,N,N',N'-tetracetic acid [22]; Palmitoyl-CoA [23, 32]; Phosphatidylethanolamine [26]; Cholesterol [26]; Myristic acid [26]; Phosphatidic acid [26]; Lysophosphatidylserine [26]; DH-990 (hypolidemic drug) [20]; Dipalmitoyl-phosphatidylethanolamine [29]; Dipalmitoylphosphatidylcholine [29]; ATP (alone, slight [32], + pantetheine: inhibition of Mn^{2+}- and Mg^{2+}-activated enzyme, + CoA: inhibition of Mn^{2+}-activated enzyme, slight stimulation of Mg^{2+}-activated enzyme, ATP alone: inhibition of Mg^{2+}- and Mn^{2+}-activated enzyme [31]) [31, 32]; N-Ethylmaleimide [32]; Reduced glutathione [32]

Cofactor(s)/prosthetic group(s)/activating agents

Phospholipid (absolute requirement) [6, 12]; Diolein (stimulates, but has no effect on glyceryl ether content of phosphatidylethanolamine) [11]; Triton X-100 (slightly stimulates with Mn^{2+}, but not with Mg^{2+} as cofactor) [22]; EGTA (stimulates) [23]; Dipalmitin (stimulates, but has no effect on glyceryl ether content of phosphatidylethanolamine) [11]; CHAPS (activates) [13]; Octyl glucoside (activates at low concentration, inhibits at higher concentration) [13]; Taurocholate [13]; Deoxycholate (1.25 mM, 50% increase of activity [18], activates [22]) [18, 22]; 1,2-Diacylglycerol (stimulated by exogenous 1,2-diacylglycerols [20], largest stimulation by 1,2-diolein and 1,2-diacylglycerol) [20]; Unsaturated fatty acids (very slight stimulation) [4]; Phospholipase C (10 min, 23°C, 26% stimulation) [23]; Bovine serum albumin (1 mg/ml, 2-fold increase of activity) [23]; Phosphatidylcholine (increases activity) [26]; Lysophosphatidylcholine (increases activity) [26]; Phosphatidylserine (increases activity) [26]; More (the rate of incorporation of CMP into CDPethanolamine is increased by increasing the concentration of phosphatidylethanolamine in detergent-phospholipid micellar system) [14]

Metal compounds/salts

Mn^{2+} (required [1, 2, 14, 20], absolute requirement for a divalent metal ion [6], requires Mn^{2+}, Mg^{2+} or Co^{2+} [9], Mn^{2+} or Mg^{2+} required [13, 14, 23], can partially replace Mg^{2+} in activation [32], much more effective than Mg^{2+} [1, 13, 22, 23, 31], Mg^{2+} (20 mM) or Mn^{2+} (1 mM) required for optimum activity [31], Mg^{2+} is less effective than Mn^{2+} as cofactor for the reverse reaction [14], K_a: 2.6 mM [6], maximal activity at: 5 mM [9], 0.7 mM [32]) [1, 2, 6, 9, 13, 14, 20, 22, 23, 31, 32]; Mg^{2+} (absolutely required [32], completely dependent on $MgCl_2$ [23], can partially replace Mn^{2+} in activation (50% [22]) [1, 13, 22], requires Mg^{2+}, Mn^{2+} or Co^{2+} [9], Mn^{2+} or Mg^{2+} required [13, 14, 23], Mg^{2+} is less effective than Mn^{2+} as cofactor for the reverse reaction [14], Mg^{2+} (20 mM) or Mn^{2+} (1 mM) required for optimum activity [31], maximal activity at: 10 mM [9], 3 mM [32]) [1, 9, 13, 14, 22, 23, 25, 31, 32]; Co^{2+} (enzyme requires Mg^{2+}, Mn^{2+} or Co^{2+}, maximal activity: 5 mM [9], negligible effect [22]) [9]

Turnover number (min^{-1})

Specific activity (U/mg)

12.1 [22]; 1.705 [29]; 0.0133 [26]; More [1]

K_m-value (mM)

0.0016 (CDPethanolamine) [14]; 0.0083 (CDPethanolamine) [2]; 0.0118 (1,2-dioleoylglycerol) [23]; 0.0183 (CDPethanolamine (+ 1,2-dioleoyl-sn-glycerol)) [23]; 0.022 (CDPethanolamine) [1, 6]; 0.04 (CMP) [8]; 0.063 (diacylglycerol) [1]; 0.083 (CDPethanolamine) [20]; 0.111 (1-dodecanoyl-2-octadecenoylglycerol) [25]; 0.114 (1-pentadecanoyl-2-octadecenoylglycerol) [25]; 0.120 (1-heptadecanoyl-2-octadecenoylglycerol) [25]; 0.128 (1-nonadecanoyl-2-octadecenoylglycerol) [25]; 0.14 (CMP) [1]; 0.147 (1,2-dioleoyl-glycerol) [25]; 0.167 (1-octadecanoyl-2-octadecenoylglycerol) [25]; 0.182 (1-arachidoyl-2-octadecenoylglycerol) [25]; 0.28 (CDPethanolamine) [5]; 1.9 (1-alkyl-2-acyl-sn-glycerol) [5]; More (3.3 mol% dioleoylglycerol [6], K_m for CDPethanolamine depends on composition of the lipid mixture utilized for reconstitution of solubilized enzyme [29]) [6, 12, 15, 19, 22, 25, 29–32]

pH-optimum
6.5 [32]; 8.0 [20]; 8.0–8.5 [1, 22]; 8.5 (broad) [13]; 8.5–9.3 [23]

pH-range

Temperature optimum (°C)
More (temperature dependence) [1]; 30–37 [20]; 37 [13, 29]

Temperature range (°C)

---

3 ENZYME STRUCTURE

Molecular weight

Subunits

Glycoprotein/Lipoprotein
–

---

4 ISOLATION/PREPARATION

Source organism
Rabbit [9, 28]; Glycine max [10]; Pisum sativum [10]; Rat [1, 3, 5, 8, 14–17, 19, 21–27, 29–31]; Bovine [2]; Saccharomyces cerevisiae [6, 12, 20]; Chicken [4, 18]; Solanum tuberosum [7]; Tetrahymena thermophila [11]; Hamster [13]; Ricinus communis [32]

Source tissue
Platelets (membrane) [9]; Liver [1, 4, 13, 15–17, 22–25, 30, 31]; Cerebellar cortex (glomeruli) [2]; Brain [3, 5, 8, 14, 18, 19, 21, 23, 26, 29]; Seeds [10]; Myocardium [28]; Fat cells [23]; Tuber [7]; Leaf [10]; Platelets (membrane) [9]; Intestinal mucosa [23]; Lung [27]; Endosperm [32]

Localization in source
Microsomes (cytoplasmic surface [1], membrane [7, 30]) [1, 3, 5, 7, 8, 10, 13–18, 21–31]; Membrane [20]; Endoplasmic reticulum [32]

Purification
Rat (partial [22, 26], solubilization [30]) [1, 15, 22, 26, 29, 30]; Hamster (partial) [13]

Crystallization
–

Cloned
–

Renatured
–

---

5 STABILITY

pH

Temperature (°C)
    37 (stable for 19 min, heating of microsomes) [23]; 49 (4 min, more than
    50% loss of activity) [23]; 50 ($t_{1/2}$: 8 min) [20]; 55 (1 min, 90% loss of activity)
    [13]

Oxidation

Organic solvent

General stability information
    Enzyme in rat liver microsomes remains unaffected even if over 90% and al-
    most 100% of microsomal phosphatidylcholine and phosphatidylethanola-
    mine is hydrolyzed by snake venom phospholipase A_2 [17]; Stable if treated
    with 0.5% Triton X-100 [1]; Phospholipase A_2 treatment of microsomes de-
    creases activity probably due to disruption of membrane structure [18]; Tri-
    ton X-100, stable to [22]; Trypsin, 70%, 0.9 min, activity remains stable [23];
    Glycerol, diacylglycerol, phosphatidylcholine or lysophosphatidylcholine
    stabilizes [26]; Lyophilization inhibits [32]

Storage
    At below –20°C, stable for more than 1 month, microsomal preparation [1];
    –20°C, stable for more than 2 weeks at any stage of purification [28]; –18°C,
    solubilized enzyme is stable for 3 weeks [26]; 4°C, solubilized enzyme is
    stable for 5 days [26]; Solubilized enzyme is stable for long periods of time
    [30]

---

6 CROSSREFERENCES TO STRUCTURE DATABANKS

PIR/MIPS code
    PIR2:S48967 (yeast (Saccharomyces cerevisiae))

Brookhaven code

7 LITERATURE REFERENCES

[1] Kanoh, H., Ohno, K.: Methods Enzymol.,71,536–546 (1981) (Review)
[2] Dorman, R.V., Bischoff, S.B., Terrian, D.M.: Neurochem. Res.,11,1167–1179 (1986)
[3] Binaglia, L., Roberti, R., Vecchini, A., De Meo, G., Porcellati, G.: Ital. J. Biochem., 29,43–45 (1980)
[4] Sribney, M., Lyman, E.M.: Can. J. Biochem.,51,1479–1486 (1973)
[5] Radominska-Pyrek, A., Horrocks, L.A.: J. Lipid Res.,13,580–587 (1972)
[6] Hjelmstad, R.H., Bell, R.M.: Methods Enzymol.,209,272–279 (1992) (Review)
[7] Justin, A.M., Demandre, C., Tremolieres, A., Mazliak, P.: Biochim. Biophys. Acta, 836,1–7 (1985)
[8] Goracci, G., Francescangeli, E., Horrocks, L.A., Porcellati, G.: Biochim. Biophys. Acta,876,387–391 (1986)
[9] Taniguchi, S., Morikawa, S., Hayashi, H., Fujii, K., Mori, H., Fujiwara, M., Fujiwara, M.: J. Biochem.,100,485–491 (1986)
[10] Justin, A.M., Demandre, C., Mazliak, P.: Biochim. Biophys. Acta,922,364–371 (1987)
[11] Smith, J.D.: J. Biol. Chem.,260,2064–2068 (1985)
[12] Hjelmstad, R.H., Bell, R.M.: J. Biol. Chem.,266,4357–4365 (1991)
[13] O, K.-M., Siow, Y.L., Choy, P.C.: Biochem. Cell Biol.,67,680–686 (1989)
[14] Roberti, R., Mancini, A., Freysz, L., Binaglia, L.: Biochim. Biophys. Acta,1165, 183–188 (1992)
[15] Kanoh, H., Ohno, K.: Biochim. Biophys. Acta,306,203–217 (1973)
[16] Kanoh, H., Ohno, K.: Biochim. Biophys. Acta,380,199–207 (1975)
[17] Morimoto, K., Kanoh, H.: Biochim. Biophys. Acta,531,16–24 (1978)
[18] Freysz, L., Horrocks, L.A., Mandel, P.: Biochim. Biophys. Acta,489,431–439 (1977)
[19] Strosznajder, J., Radominska-Pyrek, A., Horrocks, L.A.: Biochim. Biophys. Acta, 574,48–56 (1979)
[20] Percy, A.K., Carson, M.A., Moore, J.F., Waechter, C.J.: Arch. Biochem. Biophys., 230,69–81 (1984)
[21] Goracci, G., Horrocks, L.A., Porcellati, G.: FEBS Lett.,80,41–44 (1977)
[22] Kanoh, H., Ohno, K.: Eur. J. Biochem.,66,201–210 (1976)
[23] Coleman, R., Bell, R.M.: J. Biol. Chem.,252,3050–3056 (1977)
[24] Holub, B.J.: J. Biol. Chem.,253,691–696 (1978)
[25] Morimoto, K., Kanoh, H.: J. Biol. Chem.,253,5056–5060 (1978)
[26] Vecchini, A., Roberti, R., Freysz, L., Binaglia, L.: Biochim. Biophys. Acta,918,40–47 (1987)
[27] Ide, H., Miller, J.C., Weinhold, P.A: Biochim. Biophys. Acta,960,119–124 (1988)
[28] Ford, D.A., Rosenbloom, K.B., Gross, R.W.: J. Biol. Chem.,267,11222–11228 (1992)
[29] Roberti, R., Vecchini, A., Freysz, L., Masoom, M., Binaglia, L.: Biochim. Biophys. Acta,1004,80–88 (1989)
[30] Radominska-Pyrek, A.: Biochem. Biophys. Res. Commun.,85,1074–1081 (1978)
[31] Liteplo, R.G., Sribney, M.: Can. J. Biochem.,55,1049–1056 (1977)
[32] Sparace, S.A., Wagner, L.K., Moore, T.S.: Plant Physiol.,67,922–925 (1981)

1 NOMENCLATURE

EC number
2.7.8.2

Systematic name
CDPcholine:1,2-diacylglycerol cholinephosphotransferase

Recommended name
Diacylglycerol cholinephosphotransferase

Synonyms
Phosphorylcholine-glyceride transferase
Alkylacylglycerol cholinephosphotransferase
1-Alkyl-2-acetylglycerol cholinephosphotransferase
Cholinephosphotransferase
CPT [8]
Alkylacylglycerol choline phosphotransferase [1]
Diacylglycerol choline phosphotransferase [1]
1-Alkyl-2-acetyl-rn-glycerol:CDPcholine choline phosphotransferase [3]
Cholinephosphotransferase, diacylglycerol
CDP-choline diglyceride phosphotransferase
Cytidine diphosphocholine glyceride transferase
Cytidine diphosphorylcholine diglyceride transferase
Phosphocholine diacylglyceroltransferase
Sn-1,2-Diacylglycerol cholinephosphotransferase
Cholinephosphotransferase, 1-alkyl-2-acetylglycerol
1-Alkyl-2-acetyl-sn-glycerol cholinephosphotransferase
EC 2.7.8.16 (activity with 1-alkyl-2-acylglycerol as acceptor was previously
listed as EC 2.7.8.16)

CAS Reg. No.
77237-98-0; 9026-13-5

2 REACTION AND SPECIFICITY

Catalyzed reaction
CDPcholine + 1,2-diacylglycerol →
→ CMP + a phosphatidylcholine (sequential kinetic mechanism [15], bi-bi
sequential mechanism involving a direct nucleophilic attack of diacylglycer-
ol on CDPcholine during the reaction [18]);
CDPcholine + 1-alkyl-2-acylglycerol →
→ CMP + 1-alkyl-2-acylglycero-3-phosphocholine

Reaction type
Substituted phospho group transfer

Natural substrates
CDPcholine + sn-1,2-diacylglycerol (DTT-sensitive activity [7], final reaction in synthesis of phosphatidylcholine) [7, 8, 14, 19, 33]

CMP + phosphatidylcholine (principal pathway for degradation of phosphatidylcholine, particularly during brain ischemia, followed by hydrolysis of diacylglycerols by the lipase [17]) [17, 35]

More (DTT-insensitive activity [7, 10–12]: final step of the biosynthesis of platelet activating factor (PAF) in the de novo pathway [7, 9–12], renal DTT-insensitive enzyme could be a potentially important enzyme in the regulation of systemic blood presure [10], last step in de novo synthesis of diacylglycerophosphocholine [3], the reversibility of phosphoethanolamine transferase and phosphocholine transferase permits the interconversion of diacylglycerol moieties of choline and ethanolamine glycerophospholipids [28], significant selectivity which may be of considerable importance in maintaining the characteristic composition of fatty acyl chains in membrane phospholipids [44]) [3, 7, 9–12, 20, 28, 44]

Substrate spectrum
1 CDPcholine + 1,2-diacylglycerol (r [8, 17, 22, 28, 34, 35, 40]) [1–44]
2 dCDPcholine + 1,2-diacylglycerol [18]
3 CDPcholine + 1-hexadecanoyl-2-octadecanoyl-sn-glycerol [1]
4 CDPcholine + 1-hexadecyl-2-octadecanoyl-sn-glycerol [1]
5 CDPcholine + 1,2-dioleoyl-sn-glycerol [2]
6 CDPcholine + 1-chimyl-2-acetyl-sn-glycerol [3]
7 CDPcholine + 1-batyl-2-acetyl-sn-glycerol [3]
8 CDPcholine + 1-selachyl-2-acetyl-sn-glycerol [3]
9 CDPcholine + 1-hexadecyl-2-acetyl-sn-glycerol [7]
10 CDPcholine + 1,2-dipalmitoyl-sn-glycerol (preferred substrate over other disaturated species [15], poor substrate [22]) [15, 22]
11 CDPcholine + dihexanoylglycerol (at 5–10% of the activity with dioleoyl-glycerol or egg diglycerides) [16]
12 CDPcholine + dioctanoylglycerol (at 5–10% of the activity with dioleoyl-glycerol or egg diglycerides) [16]
13 CDPcholine + didecanoylglycerol (at 5–10% of the activity with dioleoyl-glycerol or egg diglycerides) [16]
14 CDPcholine + 1-palmitoyl-2-linoleoylglycerol (preferred substrate) [30]
15 More (short chain esters at the sn-2 position (acetate or propionate) are utilized [10], substrates are 1,2-diacylglycerols containing fatty acids 6, 12 and 14 carbons in length [2, 43], significant preference for unsaturated diacylglycerols over saturated sialylglycerols [4], higher activity with 16:0 and 18:1 substrates than with 18:0 substrates [7], DTT-insensitive enzyme prefers a lipid substrate with 16:0 or 18:1 sn-1-alkyl chains [10],

1-acyl-2-oleoylglycerols, ranging from 1-lauroyl to 1-heptadecanoyl species are equally well utilized. Fully saturated diacylglycerols like dipalmitoylglycerol are poorly utilized. Several 1-oleoyl-2-saturated types of
diacylglycerol are effectively used [22], no selectivity towards the molecular species of the endogenous pool of diacylglycerols [27], lacks specificity for the type of diglyceride [38], specificity overview [32, 41, 43,
44], a marked preference of the enzyme for the 1-palmitoyl over the
1-stearoyl homologue is observed with all 4 unsaturation classes of
diacylglycerols [24], influence of saturated fatty acids on the enzyme
activities depends on their location at the C-1 or C-2 position of glycerol
[41], utilizes without marked selectivity the endogenous 1,2-diacylglyc-
erol species differing in the degree of unsaturation [16], utilizes 1-myri-
styl phosphatidylcholine most rapidly and, in decreasing order the
1-palmitoyl and 1-stearyl species [16], no substrate: AMP [17], UMP
[17], 1,3-diC$_{18:1}$glycerol ether [2], 1,2-diC$_{18:1}$glycerol ether [2], analogs
with acetamide or methoxy substituents at the sn-2-position [10]) [1, 2, 4,
7, 10, 16, 17, 22, 24, 27, 32, 38, 41, 43, 44]

Product spectrum

 1 CMP + a phosphatidylcholine [1–44]
 2 CMP + 1,2-diacylglycero-3-phosphocholine
 3 CMP + 1-hexadecanoyl-2-octadecanoyl-sn-glycero-3-phosphocholine
 4 CMP + 1-hexadecyl-2-octadecanoyl-sn-glycero-3-phosphocholine
 5 CMP + 1,2-dioleoyl-sn-glycero-3-phosphocholine
 6 CMP + 1-chimyl-2-acetyl-sn-glycero-3-phosphocholine
 7 CMP + 1-batyl-2-acetyl-sn-glycero-3-phosphocholine
 8 CMP + 1-selachyl-2-acetyl-sn-glycero-3-phosphocholine
 9 CMP + 1-hexadecyl-2-acetyl-sn-glycero-3-phosphocholine
10 CMP + 1,2-dipalmitoyl-sn-glycero-3-phosphocholine
11 CMP + dihexanoylglycerophosphocholine
12 CMP + dioctanoylglycerophosphocholine
13 CMP + didecanoylglycerophosphocholine
14 CMP + 1-palmitoyl-2-linoleoylglycero-3-phosphocholine
15 ?

Inhibitor(s)

Tween-20 (at high concentrations [1], stimulates at low subsolubilizing concentrations, membrane-solubilizing concentrations lead to nearly complete
inactivation [8], inactivation at solubilization concentration, full recovery of
activity after reconstituting the membrane by adding excess lipid (soybean)
and removing detergent by gel filtration, dialysis or absorption to Bio-Beads
[19]) [1, 8, 19, 33, 35, 43]; ATP [39]; cAMP [39]; Acetylcholine [39]; Norepinephrine [39]; 5-Hydroxytryptamine [39]; Acetylcholine [39]; DH-990 (hypolidemic drug) [40]; 1-Butanol [43]; Methanol [43]; Acetone [43]; Trypsin [43];
Palmitoyl-CoA [2, 43]; Ethanol (at high concentrations [1], 2.5% [7]) [1, 7];

Mn^{2+} (2 mM, 70–80% inhibition) [1]; Ca^{2+} (above 0.01 mM [7, 8], competitive with Mg^{2+} or Mn^{2+} [8], 0.2 mM, 90% inhibition in presence of 10 mM Mg^{2+} [20], half-maximal inhibition: 0.15 mM (in presence of 5 mM Mg^{2+}), 5 mM (in presence of 5 mM Mn^{2+}) [29]) [7, 8, 10, 20, 23, 29, 35, 43]; Fatty acid esters [25]; DTT (slight enhancement of activity towards 1-alkyl-2-acetyl-sn-glycerol, inhibition of activity towards diacylglycerol [3]) [1, 3, 4, 43]; CDP (5'- [40]) [2, 40, 43]; CDPethanolamine [2, 22, 42, 43]; Phospholipase A_2 (rapid inactivation, microsomal total phospholipids partially reactivate [37]) [37, 38]; Trypsin [2, 4]; N,N-Dimethylaminoethyl p-chlorophenoxyacetate (centrophenoxine) [6]; N,N-Dimethylaminoethanol (less inhibitory than centrophenoxine) [6]; p-Chlorophenoxyacetic acid (less inhibitory than centrophenoxine) [6]; Triton (WR 1339 [43], X-100 [13, 19, 22, 42, 43], X-200 [33], irreversible inactivation at the solubilization step [13], inactivation at solubilization concentration, full recovery of activity after reconstituting the membrane by adding excess lipid (soybean) and removing detergent by gel filtration, dialysis or absorption to Bio-Beads [19], activity assayed with Mg^{2+} is more labile than that assayed with Mn^{2+} [22]) [8, 13, 19, 22, 33, 42, 43]; p-Hydroxymercuribenzoate [40]; Deoxycholate (stimulates at low subsolubilizing concentrations, membrane-solubilizing concentrations lead to nearly complete inactivation [8], stimulates, inhibition above 2 mM [22]) [8, 22, 38]; n-Octylglucoside (irreversible inactivation at the solubilization step [13], inactivation at solubilization concentration, full recovery of activity after reconstituting the membrane by adding excess lipid (soybean) and removing detergent by gel filtration, dialysis or absorption to Bio-Beads, if membrane is solubilized with octylglucoside or cholate at weight ratios of detergent:membrane protein of at least 10, the activity is irreversibly lost unless stabilizers are added with detergent, diacylglycerol and glycerol are effective stabilizers [19]) [13, 19, 33]; CHAPS (irreversible inactivation at the solubilization step [13]) [13, 33]; CHAPSO [33]; Cholate (inactivation at solubilization concentration, full recovery of activity after reconstituting the membrane by adding excess lipid (soybean) and removing detergent by gel filtration, dialysis or absorption to Bio-Beads, if membrane is solubilized with octylglucoside or cholate at weight ratios of detergent:membrane protein of at least 10, the activity is irreversibly lost unless stabilizers are added with detergent, diacylglycerol and glycerol are effective stabilizers) [19]; SDS (inactivation at solubilization concentration, full recovery of activity after reconstituting the membrane by adding excess lipid (soybean) and removing detergent by gel filtration, dialysis or absorption to Bio-Beads) [19]; CMP (5'- [40], CDPcholine + 1,2-diacylglycerol [8]) [8, 26, 40]; 5'-CTP [40]; CMPcholine (slight [18], product inhibition [22]) [18, 22]; dAcCDPcholine [18]; Phospholipids (phosphatidylcholine, phosphatidylethanolamine or lysophosphatidylethanolamine activates, lysophosphatidylcholine inhibits) [33]; 5'-AMP (little effect) [40]; 5'-UMP (little effect) [40]; 5'-GMP (little effect) [40]; More (heat-labile, nondialyzable endogenous inhibitor may act at the binding step of the enzyme to its lipid substrate) [13]

Cofactor(s)/prosthetic group(s)/activating agents

DTT (DTT-insensitive enzyme [7, 10–12], slight but significant stimulation [7], required [17], slight enhancement of activity towards 1-alkyl-2-ace-tyl-sn-glycerol, inhibition of activity towards diacylglycerol [3]) [3, 7, 10–12, 17]; Tween-20 (stimulates at low subsolubilizing concentrations, membrane-solubilizing concentrations lead to nearly complete inactivation) [8]; Lysolecithin (stimulates at low subsolubilizing concentrations, membrane-solubilizing concentrations lead to nearly complete inactivation) [8]; Phospholipase C (23°C, 10 min, slight stimulation) [43]; Triton (stimulates at low subsolubilizing concentrations, membrane-solubilizing concentrations lead to nearly complete inactivation) [8]; Deoxycholate (stimulates at low subsolubilizing concentrations, membrane-solubilizing concentrations lead to nearly complete inactivation [8], stimulates, inhibition above 2 mM [22], stimulates [42]) [8, 22, 42]; EGTA [43]; Phorbol 12-myristate 13-acetate (activates) [11]; Phosphatidylcholine (activates) [14, 16, 22]; Phosphatidyletha-nolamine (activates) [14]; Ethylene glycol bis(beta-aminoethyl ether)-N,N,N',N'-tetraacetic acid (stimulates) [43]; 1,2-Diolein (highest stimulation of diacyl-glycerols tested) [40]; 1,2-Dilaurin [40]; Cytidine nucleotides (stimulate) [23]; Phospholipids (phosphatidylcholine, phosphatidylethanolamine or lysophos-phatidylethanolamine activates, lysophosphatidylcholine inhibits [33], absolute requirement [32], total microsomal phospholipids stimulate [22], microsomal phospholipids required for maximal activity [42], activation [26]) [22, 26, 32, 33, 42]; Dioleoylphosphatidylcholine (stimulates) [16]; EGTA (0.5 mM, stimulates) [2]; Taurocholate (stimulates) [22]; Unsaturated fatty acids (a number of unsaturated fatty acids markedly stimulate, 0.8 mM oleate activates if a mixed diglyceride such as 1-palmitoyl-2-oleoyl-sn-glycerol is used as a substrate, dipalmitin or diolein incorporation into lecithin is not stimulated) [25]; More (cholinephosphotransferase requires a lipidic boundary for full activation, no activation by substrate) [16]

Metal compounds/salts

Mg^{2+} (required [2, 3, 7, 17, 20, 22, 23, 42, 43], enzyme separated into Mg^{2+}-requiring and Mn^{2+}-requiring components [42], Mn^{2+}, Mg^{2+} or Co^{2+} required [29], absolute requirement for Mg^{2+} or Mn^{2+} [14, 33, 34], Mg^{2+} is less effective than Mn^{2+} as cofactor in the reverse reaction [34], requires Mg^{2+} rather than Mn^{2+}, particularly at physiological concentrations less than 5 mM [13], activates [5], maximal activation: 10–20 mM [7], 10 mM [14, 33], 5 mM [20, 29], 5–10 mM [8], K_a: 0.7 mM [32]) [2, 3, 5, 7, 8, 13, 14, 17, 20, 22, 23, 29, 32–34, 42, 43]; Mn^{2+} (required [40], can partially replace Mg^{2+} in activation [7, 8, 22, 43], Mn^{2+}, Mg^{2+} or Co^{2+} required [29], absolute requirement for Mg^{2+} or Mn^{2+} [14, 33, 34], Mg^{2+} is less effective than Mn^{2+} as cofactor in the reverse reaction [34], maximal activation: 5 mM [29], 2–5 mM [40], enzyme separated into Mg^{2+}-requiring and Mn^{2+}-requiring components [42], cannot replace Mg^{2+} in activation [5]) [7, 8, 14, 22, 29, 33, 34, 40, 42, 43]; Co^{2+} (Mn^{2+}, Mg^{2+} or Co^{2+} required, maximal activation: 10 mM) [29]

Turnover number (min^{-1})

Specific activity (U/mg)
  0.13 [13]; 0.0132 [16]; More [22]

K$_m$-value (mM)
  More (K$_m$ for CDPcholine, depends on diacylglycerol structure [5]) [4, 5, 10,
  14, 15, 17, 21–23, 26, 28, 32, 35, 40–43]; 0.0057 (didodecanoylglycerol) [2];
  0.0141 (didecanoylglycerol) [2]; 0.0143 (1-hexadecanoyl-2-octadecanoyl-
  sn-glycerol) [1]; 0.0239 (CDPcholine (+ 1,2-dioleoyl-sn-glycerol)) [2]; 0.0262
  (1-hexadecyl-2-octadecenoyl-sn-glycerol) [1]; 0.0314 (dihexadecanoylglyc-
  erol) [2]; 0.0405 (bacterial diacylglycerol) [2]; 0.048 (CDPcholine (+ 1-hexa-
  decanoyl-2-octadecanoyl-sn-glycerol)) [1]; 0.0501 (dioctadecenoylglycerol)
  [2]; 0.0509 (1,2-dioleoylglycerol) [2]; 0.053 (CDPcholine (+ 1-hexadecyl-
  2-octadecenoyl-sn-glycerol)) [1]; 0.0681 (dihexanoylglycerol) [2]; 0.104
  (1-myristoyl-2-oleoyl-sn-glycerol) [41]; 0.109 (1-pentadecanoyl-2-oleo-
  yl-sn-glycerol) [41]; 0.116 (1-palmitoyl-2-oleoyl-sn-glycerol) [41]; 0.118 (di-
  myristoyl-sn-glycerol) [2]; 0.120 (1,2-dioleoyl-sn-glycerol) [41]; 0.122 (1-tride-
  canoyl-2-oleoyl-sn-glycerol) [41]; 0.125–0.126 (1-stearoyl-2-oleoyl-sn-glycer-
  ol) [2, 41]; 0.133 (1-lauroyl-2-oleoyl-sn-glycerol) [41]; 0.16 (1-arachidoyl-2-
  oleoyl-sn-glycerol) [41]; 0.17 (1-heptadecanoyl-2-oleoyl-sn-glycerol) [41];
  0.18–0.35 (CMP) [8]; 0.21 (1-nonadecanoyl-2-oleoyl-sn-glycerol) [41]; 0.53
  (CDPcholine) [5]

pH-optimum
  7.0 [40]; 7.5 [5]; 8 (1-alkyl-2-acetyl-sn-glycero-3-phosphocholine [3]) [3, 7];
  8.0–8.5 [22, 42]; 8.0–9.0 [1]; 8.5 [20]; 8.5–9.3 [2, 43]

pH-range

Temperature optimum (°C)
  22 [12]; 30 (assay at) [1]; 30–45 [40]; 37 (assay at) [3, 7, 10, 13]; More (tem-
  perature-dependence) [24]

Temperature range (°C)

---

3 ENZYME STRUCTURE

Molecular weight

Subunits

Glycoprotein/Lipoprotein
  –

4 ISOLATION/PREPARATION

Source organism

Tetrahymena pyriformis [12]; Rabbit [7, 19–21, 29]; Chicken [7, 25, 38];
Tetrahymena thermophila [31]; Mouse [15]; Pig [16]; Fusarium oxysporum (f.
sp. Lycopersici) [5]; Rat [1–3, 6–10, 13, 17, 18, 22, 24, 28, 34–37, 39,
41–44]; Solanum tuberosum [27]; Pisum sativum [30]; Glycine max [30];
Guinea pig [4]; Human [7, 11]; Saccharomyces cerevisiae [26, 32, 40];
Hamster [14, 33]; Bovine [23]

Source tissue

Glomeruli (mesangial cell culture) [9]; Inner medulla [10]; Umbilical vein en-
dothelial cells [11]; Heart [14]; Skeletal muscle [19]; Bacillus Calmette-Gu-
erin induced rabbit alveolar macrophages [20]; Mycelium [5]; Neutrophils
[7]; Liver [1, 3, 6, 13–16, 18, 22, 25, 33, 35, 36, 37, 41–44]; Fat cells [2, 43];
Tuber [27]; Platelets [29]; Leaf [30]; Seeds [30]; Intestinal mucosa [43];
Spleen [3, 7]; Lung (fetal) [3, 4, 43]; Kidney (inner medulla [10]) [3, 10];
Brain (cerebral cortex [21, 23], immature [21], glomeruli [23], lysed synapto-
somes [39]) [7, 17, 21–24, 28, 34, 38, 39, 43]; Retina [7]

Localization in source

Microsomes (cytoplasmic face of microsomal versicles [10, 22]) [1–4, 6, 7,
10, 12–18, 21, 22, 24, 25, 27, 28, 30, 33–38, 41–44]; Membrane (platelets)
[29]; Mitochondria (located on outside of membrane [4]) [4, 8, 12]; Endoplasmic reticulum (located on outside of membrane [4, 14]) [4, 8, 14, 20];
Golgi apparatus [8]; Membrane (integral membrane protein) [8]; Nuclear
membrane [8]; Pellicles (low activity) [12]; Sarcoplasmic reticulum [19];
Plasma membrane [20]; More (no activity in cytosol) [12]

Purification

Rat (partial [8]) [8, 13, 22, 42]; Pig (partial) [16]; Hamster (partial) [33]

Crystallization

–

Cloned

–

Renatured

–

5 STABILITY

pH

Temperature (°C)
37 (19 min, stable, microsomal preparation) [2, 43]; 49 (15 min, 50% loss of activity, microsomal preparation) [2, 43]; 50 ($t_{1/2}$: 1 min) [40]; 55 (1 min, 90% loss of activity, whole microsomes, less than 60% loss of activity, partially purified enzyme) [33]

Oxidation

Organic solvent

General stability information
Soja phosphatidylcholine protects against detergent inactivation [16]; DTT stabilizes [3]; No loss of activity by freezing and thawing one time [3]; Freezing and thawing the microsomal preparation 5times destroys activity by 50% [43]

Storage
4°C, 24 h, 10–15% loss of activity, partially purified liver enzyme [14]

---

6 CROSSREFERENCES TO STRUCTURE DATABANKS

PIR/MIPS code
PIR2:S63075 (yeast (Saccharomyces cerevisiae))

Brookhaven code

---

7 LITERATURE REFERENCES

[1] Lee, T.-C., Blank, M.L., Fitzgerald, V., Snyder, F.: Biochim. Biophys. Acta,713, 479–483 (1982)
[2] Coleman, R., Bell, R.M.: J. Biol. Chem.,252,3050–3056 (1977)
[3] Renooij, W., Snyder, F.: Biochim. Biophys. Acta,663,545–556 (1981)
[4] Ghosh, S., Oten, P.W., Mukherjee, S., Das, S.K.: Mol. Cell. Biochem.,101,157–166 (1991)
[5] Wilson, A.C., Barran, L.R.: Trans. Br. Mycol. Soc.,85,141–144 (1985)
[6] Parthasarathy, S., Cady, R.K., Kraushaar, D.S., Sladek, N.E., Bauman, W.J.: Lipids, 13,161–164 (1978)
[7] Lee, T.-C., Snyder, F.: Methods Enzymol.,209,279–283 (1992) (Review)
[8] Cornell, R.B.: Methods Enzymol.,209,267–272 (1992) (Review)
[9] Lianos, E.A., Zanglis, A.: J. Biol. Chem.,262,8990–8993 (1987)
[10] Woodard, D.S., Lee, T.-c., Snyder, F.: J. Biol. Chem.,262,2520–2527 (1987)
[11] Heller, R., Bussolino, F., Ghigo, D., Garbarino, G., Pescarmona, G., Till, U., Bosia, A.: J. Biol. Chem.,266,21358–21361 (1991)
[12] Tsoukatos, D.C., Tselepis, A.D., Lekka, M.E.: Biochim. Biophys. Acta,1170,258–264 (1993)

[13] Ishidate, K., Matsuo, R., Nakazawa, Y.: Lipids,28,89–96 (1993)
[14] O, K., Choy, P.C.: Lipids,25,122–124 (1990)
[15] Mantel, C.R., Schulz, A.R., Miyazawa, K., Broxmeyer, H.E.: Biochem. J.,289, 815–820 (1993)
[16] Bru, R., Blöchliger, E., Luisi, P.L.: Arch. Biochem. Biophys.,307,295–303 (1993)
[17] Goracci, G., Francescangeli, E., Horrocks, L.A., Porcellati, G.: Biochim. Biophys. Acta,664,373–379 (1981)
[18] Pontoni, G., Manna, C., Salluzzo, A., Del Piano, L., Gallett, P., De Rosa, M., Zappia, V.: Biochim. Biophys. Acta,836,222–232 (1985)
[19] Cornell, R., MaxLennan, D.H.: Biochim. Biophys. Acta,821,97–105 (1985)
[20] Wang, P., Dechatelet, L.R., Waite, M.: Biochim. Biophys. Acta,450,311–321 (1976)
[21] Baker, R.R., Chang, H.-Y.: Can. J. Biochem.,60,724–733 (1982)
[22] Kanoh, H., Ohno, K.: Methods Enzymol.,71,536–546 (1981) (Review)
[23] Dorman, R.V., Bischoff, S.B., Terrian, D. M.: Neurochem. Res.,11,1167–1179 (1986)
[24] Binaglia, L., Roberti, R., Vecchini, A., De Meo, G., Porcellati, G.: Ital. J. Biochem., 29,43–45 (1980)
[25] Sribney, M., Lyman, E.M.: Can. J. Biochem.,51,1479–1486 (1973)
[26] Hjelmstad, R.H., Bell, R.M.: Methods Enzymol.,209,272–279 (1992) (Review)
[27] Justin, A.M., Demandre, C., Tremolieres, A., Mazliak, P.: Biochim. Biophys. Acta,836,1–7 (1985)
[28] Goracci, G., Francescangeli, E., Horrocks, L.A., Porcellati, G.: Biochim. Biophys. Acta,876,387–391 (1986)
[29] Taniguchi, S., Morikawa, S., Hayashi, H., Fujii, K., Mori, H., Fujiwara, M., Fujiwara, M.: J. Biochem.,100,485–491 (1986)
[30] Justin, A.M., Demandre, C., Mazliak, P.: Biochim. Biophys. Acta,922,364–371 (1987)
[31] Smith, J.D.: J. Biol. Chem.,260,2064–2068 (1985)
[32] Hjelmstad, R.H., Bell, R.M.: J. Biol. Chem.,266,4357–4365 (1991)
[33] O, K.-M., Siow, Y.L., Choy, P.C.: Biochem. Cell Biol.,67,680–686 (1989)
[34] Roberti, R., Mancini, A., Freysz, L., Binaglia, L.: Biochim. Biophys. Acta,1165, 183–188 (1992)
[35] Kanoh, H., Ohno, K.: Biochim. Biophys. Acta,306,203–217 (1973)
[36] Kanoh, H., Ohno, K.: Biochim. Biophys. Acta,380,199–207 (1975)
[37] Morimoto, K., Kanoh, H.: Biochim. Biophys. Acta,531,16–24 (1978)
[38] Freysz, L., Horrocks, L.A., Mandel, P.: Biochim. Biophys. Acta,489,431–439 (1977)
[39] Strosznajder, J., Radominska-Pyrek, A., Horrocks, L.A.: Biochim. Biophys. Acta, 574,48–56 (1979)
[40] Percy, A.K., Carson, M.A., Moore, J.F., Waechter, C.J.: Arch. Biochem. Biophys., 230,69–81 (1984)
[41] Morimoto, K., Kanoh, H.: J. Biol. Chem.,253,5056–5060 (1978)
[42] Kanoh, H., Ohno, K.: Eur. J. Biochem.,66,201–210 (1976)
[43] Coleman, R., Bell, R.M.: J. Biol. Chem.,252,3050–3056 (1977)
[44] Holub, B.J.: J. Biol. Chem.,253,691–696 (1978)

1 NOMENCLATURE

EC number
2.7.8.3

Systematic name
CDPcholine:N-acylsphingosine cholinephosphotransferase

Recommended name
Ceramide cholinephosphotransferase

Synonyms
Cholinephosphotransferase, ceramide
Phosphorylcholine-ceramide transferase

CAS Reg. No.
9026-14-6

2 REACTION AND SPECIFICITY

Catalyzed reaction
CDPcholine + N-acylsphingosine →
→ CMP + sphingomyelin

Reaction type
Substituted phospho group transfer

Natural substrates
More (in vivo the proximal donor of the phosphocholine moiety of sphingomyelin is not CDPcholine but most probably phosphatidylcholine [4],
sphingomyelin is synthesized predominantly via direct transfer of the phosphorylcholine group from phosphatidylcholine, synthesis of sphingomyelin
by the action of CDPcholine:ceramide cholinephosphotransferase occurs
only to an extent of 2% in vivo [5]) [4, 5]

Substrate spectrum
1 CDPcholine + N-acyl-threo-trans-sphingosine (CDPcholine + ceramide
[2], highly specific for CDPcholine [1, 2], active with ceramides containing isomers of sphingosine with the threo configuration of the substituents
at carbon 2 and 3 and with a trans double bond or triple bond at carbon
4, much less active if the spingosine moiety has the erythro configuration
or a cis double bond, the fatty acid moiety of the threo-trans ceramides
may be varied widely [1], sphingosine of active ceramides must have the
trans configuration of the double bond and the hydroxyl group on carbon
3 must have the threo relationship to the amino group on carbon 2.
Ceramides of dihydrosphingosine are inactive but derivatives of sphingosine containing a triple bond rather than a double bond at carbon 4 are
active if the hydroxyl group on carbon 3 is threo [2]) [1, 2]

Product spectrum
    1 CMP + sphingomyelin [1]

Inhibitor(s)
    Ca^{2+} [1]; Ba^{2+} [1]

Cofactor(s)/prosthetic group(s)/activating agents

Metal compounds/salts
    Mn^{2+} (activates) [1]; Mg^{2+} (activation less effective than with Mn^{2+}) [1]

Turnover number (min^{-1})

Specific activity (U/mg)

K_m-value (mM)

pH-optimum
    7.5–8.0 [2]; 7.8 [1]

pH-range
    7–8.5 (7: about 50% of activity maximum, 8.5: about 45% of activity maxi-
    mum) [2]

Temperature optimum (°C)
    37 (assay at) [1]; 45 [2]

Temperature range (°C)
    25–55 (25°C: about 60% of activity maximum, 55°C: about 70% of activity
    maximum) [2]

3 ENZYME STRUCTURE

Molecular weight

Subunits

Glycoprotein/Lipoprotein
    –

4 ISOLATION/PREPARATION

Source organism
    Pig (hog) [2]; Chicken [1, 2]; Rat [2]; Guinea pig [2]; Ascaridia galli [3];
    Hamster [4]; Mouse [5]

Source tissue
    Kidney (baby hamster kidney cells (BHK 21) [4]) [2, 4]; Spleen [2]; Liver [1,
    2]; Brain [2]; Fibroblasts [4]

Localization in source
   Mitochondria [1]; Microsomes [1]

Purification

Crystallization
   –

Cloned
   –

Renatured
   –

5 STABILITY

pH

Temperature (°C)

Oxidation

Organic solvent

General stability information

Storage

6 CROSSREFERENCES TO STRUCTURE DATABANKS

PIR/MIPS code

Brookhaven code

7 LITERATURE REFERENCES

[1] Kennedy, E.P.: Methods Enzymol.,5,486–488 (1962) (Review)
[2] Sribney, M., Kennedy, E.P.: J. Biol. Chem.,233,1315–1322 (1958)
[3] Bankov, I., Barrett, J.: Int. J. Parasitol.,23,1083–1085 (1993)
[4] Voelker, D.R., Kennedy, E.P.: Biochemistry,21,2753–2759 (1982)
[5] Marggraf, W.-D., Anderer, F.A.: Hoppe-Seyler's Z. Physiol. Chem.,355,803–810 (1974)

1 NOMENCLATURE

EC number
2.7.8.4

Systematic name
CDPethanolamine:L-serine ethanolaminephosphotransferase

Recommended name
Serine-phosphoethanolamine synthase

Synonyms
Ethanolaminephosphotransferase, serine
Serine ethanolamine phosphate synthetase
Serine ethanolamine phosphodiester synthase
Serine ethanolaminephosphotransferase
Serine-phosphinico-ethanolamine synthase
Serinephosphoethanolamine synthase

CAS Reg. No.
9023-23-8

2 REACTION AND SPECIFICITY

Catalyzed reaction
CDPethanolamine + L-serine →
→ CMP + L-serine-phosphoethanolamine

Reaction type
Substituted phospho group transfer

Natural substrates

Substrate spectrum
1 CDPethanolamine + L-serine [1, 2]
2 CDP-2-amino-2-methylpropanol + L-serine [1]
3 CMPaminoethylphosphonate + L-serine [1]
4 CDPethanolamine + DL-alpha-methylserine (can replace L-serine) [1]
5 More (CDPcholine or CDPserine cannot replace CDPethanolamine,
N-acetyl-L-serine, L-homoserine, 4-hydroxy-L-proline, 5-hydroxy-DL-lysine,
3-hydroxy-DL-glutamic acid, L-threonine, ethanolamine or 3-hydroxyprop-
ionic acid cannot replace L-serine) [1]

Product spectrum

1 CMP + L-serineethanolamine phosphate [1]
2 ?
3 ?
4 ?
5 ?

Inhibitor(s)

Ca^{2+} [1]; L-Threonine (10 mM: 27–34% inhibition) [1]; L-Homoserine (10 mM: 27–34% inhibiton) [1]; L-Alanine (10 mM: 27–34% inhibition) [1]; p-Hydroxy-mercuribenzoate (0.1 mM: 2% remaining activity) [1]; Methylmercuric iodide (0.1 mM: 3% remaining activity) [1]; Methylmercuric bromide (0.1 mM: 6% remaining activity) [1]; Phenylmercuric acetate (0.1 mM: 2% remaining activity) [1]; N-Ethylmaleimide (0.2 mM: 25% remaining activity) [1]; Iodoacetate (1 mM: 42% remaining activity) [1]; [1]; Iodoacetamide (1 mM: 6% remaining activity) [1]; N-Acetylimidazole (10 mM: 8% remaining activity) [1]; Diazotized sulfanilic acid (0.2 mM: 36% remaining activity) [1]; CMP (competitive inhibition to CDPethanolamine, non-competitive inhibition to L-serine) [1]; CDP (strong) [1]; Shell Nonidet P-40 (strong) [2]; Duponal (strong) [2]; CETAB (cationic detergent, strong) [2]

Cofactor(s)/prosthetic group(s)/activating agents

Metal compounds/salts

Mg^{2+} (10 mM, most effective bivalent cation [1], requirement [1, 2]) [1, 2]; Mn^{2+} (10 mM: 6% of activity compared to Mg^{2+}) [1]; Co^{2+} (10 mM: 10% of activity compared to Mg^{2+}) [1]

Turnover number (min⁻¹)

Specific activity (U/mg)

0.0034 [2]

K_m-value (mM)

0.011 (CMPaminoethylphosphonate) [1]; 0.085 (CDPethanolamine) [1]; 1 (L-serine) [1]

pH-optimum

7.5 [2]

pH-range

6.5–9.0 (55% of maximal activity at pH 6.5, 30% of maximal activity at pH 9.0) [1]

Temperature optimum (°C)

37 (assay at) [1, 2]

Temperature range (°C)

3 ENZYME STRUCTURE

Molecular weight

Subunits

Glycoprotein/Lipoprotein

–

4 ISOLATION/PREPARATION

Source organism
Chicken [1, 2]

Source tissue
Gut mucosa [1, 2]; Kidney (to a smaller extent than in gut mucosa) [1]

Localization in source
Microsomes [1, 2]

Purification
Chicken (partial) [1, 2]

Crystallization
–

Cloned
–

Renatured
–

5 STABILITY

pH

Temperature (°C)
37 (60 min, 8 M urea: no influence) [1]; 55 (15 min: inactivation, heat-labile [2]) [1, 2]

Oxidation

Organic solvent

General stability information
Aggregation on thawing, making withdrawals of homogeneous samples difficult [1]

Storage
4°C, crude extract, stable [1]; 10°C, storage of whole microsomes [2]

6 CROSSREFERENCES TO STRUCTURE DATABANKS

PIR/MIPS code

Brookhaven code

---

7 LITERATURE REFERENCES

[1] Allen, A.K., Rosenberg, H.: Biochim. Biophys. Acta,151,504–519 (1968)
[2] Rosenberg, H., Ennor, A.H.: Biochim. Biophys. Acta,115,23–32 (1966)

1 NOMENCLATURE

EC number
2.7.8.5

Systematic name
CDPdiacylglycerol:sn-glycerol-3-phosphate 3-phosphatidyltransferase

Recommended name
CDPdiacylglycerol-glycerol-3-phosphate 3-phosphatidyltransferase

Synonyms
Glycerophosphate phosphatidyltransferase
3-Phosphatidyl 1'-glycerol-3'-phosphate synthase
CDPdiacylglycerol:glycerol-3-phosphate phosphatidyltransferase [1]
Cytidine 5'-diphospho-1,2-diacyl-sn-glycerol (CDPdiglyceride):sn-glycer-ol-3-phosphate phosphatidyltransferase [2]
Phosphatidylglycerophosphate synthase [3]
Phosphatidylglycerolphosphate synthase [7]
PGP synthase [7]
CDPdiacylglycerol-sn-glycerol-3-phosphate 3-phosphatidyltransferase [5]
CDPdiacylglycerol:sn-glycero-3-phosphate phosphatidyltransferase [6]
Phosphatidyltransferase, glycerol phosphate
Glycerol 3-phosphate phosphatidyltransferase
Phosphatidylglycerol phosphate synthase
Phosphatidylglycerol phosphate synthetase
Phosphatidylglycerophosphate synthase
Phosphatidylglycerophosphate synthetase
sn-Glycerol-3-phosphate phosphatidyltransferase

CAS Reg. No.
9068-49-9

---

2 REACTION AND SPECIFICITY

Catalyzed reaction
CDPdiacylglycerol + sn-glycerol 3-phosphate →
→ CMP + 3(3-sn-phosphatidyl)-sn-glycerol 1-phosphate (ordered sequential Bi-Bi reaction [2, 5, 6])

Reaction type
Substituted phospho group transfer

Enzyme Handbook © Springer-Verlag Berlin Heidelberg 1997
Duplication, reproduction and storage in data banks are only
allowed with the prior permission of the publishers

Natural substrates

CDP-1,2-diacyl-sn-glycerol + glycerol 3-phosphate (involved in synthesis of phosphatidylglycerol) [7]

Substrate spectrum

1 CDPdiacylglycerol + sn-glycero-3-phosphate (r [2]) [2, 3, 5–10]
2 dCDPdiacylglycerol + sn-glycero-3-phosphate [2, 5, 6]
3 DL-2-Hexadecoxy-3-octadecoxypropylphosphonyl-O-(cytidine 5'-phosphate) + sn-glycero-3-phosphate [11]
4 DL-3,4-Dioctadecoxybutylphosphonyl-O-(cytidine 5'-phosphate) + sn-glycerol 3-phosphate [11]
5 More (cytosine-beta-D-arabinofuranoside-5'-monophosphate-dependent incorporation of glycerol 3-phosphate at pH 8.5 but not at pH 6.8 [1], cytidine 5'-monophosphate dependent exchange between glycerol 3-phosphate and phosphatidylglycerophosphate [2]) [1, 2]

Product spectrum

1 CMP + phosphatidylglycerophosphate (3-sn-phosphatidyl-1'-sn-glycerol 3'-phosphate [2], 1-(3-glycerophosphoryl)-glycerol 3-phosphate [8], phosphatidylglycerophosphate is the predominant product at pH 9.5, phosphatidylglycerol is the predominant product at pH 7.0 [3]) [2, 3, 5, 8, 9]
2 ?
3 ?
4 ?
5 ?

Inhibitor(s)

Liponucleotide (forms a dead-end complex at high concentrations inhibiting both, the forward and the reverse reaction) [2]; CDPdiacylglycerol (inhibition of CDPdiacylglycerol formation [2], uncompetitive at high concentration [6]) [2, 6]; Glycerol 3-phosphate (inhibition of CDPdiacylglycerol formation) [2]; Thioreactive agents (slight inhibition) [3]; Ca^{2+} [5]; Inositol (inhibits CMP-dependent incorporation of glycerol 3-phosphate by microsomes) [1]; Triton X-100 (0.2% inhibits, lower concentrations are required for CMP-dependent incorporation of glycerol 3-phosphate) [1]; Mg^{2+} (required, K_m: 50 mM, inhibition above 150 mM) [6]; Cd^{2+} (strong inhibition at concentrations above 0.5 mM in presence of Mg^{2+}) [9]; Zn^{2+} (strong inhibition at concentrations above 0.5 mM in presence of Mg^{2+}) [9]; Hg^{2+} (strong inhibition at concentrations above 0.5 mM in presence of Mg^{2+}) [9]; Cu^{2+} (strong inhibition at concentrations above 0.5 mM in presence of Mg^{2+}) [9]

Cofactor(s)/prosthetic group(s)/activating agents

Triton X-100 (0.5–6%, absolute requirement [2], maximal activity at pH 9.5 is dependent on Triton X-100, 0.5 mM [3], maximal activity is dependent on Triton X-100 [4, 6, 8, 10]: at 1 mM [4], 0.2% [6], maximal activity at a molar ratio of Triton X-100 to CDPdiacylglycerol of 50:1 [10]) [2–4, 6, 8, 10]; Phosphatidylethanolamine (stimulates, no stimulation by other diacyl-glycerophosphatides or lysophosphatides) [9]; Phosphatidylinositol (required for CMP-dependent incorporation of glycerol 3-phosphate) [1]

Metal compounds/salts

Mn^{2+} (required for CMP-dependent incorporation of glycerol 3-phosphate [1], at pH 9.5 maximal activity depends on 0.5 mM Mn^{2+}, 10 mM Mg^{2+} or 20 mM Co^{2+} [3], maximal activity depends on 0.1 mM Mn^{2+}, 0.3 mM Mg^{2+} or 1 mM Co^{2+} [4], cannot substitute for Mg^{2+} [8], cation requirement is relatively non-specific, Mg^{2+}, Ba^{2+} or Ca^{2+} provides maximal activation in the 10 mM range. Mn^{2+} or Co^{2+} stimulates at lower concentration, inhibition at higher concentration [9]) [1, 3, 4, 8, 9]; Mg^{2+} (at pH 9.5 maximal activity is dependent on 0.5 mM Mn^{2+}, 10 mM Mg^{2+} or 20 mM Co^{2+} [3], maximal activity is dependent on 0.1 mM Mn^{2+}, 0.3 mM Mg^{2+} or 1 mM Co^{2+} [4], 80 mM required for maximal activity [8], cation requirement is relatively nonspecific, Mg^{2+}, Ba^{2+} or Ca^{2+} provides maximal activation in the 10 mM range. Mn^{2+} or Co^{2+} stimulates at lower concentration, inhibition at higher concentration [9], activity depends on Mg^{2+}, 100 mM [10], absolute requirement for a divalent metal, Mg^{2+} required, K_m: 50 mM, inhibition above 150 mM [6]) [3, 4, 6, 8–10]; Ca^{2+} (cannot substitute for Mg^{2+} in activation [8], cation requirement is relatively nonspecific, Mg^{2+}, Ba^{2+} or Ca^{2+} provides maximal activation in the 10 mM range. Mn^{2+} or Co^{2+} stimulates at lower concentration, inhibition at higher concentration [9]) [9]; Co^{2+} (at pH 9.5 maximal activity depends on 0.5 mM Mn^{2+}, 10 mM Mg^{2+} or 20 mM Co^{2+} [3], maximal activity depends on 0.1 mM Mn^{2+}, 0.3 mM Mg^{2+} or 1 mM Co^{2+} [4], cation requirement is relatively nonspecific, Mg^{2+}, Ba^{2+} or Ca^{2+} provides maximal activation in the 10 mM range. Mn^{2+} or Co^{2+} stimulates at lower concentration, inhibition at higher concentration [9]) [3, 4, 9]; Ba^{2+} (cation requirement is relatively nonspecific, Mg^{2+}, Ba^{2+} or Ca^{2+} provides maximal activation in the 10 mM range. Mn^{2+} or Co^{2+} stimulates at lower concentration, inhibition at higher concentration) [9]

Turnover number (min^{-1})

Specific activity (U/mg)

More [2, 9]; 0.03 [10]; 0.36 [8]; 18.6 [6]; 22 [5]

K_m-value (mM)

More (anomalous kinetics for CDPdiacylglycerol) [8]; 0.020 (glycerol 3-phosphate) [7]; 0.034 (dCDPdiacylglycerol) [2, 6]; 0.04 ((d)CDPdiacyl-glycerol) [5]; 0.046 (CDPdiacylglycerol) [2, 6, 7]; 0.060 (DL-3,4-dioctadecoxybutylphosphonyl-O-(cytidine 5'-phosphate)) [11]; 0.080 (DL-2-hexadecoxy-3-octadecoxypropylphosphonyl-O-(cytidine 5'-phosphate)) [11]; 0.1 (CDPdiacylglycerol [3], sn-glycerol 3-phosphate [10]) [3, 10]; 0.15 (sn-glycerol 3-phosphate) [8]; 0.17 (glycerol 3-phosphate) [3]; 0.19 (CMP) [1]; 0.32 (sn-glycerol 3-phosphate) [2, 5, 6]

---

pH-optimum
   7.0 (2 optima: pH 7.0 and 9.5 [3]) [3, 4]; 7.4 [1]; 8.0 [10]; 9.5 (2 optima: pH
   7.0 and 9.5) [3]

pH-range

Temperature optimum (°C)

Temperature range (°C)
   30 (assay at) [1, 8, 10]; 37 (assay at) [5, 6, 9, 11]

---

3 ENZYME STRUCTURE

Molecular weight
   200000 (E. coli, gel filtration) [2]

Subunits
   ? (x × 24000, E. coli, SDS-PAGE) [2, 6]

Glycoprotein/Lipoprotein
   –

---

4 ISOLATION/PREPARATION

Source organism
   Rat [7, 9]; Rabbit [1]; E. coli [2, 5, 6, 11]; Soybean [3]; Saccharomyces cerevisiae [4]; Bacillus licheniformis [8]; Pig [9]; Clostridium perfringens [10]

Source tissue
   Liver [9]; Lung [1]; Heart [7]

Localization in source
   Microsomes [1]; Membrane (associated [2], bound [8]) [2, 8]; Mitochondria
   (membrane [9], crude mitochondrial fraction [3]) [3, 4, 7, 9]; Cell envelope
   [10]

Purification
   E. coli [2, 5, 6]; Bacillus licheniformis (partial) [8]

Crystallization
   –

Cloned
   [5]

Renatured
   –

5 STABILITY

pH

Temperature (°C)
    55 (5 min, stable, partially purified enzyme, Triton X-100 extract [2, 5], mitochondrial fraction, 26% loss of activity after 1 min, 70% loss of activity after 5 min [7]) [2, 5, 7]; 60 (5 min, 50% loss of activity, partially purified enzyme, Triton X-100 extract [2, 5], 100% stable for at least 20 min [10]) [2, 5, 10]; 65 (5 min, complete loss of activity, partially purified enzyme, Triton X-100 extract) [2, 5]

Oxidation

Organic solvent

General stability information
    Urea, 8 M, stable in presence of 0.1% Triton X-100 after 2 h at 30°C [2, 5]; SDS, 1%, 50% loss of activity after 2 h at 30°C [2, 5]; Combination of both 4 M urea and 1% SDS at 30°C completely inactivates [2, 5]

Storage
    –80°C, 50 mM Tris-HCl, pH 7.0, 10 mM $MgCl_2$, 2 mM DTT, 0.1% Triton X-100, stable for at least 3 years [5]; 4°C, 50 mM Tris-HCl, pH 7.0, 10 mM $MgCl_2$, 2 mM DTT, 0.1% Triton X-100, stable for several months [5]; 4°C, 50% loss of activity after 12 h, broken cell preparation [8]

---

6 CROSSREFERENCES TO STRUCTURE DATABANKS

PIR/MIPS code

Brookhaven code

---

7 LITERATURE REFERENCES

[1] Bleasdale, J.E., Johnston, J.M.: Biochim. Biophys. Acta,710,377–390 (1982)
[2] Hirabayashi, T., Larson, T.J., Dowhan, W.: Biochemistry,15,5205–5211 (1976)
[3] Carman, G.M., Greenberg, A.S.: J. Food Biochem.,8,321–333 (1984)
[4] Carman, G.M., Belunis, C.J.: Can. J. Biochem.,61,1452–1457 (1983)
[5] Dowhan, W.: Methods Enzymol.,209,313–321 (1992) (Review)
[6] Dowhan, W., Hirabayashi, T.: Methods Enzymol.,71,555–561 (1981) (Review)
[7] Cao, S.G., Hatch, G.M.: Lipids,29,475–480 (1994)
[8] Larson, T.J., Hirabayashi, T., Dowhan, W.: Biochemistry,15,974–979 (1976)
[9] McMurray, W.C., Jarvis, E.C.: Can. J. Biochem.,56,414–419 (1978)
[10] Carman, G.M., Wieczorek, D.S.: J. Bacteriol.,142,262–267 (1980)
[11] Tyhach, R.J., Rosenthal, A.F., Tropp, B.E.: Biochim. Biophys. Acta,388,29–37 (1975)

1 NOMENCLATURE

EC number
2.7.8.6

Systematic name
UDPgalactose:undecaprenyl-phosphate galactosephosphotransferase

Recommended name
Undecaprenyl-phosphate galactosephosphotransferase

Synonyms
Poly(isoprenol)-phosphate galactosephosphotransferase
Galactosephosphotransferase, poly(isoprenol) phosphate
Poly(isoprenyl)phosphate galactosephosphatetransferase
Undecaprenyl phosphate galactosyl-1-phosphate transferase

CAS Reg. No.
37278-29-8

2 REACTION AND SPECIFICITY

Catalyzed reaction
UDPgalactose + undecaprenyl phosphate →
→ UMP + alpha-D-galactosyl-diphosphoundecaprenol

Reaction type
Substituted phospho group transfer

Natural substrates
UDPgalactose + glycosyl carrier lipid (involved in biosynthesis of O-antigen in Salmonella typhimurium and Citrobacter strain 139 [2], involved in biosynthesis of O-antigen in Salmonella newington [1]) [1, 2]

Substrate spectrum
1 UDP-D-galactose + antigen carrier lipid-phosphate (r) [1]
2 UDPgalactose + phospholipid (i.e. glycosyl carrier lipid) (r) [2]

Product spectrum
1 UMP + galactose-diphosphate antigen carrier lipid (polyisoprenoid structure linked to sugars by a diphosphate bridge) [1]
2 UMP + galactose-1-diphospho-lipid (crude extract of cell envelope fraction) [2]

Inhibitor(s)
UMP (0.04 mM: 56% inhibition, 0.1 mM: 76% inhibition, 0.4 mM: 93% inhibition) [2]; dUMP (0.13 mM: 48% inhibition, 0.37 mM: 78% inhibition, 1 mM: 87% inhibition) [2]; UDP (0.12 mM: no inhibition, 1 mM: 40% inhibition) [2]; UTP (0.11 mM: 5% inhibition, 1 mM: 49% inhibition) [2]; TMP (1.1 mM: 12% inhibition) [2]; CMP (1 mM: 1% inhibition) [2]; GMP (1 mM: 5% inhibition) [2]; More (1 mM AMP: no inhibition) [2]

Cofactor(s)/prosthetic group(s)/activating agents

Metal compounds/salts

Turnover number (min^{-1})

Specific activity (U/mg)

K_m-value (mM)

pH-optimum
8.5 (assay at) [2]

pH-range

Temperature optimum (°C)
37 (assay at) [2]

Temperature range (°C)

---

3 ENZYME STRUCTURE

Molecular weight

Subunits

Glycoprotein/Lipoprotein
–

---

4 ISOLATION/PREPARATION

Source organism
Salmonella newington [1]; Citrobacter sp. (strain 139) [2]; Salmonella typhimurium (LT-2, mutant strain G30) [2]

Source tissue
Cell [1]; Cell envelope [2]

Localization in source

Purification
Salmonella newington (partial) [1]; Citrobacter sp. (strain 139, partial) [2];
Salmonella typhimurium (LT-2, mutant strain G30, partial) [2]

Crystallization
–

Cloned
–

Renatured
–

---

5 STABILITY

pH

Temperature (°C)

Oxidation

Organic solvent

General stability information

Storage
–18°C, stable at least 1 month [2]

---

6 CROSSREFERENCES TO STRUCTURE DATABANKS

PIR/MIPS code

Brookhaven code

---

7 LITERATURE REFERENCES

[1] Wright, A., Dankert, M,, Pennessy, P., Robbins, P.W.: Biochemistry,57,1798–1803
(1967)
[2] Osborn, M.J., Tze-Yuen,R.: J. Biol. Chem.,243,5145–5152 (1968)

1 NOMENCLATURE

EC number
2.7.8.7

Systematic name
CoA:apo-[acyl-carrier-protein] pantetheinephosphotransferase

Recommended name
Holo-[acyl-carrier-protein] synthase

Synonyms
Acyl carrier protein holoprotein (holo-ACP) synthetase [1]
Holo-ACP synthetase [1]
Coenzyme A:fatty acid synthetase apoenzyme 4'-phosphopantetheine transferase [5]
Synthase, holo-
Acyl carrier protein synthetase
Holo-ACP synthase

CAS Reg. No.
37278-30-1

---

2 REACTION AND SPECIFICITY

Catalyzed reaction
CoA + apo-[acyl-carrier protein] →
→ adenosine 3',5'-bisphosphate + holo-[acyl-carrier protein]

Reaction type
Substituted phospho group transfer

Natural substrates

Substrate spectrum
1 CoA + apo-[acyl-carrier protein] (r [1, 3], amino terminal hexapeptide of apo-ACP is important for activity [3], no substrates: dephospho-CoA [1–3], oxidized CoA [1–3]) [1–5]

Product spectrum
1 Adenosine 3',5'-bisphosphate + holo-[acyl-carrier protein] [1, 3]

Inhibitor(s)
Apo-ACP (substrate inhibition) [3]; 3',5'-ADP [4]; More (no substrate inhibition by apo-peptide(1→74)) [3]

Cofactor(s)/prosthetic group(s)/activating agents

Metal compounds/salts
   Mg^{2+} (Mg^{2+} or Mn^{2+} required [1–3], K_m: 3 mM [1, 3, 4]) [1–4]; Mn^{2+} (Mg^{2+} or
   Mn^{2+} required [1–3], saturating concentration: 0.1 mM, increasing inhibition
   up to 5 mM, increasing concentration further increase the activity, being followed by slow but extensive recovery, rate at 100 mM is about twice that at
   0.1 mM [1, 3]) [1–3]

Turnover number (min^{-1})

Specific activity (U/mg)
   More [1]; 0.0079 [3]

K_m-value (mM)
   0.0004 (apo-ACP) [1, 2]; 0.00055 (apo-ACP) [3]; 0.002 (apo-peptide(1→74)
   [3], apo-ACP [4]) [3, 4]; 0.072 (CoA) [4]

pH-optimum
   7.5–9.0 [1, 3]; 8.2 [4]

pH-range
   6.5–11 (6.5: about 20% of activity maximum, 11: about 30% of activity maximum) [1]

Temperature optimum (°C)

Temperature range (°C)

3 ENZYME STRUCTURE

Molecular weight
   500000 (E. coli, sucrose density gradient sedimentation) [3]

Subunits

Glycoprotein/Lipoprotein
   –

4 ISOLATION/PREPARATION

Source organism
   Saccharomyces cerevisiae [5]; E. coli (B [1, 3]) [1–3]; Spinacia oleracea [4];
   Ricinus communis [4]

Source tissue
   Leaf [4]; Endosperm [4]

Localization in source
Soluble [2]; Cytosol [4]

Purification
E. coli B [1, 3]; Spinacia oleracea (partial) [4]

Crystallization
–

Cloned
–

Renatured
–

---

5 STABILITY

pH

Temperature (°C)

Oxidation

Organic solvent

General stability information
Stable, if CoASH is present at half-saturating concentrations, it is necessary to include this substrate in all steps of purification beyond the 54°C treatment [3]

Storage

---

6 CROSSREFERENCES TO STRUCTURE DATABANKS

PIR/MIPS code

Brookhaven code

---

7 LITERATURE REFERENCES

[1] Elovson, J., Vagelos, P.R.: J. Biol. Chem.,243,3603–3611 (1968)
[2] Prescott, D.J., Vagelos, P.R.: Adv. Enzymol. Relat. Areas Mol. Biol.,36,269–311 (1972) (Review)
[3] Prescott, D.J., Elovson, J., Vagelos, P.R.: Methods Enzymol.,35B,95–101 (1975) (Review)
[4] Elhussein, S.A., Miernyk, J.A., Ohlrogge, J.B.: Biochem. J.,252,39–45 (1988)
[5] Werkmeister, K., Wieland, F., Schweizer, E.: Biochem. Biophys. Res. Commun., 96,483–490 (1980)

Enzyme Handbook © Springer-Verlag Berlin Heidelberg 1997
Duplication, reproduction and storage in data banks are only
allowed with the prior permission of the publishers

1 NOMENCLATURE

EC number
2.7.8.8

Systematic name
CDPdiacylglycerol:L-serine 3-O-phosphatidyltransferase

Recommended name
CDPdiacylglycerol-serine O-phosphatidyltransferase

Synonyms
PS synthase [15]
Phosphatidylserine synthase
CDPdiglyceride-serine O-phosphatidyltransferase
Cytidine 5'-diphospho-1,2-diacyl-sn-glycerol (CDPdiglyceride):L-serine
O-phosphatidyltransferase [1]
Phosphatidylserine synthetase [2]
CDP-diglyceride:L-serine phosphatidyltransferase [2]
CDPdiacylglycerol:L-serine O-phosphatidyltransferase [4]
CDPdiacylglycerol-L-serine O-phosphatidyltransferase [5]
Phosphatidyltransferase, cytidine diphosphoglyceride-serine O-
CDP-diacylglycerol:L-serine O-phosphatidyltransferase
CDP-diglyceride-L-serine phosphatidyltransferase
CDP-diglyceride:serine phosphatidyltransferase
Cytidine 5'-diphospho-1,2-diacyl-sn-glycerol:L-serine O-phosphatidyltrans-
ferase [11]

CAS Reg. No.
9068-48-8

---

2 REACTION AND SPECIFICITY

Catalyzed reaction
CDPdiacylglycerol + L-serine →
→ CMP + 3-O-sn-phosphatidyl-L-serine (bi-bi sequential reaction, mechanism [5, 8, 13, 17], ping-pong mechanism (not [8]) [1, 3, 4], reaction of E.
coli enzyme proceeds with retention of configuration at phosphorus:
two-step mechanism involving a phosphatidyl-enzyme intermediate, Saccharomyces cerevisiae enzyme catalyzes the reaction with inversion of configuration, single-displacement mechanism [14])

Reaction type
Substituted phospho group transfer

Natural substrates

More (enzyme catalyzes the first committed step in biosynthesis of phosphatidylethanolamine, the major phospholipid of E. coli [7], enzyme plays an important role in regulation of phospholipid biosynthesis in Saccharomaces cerevisiae [5], enzyme of phospholipid biosynthesis [6], reconstitution of the enzyme into phospholipid vesicles suggests that the activity is modulated by the phosphatidylinositol to phosphatidylserine ratio in the membrane, increase in the ratio of phosphatidylinositol to phosphatidylserine in vesicles results in decrease in phosphatidylserine synthase activity [15], normal function of the enzyme involves membrane association which is primarily induced by the presence of a membrane-associated substrate [16], phosphatidylserine synthesis [17]) [5–7, 15–17]

Substrate spectrum

1 CDPdiacylglycerol + serine (r [2, 5, 12, 17, 19], equilibrium strongly favours synthesis of phosphatidylserine [2], forward reaction favoured [19], reversible in presence of 5'-CMP, but not 2'-CMP, 3'-CMP or 5'-AMP [12] CDP-diacyl-L-glycerol [4, 7]) [1–20]
2 dCDPdiacylglycerol + serine [1, 8]
3 CDPdiacylglycerol + glycerol (at a very slow rate) [1]
4 CDPdiacylglycerol + sn-glycerol 3-phosphate (at a very slow rate) [1]
5 CDP-1,2-dipalmitoylglycerol + serine [4, 7]
6 CDP-1,2-dicaproylglycerol + serine [4, 7]
7 More (enzyme also catalyzes hydrolysis of CDPdiacylglycerol to form CMP and phosphatidic acid [1], also hydrolyzes phosphatidylserine and CDPdiacylglycerol at a slow rate [2], specific for the L-glycerol 3-phosphate isomer of the liponucleotide, it does not recognize the D-isomer of the 1-monoacyl derivative [7]) [1, 2, 7]

Product spectrum

1 CMP + phosphatidylserine [1–5, 13]
2 dCMP + phosphatidylserine
3 CMP + phosphatidylglycerol [1]
4 CMP + phosphatidylglycerophosphate [1]
5 CMP + 1,2-dipalmitoyl-3-phosphatidylserine
6 CMP + 1,2-dicaproyl-3-phosphatidylserine
7 ?

Inhibitor(s)

EDTA [12]; Thioreactive agents [5]; Inositol [5]; Cardiolipin [5]; Diacylglycerol [5]; Triton X-100 (inhibition at detergent concentration higher than a molar ratio of 20:1 [8], increasing levels of Triton X-100 at low molar ratios of Triton X-100 to CDPdiacylglycerol stimulate enzyme activity to a maximum and then apparently inhibit activity as the molar ratio is raised beyond the point of maximal activity [7]) [7, 8]; Phosphatidylglycerol [9]; p-Hydroxymercuribenzoate (reversed by DTT) [12]; Ca^{2+} (in presence of Mn^{2+} [12]) [12, 20]; Hg^{2+} [12]; More (phosphorylation by cAMP-dependent protein kinase results in a 60–70% reduction of enzyme activity) [5, 10]

Cofactor(s)/prosthetic group(s)/activating agents

Cardiolipin (activates) [9]; Phosphatidylethanolamine (slight activation) [9]; Detergent (required [13], nonionic detergent required [4, 8], at 0.1 mM CDPdiacylglycerol the optimum activity occurs at a Triton to substrate ratio of 8:1 [4], nonionic detergents, e.g. Triton X-100, octyl glucoside, stimulate [17], Triton X-100, 0.1–0.5% w/v stimulates [20], increasing levels of Triton X-100 at low molar ratios of Triton X-100 to CDPdiacylglycerol stimulate enzyme activity to a maximum and then apparently inhibit activity as the molar ratio is raised beyond the point of maximal activity [7], maximal activity at a molar ratio of Triton X-100 to CDPdiacylglycerol of 20:1, inhibition at higher detergent concentration [8], 2.7 mM required [19]) [4, 7, 8, 13, 17, 19, 20]; Phosphatidate (activates) [5]; Phosphatidylcholine (activates) [5]; Phosphatidylinositol (activates) [5]; Mercaptoethanol (slight stimulation) [8]; CDPdiolein (phosphatidylserine biosynthesis of particle-bound enzyme is dependent on, K_m: 0.17 mM) [12]; Phosphatidylserine (stimulates its own formation) [17]

Metal compounds/salts

Mn^{2+} (required [8, 13], activity is dependent on Mn^{2+}, 3 mM [19], Mn^{2+} or Mg^{2+} required [20], activity requires either $MnCl_2$ (0.6 mM) or $MgCl_2$ (20 mM) [5, 8], divalent metal ion required, Mn^{2+} is optimal at around 10 mM [17], maximal activity obtained with Mn^{2+} is 2-fold greater than maximal activity obtained with Mg^{2+} [8], 300-fold stimulation by 5 mM [12], activity is dependent on 0.1 mM Mn^{2+}, or 50 mM Mg^{2+} [18]) [5, 8, 12, 13, 17–20]; Mg^{2+} (Mn^{2+} or Mg^{2+} required [20], activity requires either $MnCl_2$ (0.6 mM) or $MgCl_2$ (20 mM) [5, 8], maximal activity obtained with Mn^{2+} is 2-fold greater than maximal activity obtained with Mg^{2+} [8], activity is dependent on Mn^{2+} (0.1 mM) or Mg^{2+} (50 mM) [18]) [5, 8, 18, 20]; Co^{2+} (can partially replace Mn^{2+} in activation) [12]; More (no activation with: Ca^{2+} [8, 12], Cu^{2+} [12], Hg^{2+} [12]) [8, 12]

Turnover number (min^{-1})

3000 (phosphatidylserine produced) [1]

Specific activity (U/mg)

1.31 [2]; 33.0 [1]; 39 [3, 4]; 2.3 (strain S288C) [5, 8]; 3.76 (strain VAL2C) [8]

K_m-value (mM)

More [4]; 0.060 (CDPdiacylglycerol) [8]; 0.12 (CDPdiacylglycerol) [20]; 0.13 (L-serine) [20]; 0.15 (L-serine) [18]; 0.23 (L-serine) [13]; 0.24 (CDPdiacylglycerol) [19]; 0.26 (L-serine) [19]; 0.58 (serine) [8]; 0.83 (serine) [5]; 4 (L-serine) [12]

pH-optimum

7.0 [12]; 7.0–8.5 [3]; 7.4 (assay at) [1]; 8.0 [5, 8, 18]; 8.0–8.5 [20]; 8.5 [19]

pH-range

6–9 (6: about 20% of activity maximum, 9: about 50% of activity maximum) [8]

Temperature optimum (°C)

30 (assay at) [1, 4, 5, 8, 19, 20]; 37 (assay at) [2]; 40 [19]

Temperature range (°C)

20–60 (20°C: about 30% of activity maximum, 60°C: about 60% of activity maximum) [19]

3 ENZYME STRUCTURE

Molecular weight

500000 (E. coli [3, 4], glycerol density gradient centrifugation [3]) [3, 4]

Subunits

? (x × 54000, E. coli, SDS-PAGE [3, 4], x × 23000, Saccharomyces cerevisiae, SDS-PAGE [8], x × 53000, Bacillus licheniformis, SDS-PAGE [13]) [3, 4, 8, 13]

Glycoprotein/Lipoprotein

–

4 ISOLATION/PREPARATION

Source organism

Bacillus subtilis [17]; Clostridium perfringens (ATCC 3624 [18, 19]) [11, 18, 19]; Bacillus licheniformis [13, 17]; E. coli (strain RA324 [3]) [1–7, 9, 14, 16]; Saccharomyces cerevisiae (strain S288C [5, 8], overproducing strain VAL2C [8, 14, 15], X2180–1B [20]) [5, 8, 10, 12, 14, 15, 20]

Source tissue

Cell [5, 19]

Localization in source

Ribosomes (associated with) [1, 2, 4, 5, 16]; Membrane (tight membrane association [17]) [13, 17, 19]; Microsomes [5, 8]; Cell envelope [11, 18]

Purification

Bacillus licheniformis [13]; E. coli (partial [2]) [1–4]; Saccharomyces cerevisiae (partial [20]) [5, 20]; Clostridium perfringens (solubilization [11], partial [19]) [11, 19]

Crystallization
–

Cloned
–

Renatured
–

5 STABILITY

pH

Temperature (°C)
   30 (10 min, unstable above) [19]; 40 (labile above) [5, 8]; 50 (20 min, 90%
   loss of activity) [8]; 60 (20 min, total inactivation [8], stable for at least 20
   min [18]) [8, 18]

Oxidation

Organic solvent

General stability information
   High ionic strength buffers are necessary to prevent irreversible precipitation and inactivation [1]; Presence of nonionic detergents, Triton X-100 or
   octylglucoside above their critical micelle concentration increases the
   stability [4]; Stable to one cycle of freezing and thawing [19]

Storage
   –20°C, 10% glycerol, stable for at least 1 month [1]; 4°C, 1 M NaCl, 20 mM
   sodium phosphate buffer, pH 7.0, 0.5 mM DTT, 25% loss of activity after 1
   month [1]; –80°C, 0.1 M potassium phosphate, pH 7.4, 1% Triton X-100,
   10% glycerol, 0.5 mM DTT, 0.65 M NaCl, stable for several years [4]; 4°C,
   0.1 M potassium phosphate, pH 7.4, 1% Triton X-100, 10% glycerol, 0.5 mM
   DTT, 0.65 M NaCl, stable for several months [4]; –80°C, 50 mM Tris-HCl, pH
   7.5, 3.0 mM CDPdiacylglycerol, 0.94% Triton X-100, 2.0 mM $MnCl_2$, 100%
   stable for at least 4 months [19]

6 CROSSREFERENCES TO STRUCTURE DATABANKS

PIR/MIPS code
   PIR2:A55537 (Bacillus subtilis); PIR2:JH0368 (Escherichia coli);
   PIR2:C64451 (Methanococcus jannaschii); PIR2:S00080 (yeast (Sac-
   charomyces cerevisiae)); PIR2:S59798 (yeast (Saccharomyces cerevisiae))

Brookhaven code

7 LITERATURE REFERENCES

[1] Larson, T.J., Dowhan, W.: Biochemistry,15,5212–5218 (1976)
[2] Raetz, C.R.H., Kennedy, E.P.: J. Biol. Chem.,249,5038–5045 (1974)
[3] Dowhan, W., Larson, T.: Methods Enzymol.,71,561–571 (1981) (Review)
[4] Dowhan, W.: Methods Enzymol.,209,287–298 (1992) (Review)
[5] Carman, G.M., Bae-Lee, M.: Methods Enzymol.,209,298–305 (1992) (Review)
[6] Raetz, C.R.H., Kennedy, E.P.: J. Biol. Chem.,247,2008–2014 (1972)
[7] Carman, G.M., Dowhan, W.: J. Biol. Chem.,254,8391–8397 (1979)
[8] Bae-Lee, M.S., Carman, G.M.: J. Biol. Chem.,259,10857–10862 (1984)
[9] Ishinaga, M., Kato, M., Kito, M.: FEBS Lett.,49,201–202 (1974)
[10] Kinney, A.J., Carman, G.M.: Proc. Natl. Acad. Sci. USA,85,7962–7966 (1988)
[11] Cousminer, J.J., Carman, G.M.: Can. J. Microbiol.,27,544–547 (1981)
[12] Carson, M.A., Atkinson, K.D., Waechter, C.J.: J. Biol. Chem.,257,8115–8121 (1982)
[13] Dutt, A., Dowhan, W.: Biochemistry,24,1073–1079 (1985)
[14] Raetz, C.R.H., Carman, G.M., Dowhan, W., Jiang, R.-T., Waszkuc, W., Loffredo, W.,
     Tsai, M.-D.: Biochemistry,26,4022–4027 (1987)
[15] Hromy, J.M., Carman, G.M.: J. Biol. Chem.,261,15572–15576 (1986)
[16] Louie, K., Chen, Y.-C., Dowhan, W.: J. Bacteriol.,165,805–812 (1986)
[17] Dutt, A., Dowhan, W.: J. Bacteriol.,147,535–542 (1981)
[18] Carman, G.M., Wieczorek, D.S.: J. Bacteriol.,142,262–267 (1980)
[19] Cousminer, J.J., Fischl, A.S., Carman, G.M.: J. Bacteriol.,151,1372–1379 (1982)
[20] Nikawa, J.-I., Yamashita, S.: Biochim. Biophys. Acta,665,420–426 (1981)

1 NOMENCLATURE

EC number
2.7.8.9

Systematic name
GDPmannose:phosphomannan mannosephosphotransferase

Recommended name
Phosphomannan mannosephosphotransferase

Synonyms
Mannosephosphotransferase, phosphomannan

CAS Reg. No.
37278-31-2

2 REACTION AND SPECIFICITY

Catalyzed reaction
GDPmannose + (phosphomannan)$_n$ →
→ GMP + (phosphomannan)$_{n+1}$

Reaction type
Substituted phospho group transfer

Natural substrates

Substrate spectrum
1 GDPmannose + endogenous acceptor molecule (crude extract, acceptor
may be cell wall mannan, glycoprotein or exocellular phosphomannan)
[1]
2 More (mannose 6-phosphate, GDP are ineffective) [1]

Product spectrum
1 GDP + endogenous acceptor molecule modified (mannose 1-phosphate
is transferred from GDPmannose forming a 1,6'-phosphodiester linkage
between 2 mannose residues (proposed)) [1]
2 ?

Inhibitor(s)
GMP (0.5 mM: 70% inhibition) [1]; GDP (0.5 mM: 70% inhibition) [1]

Cofactor(s)/prosthetic group(s)/activating agents

Metal compounds/salts
Mn^{2+} (10 mM: optimum, divalent metal ion required) [1]

Turnover number (min⁻¹)

Specific activity (U/mg)

K_m-value (mM)

pH-optimum
    6.5 (assay at) [1]

pH-range

Temperature optimum (°C)
    30 (assay at) [1]

Temperature range (°C)

---

3 ENZYME STRUCTURE

Molecular weight

Subunits

Glycoprotein/Lipoprotein
    –

---

4 ISOLATION/PREPARATION

Source organism
    Hansenula holstii (NRRL-Y 2448) [1]

Source tissue
    Cell [1]

Localization in source

Purification
    Hansenula holstii (NRRL-Y 2448, partial) [1]

Crystallization
    –

Cloned
    –

Renatured
    –

5 STABILITY

pH

Temperature (°C)

Oxidation

Organic solvent

General stability information

Storage
Liquid N_2 [1]

6 CROSSREFERENCES TO STRUCTURE DATABANKS

PIR/MIPS code

Brookhaven code

7 LITERATURE REFERENCES

[1] Bretthauer, R.K., Kozak, L.P., Irwin, W.E.: Biochem. Biophys. Res. Commun.,37, 820–827 (1969)

3

1 NOMENCLATURE

EC number
2.7.8.10

Systematic name
CDPcholine:sphingosine cholinephosphotransferase

Recommended name
Sphingosine cholinephosphotransferase

Synonyms
CDP-choline-sphingosine cholinephosphotransferase [1]
Phosphorylcholine-sphingosine transferase [1]
Cholinephosphotransferase, sphingosine
Cytidine diphosphocholine-sphingosine cholinephosphotransferase

CAS Reg. No.
9027-12-7

2 REACTION AND SPECIFICITY

Catalyzed reaction
CDPcholine + sphingosine →
→ CMP + sphingosyl-phosphocholine

Reaction type
Substituted phospho group transfer

Natural substrates

Substrate spectrum
1 CDPcholine + sphingosine [1]

Product spectrum
1 CMP + sphingosylphosphorylcholine [1]

Inhibitor(s)

Cofactor(s)/prosthetic group(s)/activating agents

Metal compounds/salts

Turnover number (min^{-1})

Specific activity (U/mg)

K_m-value (mM)

pH-optimum

pH-range

Temperature optimum (°C)

Temperature range (°C)

3 ENZYME STRUCTURE

Molecular weight

Subunits

Glycoprotein/Lipoprotein
  –

4 ISOLATION/PREPARATION

Source organism
  Chicken [1]

Source tissue
  Liver [1]

Localization in source
  Particulate [1]

Purification

Crystallization
  –

Cloned
  –

Renatured
  –

5 STABILITY

pH

Temperature (°C)

Oxidation

Organic solvent

General stability information

Storage

6 CROSSREFERENCES TO STRUCTURE DATABANKS

PIR/MIPS code

Brookhaven code

7 LITERATURE REFERENCES

[1] Fujino, Y., Negishi, T., Ito, S.: Biochem. J.,109,310–311 (1968)

1 NOMENCLATURE

EC number
2.7.8.11

Systematic name
CDPdiacylglycerol:myo-inositol 3-phosphatidyltransferase

Recommended name
CDPdiacylglycerol-inositol 3-phosphatidyltransferase

Synonyms
CDPdiglyceride-inositol phosphatidyltransferase
Phosphatidylinositol synthase
CDP-diacylglycerol-inositol phosphatidyltransferase [1]
CDP-diglyceride:inositol transferase [3]
Cytidine 5'-diphospho-1,2-diacyl-sn-glycerol:myo-inositol 3-phosphatidyl-
transferase [8]
CDP-DG:inositol transferase [13]
Phosphatidyltransferase, cytidine diphosphodiglyceride-inositol
CDP-diacylglycerol:myo-inositol-3-phosphatidyltransferase
CDP-diglyceride-inositol transferase
Cytidine diphosphoglyceride-inositol phosphatidyltransferase
Cytidine diphosphoglyceride-inositol transferase

CAS Reg. No.
9027-01-4

---

2 REACTION AND SPECIFICITY

Catalyzed reaction
CDPdiacylglycerol + myo-inositol →
→ CMP + phosphatidyl-1D-myo-inositol (bi-bi sequential reaction [6])

Reaction type
Substituted phospho group transfer

Natural substrates
More (enzyme catalyzes the final step of de novo synthesis of phosphatidylinositol [3, 5, 10, 20], enzyme plays a role in resynthesis of phosphatidylinositol during agonist-stimulated inositol-phospholipid metabolism [19]) [3, 5,
10, 19, 20]

Substrate spectrum
1 CDPdiacylglycerol + myo-inositol (r [16]) [1–21]
2 CDPdioleoylglycerol + myo-inositol (r [1]) [1, 5]
3 CDPdidecanoyl-sn-glycerol + myo-inositol (preferred substrate) [5]
4 CDPdipalmitoyl-sn-glycerol + myo-inositol (70% of the activity with CDP-1-stearoyl-2-oleoylglycerol [7]) [5, 7, 21]
5 CDP-1-stearoyl-2-oleoylglycerol + myo-inositol [7]
6 CDPdistearoylglycerol + myo-inositol (38% of the activity with CDP-1-stearoyl-2-oleoylglycerol) [7]
7 CDPdiarachidonoylglycerol + myo-inositol (9% of the activity with CDP-1-stearoyl-2-oleoylglycerol) [7]
8 CDP-1-arachidonoyl-2-stearoylglycerol + myo-inositol (6% of the activity with CDP-1-stearoyl-2-oleoylglycerol) [7]
9 CDP-1-stearoyl-2-arachidonoylglycerol + myo-inositol (4% of the activity with CDP-1-stearoyl-2-oleoylglycerol) [7]
10 1-Stearoyl-2-arachidonoylphosphatidylinositol + CMP [9]
11 More (not: CDP [9], CTP [9], UMP [9], AMP [9], GMP [9], enzyme also catalyzes diacylglycerol-independent exchange reaction between phosphatidylinositol and inositol [16, 21]) [9, 16, 21]

Product spectrum
1 CMP + phosphatidyl-1D-myo-inositol [1]
2 CMP + dioleoylphosphatidylinositol [1]
3 CMP + didecanoylphosphatidylinositol
4 CMP + dipalmitoylphosphatidylinositol
5 CMP + 1-stearoyl-2-oleoylphosphatidylinositol
6 CMP + distearoylphosphatidylinositol
7 CMP + diarachidonoylphosphatidylinositol
8 CMP + 1-arachidonoyl-2-stearoylphosphtidylinositol
9 CMP + 1-stearoyl-2-arachidonoylphosphatidylinositol
10 CDP-1-stearoyl-2-arachidonoylglycerol + myo-inositol [9]
11 ?

Inhibitor(s)
CTP [12]; UDP [12]; UTP [12]; TTP [12]; ADP [12]; ATP [12]; GTP [12]; Inorganic phosphate [12]; Inorganic diphosphate [12]; Inostamycin [13]; Hexachlorocyclohexanes (delta > gamma > alpha > beta) [13]; Mn^{2+} (maximal activation at 0.5–2 mM, inhibition at higher concentration) [12]; Cetyltrimethylammonium bromide [10]; Ca^{2+} (2 mM [5], in presence of Mg^{2+}, ineffective alone [4], $CaCl_2$ [12]) [1, 4, 5, 12, 15, 18]; Detergents (activity of salt-extracted enzyme is maximized by 0.5 mM CHAPS, 0.1 mM Triton X-100 or a phospholipid mixture, 0.05 mg/ml, higher concentrations are inhibitory [15], n-octyl-beta-D-glucopyranoside activates at 15–20 mM, decrease in activity at higher concentrations [19], Brij W-1-solubilized enzyme requires Triton X-100, 3.6 mM [20], deoxycholate, 0.1% w/v stimulates [21], Triton X-100,

0.1%, 15-fold increase of activity [11], 6-O-(N-heptylcarbamoyl)-methyl-D-glucopyranoside: 0.7%, 5-fold increase of activity [11]) [1, 11, 15, 19–21]; Hg^{2+} [1, 8]; Thioreactive agents [3, 6, 8]; Inosose-2 (competitive to inositol) [3]; p-Chloromercuribenzene sulfonate [3, 8, 10]; NEM [3]; Zn^{2+} ($ZnCl_2$ [12]) [8, 12]; Cd^{2+} [8]; Inositol (product inhibition) [9]; Iodoacetate [10]; Iodoacetamide [10]; Bovine serum albumin (slight) [10]; NaF (slight) [10]; CDPdiacylglycerol (above 0.4 mM [12]) [12, 9]; CDP [12]

Cofactor(s)/prosthetic group(s)/activating agents

Phospholipids (various phospholipid classes activate the enzyme rather nonspecifically [3], 10–18% activation when microsomal lipids are added back to enzyme in lipid-depleted microsomes [10]) [3, 10]; Detergents (activity of salt-extracted enzyme is maximized by 0.5 mM CHAPS, 0.1 mM Triton X-100 or a phospholipid mixture, 0.05 mg/ml, higher concentrations are inhibitory [15], n-octyl-beta-D-glucopyranoside activates at 15–20 mM, decrease in activity at higher concentrations [19], 6-O-(N-heptyl-carbamoyl)-methyl-D-glucopyranoside: 0.7%, 5fold increase of activity [11]) [11, 15, 19]; CMP (stimulates) [16]

Metal compounds/salts

Mn^{2+} (required [9, 21], activity is dependent on Mg^{2+} or Mn^{2+} [1–3, 6, 8, 12, 15, 19, 20], divalent cation required [4], maximal activity at: 2 mM [6, 8], 0.5 mM [3], 100 mM [4], 7.5 mM [2], 5 mM (Brij W-1-solubilized enzyme [20]) [15, 20], 1.5 mM $MnCl_2$ [21], 0.5–2 mM (inhibition at higher concentration) [12], maximal activity with $MgCl_2$ is approximately 4times higher [12], twice that [2] of maximal activity with $MnCl_2$) [1–4, 6, 8, 9, 12, 15, 19–21]; Mg^{2+} (activity is dependent on Mg^{2+} or Mn^{2+} [1–3, 6, 8, 12, 15, 19, 20], maximal activity at: 20 mM [3, 6, 8], 0.8 mM [2], 30 mM (Brij W-1-solubilized enzyme) [20], 0.1 mM [15], maximal activity with $MgCl_2$ is approximately 4times higher [12], twice that [2] of maximal activity with $MnCl_2$, K_a: 0.042 mM [19], cannot replace Mn^{2+} in activation [9]) [1–3, 6, 8, 12, 15, 19, 20]

Turnover number (min^{-1})

Specific activity (U/mg)
0.0104 [3]; 2.788 [11]; 0.8 [6, 8]; 34.0 [12]; 0.29 [19]

K_m-value (mM)
0.0095 (CDPdiacylglycerol) [11]; 0.022 (CMP) [12]; 0.030 (CDP-1-stearoyl-2-oleoylglycerol) [7]; 0.045 (1-stearoyl-2-arachidonoylglycerol) [7]; 0.050 (dipalmitoylglycerol) [7]; 0.06 (CDP-didecanoyl-sn-glycerol) [5]; 0.071 (myo-inositol) [5]; 0.1 (myo-inositol) [1]; 0.16 (distearoylglycerol [7], CDPdiacylglycerol [4]) [4, 7]; 0.17 (CDPdiglyceride) [3]; 0.18 (CDPdioleoylglycerol) [1]; 0.21 (inositol) [6]; 0.28 (myo-inositol) [12]; 1 (inositol) [1]; 1.3 (inositol) [11]; 2.5 (myo-inositol) [3]; 2.8 (CMP) [1]; More [15, 16, 19–21]

pH-optimum

6.2 (CMP + phosphatidylinositol) [1]; 7.0 (diacylglycerol-independent exchange reaction between phosphatidylinositol and inositol) [16]; 7.0–7.5 [15]; 7.5 [2]; 7.7 (CTP + phosphatidic acid) [5]; 8 (Tris-HCl buffer [12], inositol + CDPdiacylglycerol [16]) [4, 6, 8, 12, 16]; 8.0–9.0 [18]; 8.5 (CMP + phosphatidyl myo-inositol [16]) [16, 21]; 8.5–9.0 (CDP-diacyl-sn-glycerol [5]) [5, 19]; 8.6 [3]; 8.8–9.4 (CDPdioleoylglycerol + myo-inositol) [1]; 9.0 (glycylglycine/NaOH buffer, glycine/NaOH buffer) [12]

pH-range

Temperature optimum (°C)

30 (assay at) [8, 16, 20]; 35 [8]; 37 (assay at) [1, 10–12, 21]; 50 [12]

Temperature range (°C)

20–50 (20°C: about 20% of activity maximum, 50°C: about 35% of activity maximum) [8]; 30–60 (30°C: about 40% of activity maximum, 60°C: about 55% of activity maximum) [12]

---

3 ENZYME STRUCTURE

Molecular weight

150000–200000 (rat, gel filtration) [11]
300000 (rat, ultrafiltration, partially purified enzyme) [10]

Subunits

? (x × 60000, rat, SDS-PAGE [3], x × 34000, Saccharomyces cerevisiae, SDS-PAGE [8], x × 21000, rat, SDS-PAGE [11], x × 24000, human, SDS-PAGE [12]) [3, 8, 11, 12]

Glycoprotein/Lipoprotein

–

---

4 ISOLATION/PREPARATION

Source organism

Glycine max (solubilization) [20]; Ricinus communis [21]; Rat (two forms of membrane-associated enzyme namely salt-extractable and salt-resistant with different intracellular localization [15]) [3, 10, 11, 15]; Plasmodium knowlesi (erythrocytes infected with) [4]; Saccharomyces cerevisiae (S288C [6], expression in E. coli [16]) [6, 8, 16, 17]; Dog [14, 18, 19]; Bovine [5]; Mouse [9]; Human [12, 13]; Rabbit [1]; Guinea pig [2, 7]

Source tissue

Brain [7, 10]; Human epidermoid carcinoma A431 cells [13]; Placenta [12]; Pituitary tumors [15]; Ventricle [18]; Seed [20]; Endosperm [21]; Liver [3, 11]; Lung [1]; Pancreas [2, 9, 14, 19]; Erythrocytes (infected with Plasmodium knowlesi) [4]; Mammary tissue [5]

Localization in source
Membranes [15]; Sarcoplasmic reticulum [18]; Endoplasmic reticulum [20, 21]; Microsomes (membrane [14, 19], associated [8]) [1–3, 5, 6, 8–11, 14, 19, 20]

Purification
Rat (partial [10]) [3, 10, 11]; Plasmodium knowlesi (erythrocytes infected with, solubilization) [4]; Saccharomyces cerevisiae [6, 8]; Human [12]; Dog (partial) [19]

Crystallization
–

Cloned
[16]

Renatured
–

---

5 STABILITY

pH

Temperature (°C)
30 (unstable above) [20]; 50 (half-life: 10 min, 100% stable) [8]; 60 (labile above [6], 10 min, 35% loss of activity [8]) [6, 8]; 70 (10 min, 100% loss of activity) [8]

Oxidation

Organic solvent

General stability information
Solubilized enzyme stable for at least 2 cycles of freezing and thawing [20]; Inactivation caused by repeated freezing and thawing or by contact with detergent is prevented by addition of DTT and CDPdiacyl-sn-glycerol [10]; Mn^{2+}, absolute requirement for stabilization of the n-octyl glucopyranoside-solubilized enzyme with half-maximal stabilization at 0.04 mM, other metal ions: Co^{2+}, Fe^{2+}, Fe^{3+}, Ag^{2+}, Cu^{2+}, Ni^{2+}, Zn^{2+}, Al^{3+} and Li^+ at 1 mM, less than 4% stabilization [19]; A mixture of phosphatidylinositol, 0.5 mM and crude egg yolk phospholipids, 0.005 mg/ml phospholipid-phosphate/ml stabilizes enzyme during chromatography [19]; Addition of phospholipid during the purification and assay procedure prevents irreversible loss of activity to some extent [3]; Activity stable for at least 4 cycles of freezing and thawing [8]

Storage

–80°C, stable for a few days without appreciable loss of activity [3]; –80°C, completely stable for at least 6 months [8]; –80°C, stable for up to 2 months [12]; –80°C, solubilized enzyme is stable for at least 3 months [20]

6 CROSSREFERENCES TO STRUCTURE DATABANKS

PIR/MIPS code

Brookhaven code

7 LITERATURE REFERENCES

[1] Bleasdale, J.E., Wallis, P., MacDonald, P.C., Johnston, J.M.: Biochim. Biophys. Acta, 575,135–147 (1979)
[2] Prottey, C., Hawthorne, J.N.: Biochem. J.,105,379–392 (1967)
[3] Takenawa, T., Egawa, K.: J. Biol. Chem.,252,5419–5423 (1977)
[4] Elabbadi, N., Ancelin, M.L., Vial, H.J.: Mol. Biochem. Parasitol.,63,179–192 (1994)
[5] Wootton, J.A., Kinsella, J.E.: Int. J. Biochem.,8,449–456 (1977)
[6] Carman, G.M., Fischl, A.S.: Methods Enzymol.,209,305–312 (1992) (Review)
[7] Murthy, P.P.N., Agranoff, B.W.: Biochim. Biophys. Acta,712,473–483 (1982)
[8] Fischl, A.S., Carman, G.M.: J. Bacteriol.,154,304–311 (1983)
[9] Hokin-Neaverson, M., Sadeghian, K., Harris, D.W., Merrin, J.S.: Biochim. Biophys. Res. Commun.,78,364–371 (1977)
[10] Rao, R.H., Strickland, K.P.: Biochim. Biophys. Acta,348,306–314 (1974)
[11] Monaco, M.E., Feldman, M., Kleinberg, D.L.: Biochem. J.,304,301–305 (1994)
[12] Antonsson, B.E.: Biochem. J.,297,517–522 (1994)
[13] Imoto, M., Taniguchi, Y., Umezawa, K.: J. Biochem.,112,299–302 (1992)
[14] Parries, G.S., Hokin-Neaverson, M.: J. Biol. Chem.,260,2687–2693 (1985)
[15] Cubitt, A.B., Gershengorn, M.C.: Biochem. J.,257,639–644 (1989)
[16] Klezovitch, O., Brandenburger, Y., Geindre, M., Deshusses, J.: FEBS Lett., 320,256–260 (1993)
[17] Fischl, A.S., Homann, M.J., Poole, M.A., Carman, G.M.: J. Biol. Chem.,261, 3178–3183 (1986)
[18] Kasinathan, C., Kirchberger, M.A.: Biochemistry,27,2834–2839 (1988)
[19] Parries, G.S., Hokin-Neaverson, M.: Biochemistry,23,4785–4791 (1984)
[20] Robinson, M.L., Carman, G.M.: Plant Physiol.,69,146–149 (1982)
[21] Sexton, J.S., Moore, T.S.: Plant Physiol.,62,978–980 (1978)

1 NOMENCLATURE

EC number
2.7.8.12

Systematic name
CDPglycerol:poly(glycerophosphate) glycerophosphotransferase

Recommended name
CDPglycerol glycerophosphotransferase

Synonyms
Teichoic-acid synthase
Glycerophosphotransferase, cytidine diphosphoglycerol
Poly(glycerol phosphate) polymerase
Teichoic acid glycerol transferase
Glycerophosphate synthetase [1]
CGPTase [2]

CAS Reg. No.
9076-71-5

---

2 REACTION AND SPECIFICITY

Catalyzed reaction
CDPglycerol + (glycerophosphate)$_n$ →
→ CMP + (glycerophosphate)$_{n+1}$

Reaction type
Substituted phospho group transfer

Natural substrates
More (may be involved in glycerol teichoic acid synthesis in Bacillus subtilis
[1, 2] and Bacillus licheniformis [1]) [1, 2]

Substrate spectrum
1 CDPglycerol + (glycerophosphate)$_n$ [1, 2]

Product spectrum
1 CMP + (glycerophosphate)$_{n+1}$ (crude extract of protoplast membrane
preparation, CMP postulated [1], crude extract of membrane preparation
[2]) [1, 2]

Inhibitor(s)

Mn^{2+} (slight stimulation, in presence of Mg^{2+} inhibition) [1]; Bacitracin (94% of maximal activity) [1]; Cetylpyridinium chloride (47% of maximal activity) [1]; Crystal violet (22% of maximal activity) [1]; Novobiocin (complete inhibition) [1]; Penicillin (27% of maximal activity) [1]; Ristocetin (90% of maximal activity) [1]; Spermidine (78% of maximal activity) [1]; Streptomycin (89% of maximal activity) [1]; Vancomycin (31% of maximal activity) [1]

Cofactor(s)/prosthetic group(s)/activating agents

Metal compounds/salts

Ca^{2+} (0.01 M: optimal, requirement for a divalent cation) [1]; Mg^{2+} (0.04 M: optimal, requirement for a divalent cation) [1]; Mn^{2+} (slight stimulation, in presence of Mg^{2+} inhibition) [1]

Turnover number (min^{-1})

Specific activity (U/mg)

K_m-value (mM)

0.00083 (CDPglycerol) [1]

pH-optimum

7.0 (and pH 9.0, 2 maxima) [1]; 9.0 (and pH 7.0, 2 maxima) [1]

pH-range

6.5–9.5 (50% of maximal activity at pH 6.5, 25% of maximal activity at pH 8.0, 70% of maximal activity at pH 9.5) [1]

Temperature optimum (°C)

30 (assay at) [2]; 37 (assay at) [1]

Temperature range (°C)

3 ENZYME STRUCTURE

Molecular weight

Subunits

Glycoprotein/Lipoprotein

–

4 ISOLATION/PREPARATION

Source organism

Bacillus subtilis (ATCC 6051 [1], NCTC 3610 [1], strain 168 (wild type) [2]) [1, 2]; Bacillus licheniformis (ATCC 9945) [1]

Source tissue
Cell [1]; Protoplast [1]; Membranes of cell or protoplast [2]

Localization in source
Membrane-bound (protoplast membrane [1]) [1, 2]

Purification
Bacillus subtilis (ATCC 6051, partial [1], NCTC 3610, partial [1], strain 168, partial [2]) [1, 2]; Bacillus licheniformis (ATCC 9945, partial) [1]

Crystallization
–

Cloned
[2]

Renatured
–

---

5 STABILITY

pH

Temperature (°C)

Oxidation

Organic solvent

General stability information

Storage
–70°C, little loss of activity after 6 months [2]; Frozen, no essential loss of activity within several weeks [1]

---

6 CROSSREFERENCES TO STRUCTURE DATABANKS

PIR/MIPS code

Brookhaven code

---

7 LITERATURE REFERENCES

[1] Burger, M.M., Glaser, L.: J. Biol. Chem.,239,3168–3177 (1964)
[2] Pooley, H.M., Abellan, F.-X., Karamata, D.: J. Bacteriol.,1,646–649 (1992)

---

1 NOMENCLATURE

EC number
2.7.8.13

Systematic name
UDPMurAc(oyl-L-Ala-D-gamma-Glu-L-Lys-D-Ala-D-Ala):undecaprenyl-phos-
phate phospho-N-acetylmuramoyl-pentapeptide-transferase

Recommended name
Phospho-N-acetylmuramoyl-pentapeptide-transferase

Synonyms
Phospho-NAc-muramoyl-pentapeptide translocase (UMP) [1]
UDP-MurNAc-L-Ala-D-gamma-Glu-L-Lys-D-Ala-D-Ala:C_{55}-isoprenoid alcohol
transferase [2]
UDP-MurNAc-Ala-gamma-D-Glu-Lys-DAla-DAla:undecaprenylphosphate
transferase
Phospho-MurNAc-pentapeptide transferase [5]
Phosphoacetylmuramoylpentapeptidetransferase
Phospho-N-acetylmuramoylpentapeptidetransferase
Phospho-N-acetylmuramoyl pentapeptide translocase
Phosphoacetylmuramoylpentapeptide translocase

CAS Reg. No.
9068-50-2

---

2 REACTION AND SPECIFICITY

Catalyzed reaction
UDPMurAc(oyl-L-Ala-D-gamma-Glu-L-Lys-D-Ala-D-Ala) + undecaprenyl phosphate →
→ UMP + MurAc(oyl-L-Ala-D-gamma-Glu-L-Lys-D-Ala-D-Ala)-diphosphounde-
caprenol

Reaction type
Substituted phospho group transfer

Natural substrates
More (enzyme has a key role in selecting analogs of UDP-MurNAc-
Ala-gamma-D-Glu-Lys-D-Ala-D-Ala for peptidoglycan synthesis [5], biosynthesis of peptidoglycan [1, 4, 5, 9]: catalyzes the initial membrane reactions in the enzyme which is involved in the transfer of phospho-NAc-muramoyl-pentapeptide from the uridine monophosphate moiety to the membrane acceptor [5], first step in the lipid cycle reactions in biosynthesis of
bacterial membrane reactions [9]) [1, 4, 5, 9]

Substrate spectrum

1 UDP-NAc-muramoyl-pentapeptide + acceptor (r, equilibrium constant: 0.25) [1]
2 UDP-N-acetylmuramoyl-pentapeptide + [^{3}H]UMP (UDP-MurNAc-L-Ala-D-gamma-Glu-L-Lys-D-Ala-D-Ala, exchange reaction [5]) [2, 3, 5–7, 9, 10]
3 UDP-MurNAc-pentapeptide + undecaprenoid-1-ol-phosphate (r [3], i.e. UDP-MurNAc-L-Ala-D-gamma-Glu-L-Lys-D-Ala-D-Ala + lipid phosphate, specificity profile towards peptide subunit [5]) [3, 5, 6, 8–10]
4 UDP-MurNAc-pentapeptide + C_{55}-isoprenyl phosphate [7, 10]

Product spectrum

1 Acceptor-phospho-NAc-muramoyl-pentapeptide + UMP [1]
2 UMP + [^{3}H]UDP-NAc-muramoyl-pentapeptide [2, 3]
3 Lipid-P-P-MurNAc-pentapeptide + UMP [3]
4 C_{55}-isoprenyl-diphosphate-MurNAc-pentapeptide [7]

Inhibitor(s)

Tunicamycin [8]; Ristomycin [1]; Vancomycin [1]; Amphomycin [3]; NH_4^+ [3]; Cs^+ [3]; Liposidomycin ($C_{24}H_{67}N_5O_{21}S$, MW 1009) [4]; Triton X-100 (activity can be restored by addition of lipid fractions to the assay) [7]

Cofactor(s)/prosthetic group(s)/activating agents

Phospholipids (phospholipids sensitive to phospholipase necessary for enzymatic activity) [10]; Moenomycin (activates partially purified enzyme, maximal activation at 1 mg moenomycin/mg protein) [10]; Undecaprenyl-phosphate (stimulates exchange reaction) [6]; Undecaprenyl-diphosphate-MurNAc-pentapeptide (stimulates exchange reaction) [6]; Neutral lipid (stimulates synthesis of C_{55}-isoprenyl-P-P-MurNAc-pentapeptide from UDPMurNAc-pentapeptide, no effect on exchange reaction of UDP-MurNAc-pentapeptide with UMP) [7]; Polar lipid fraction (required by exchange reaction of UDP-MurNAc-pentapeptide with UMP) [7]; More (the physical state of the lipid microenvironment of the enzyme has a significant effect on the catalytic activity) [9]

Metal compounds/salts

Mg^{2+} (required [3, 9], maximal activation at 25 mM [3]) [3, 9]

Turnover number (min^{-1})

Specific activity (U/mg)

K_m-value (mM)

0.0018 (UDP-NAc-muramoyl-pentapeptide) [1]; 0.016 (UDP-MurNAc-Ala-D-Glu-Orn-D-Ala-D-Ala, exchange reaction) [5]; 0.02 (UDP-MurNAc-Ala-D-Glu-Lys-D-Ala-D-Ala, exchange reaction) [5]; 0.022 (UDP-MurNAc-Ala-D-Glu-Dap-D-Ala-D-Ala, exchange reaction) [5]; 0.027 (UMP) [1]; 0.029 (UDP-MurNAc-Ala-D-Glu-Lys-D-Ala-Gly, exchange reaction) [5]; 0.044

(UDP-MurNAc-Gly-D-Glu-Lys-D-Ala-D-Ala, exchange reaction) [5]; 0.055
(UDP-MurNAc-Ala-D-Glu-Lys-D-Ala-D-Ala, transfer reaction) [5]; 0.058
(UDP-MurNAc-Ala-D-Glu-Lys-D-Ala, exchange reaction) [5]; 0.063 (UDP-Mur-
NAc-Ala-D-Glu-Lys-Gly-D-Ala, exchange reaction) [5]; 0.067 (5'-UMP) [10];
0.080 (UDP-MurNAc-Gly-D-Glu-Lys-D-Ala-D-Ala, exchange reaction) [5];
0.083 (UDP-MurNAc-Ala-D-Glu-D-Ala-Glu, transfer reaction) [5]; 0.093
(UDP-MurNAc-pentapeptide, not purified enzyme) [3]; 0.1 (UDP-Mur-
NAc-Ala-D-Glu-mDap-D-Ala-D-Ala, transfer reaction) [5]; 0.18 (UDP-Mur-
NAc-Ala-D-Glu-Lys) [5]; 0.27 (UDP-MurNAc-pentapeptide) [10]; 0.4
(UDP-MurNAc-Ala-D-Glu-Lys-D-Ala, transfer reaction) [5]; 0.44 (UDP-Mur-
NAc-Ala-D-Glu-Lys-Gly-D-Ala, transfer reaction) [5]; 0.5 (UDP-Mur-
NAc-Gly-D-Glu-Lys-D-Ala-D-Ala, transfer reaction) [5]; 4.4 (UDP-Mur-
NAc-Ala-Glu-Lys, transfer reaction) [5]

pH-optimum
    8.5 [3]

pH-range
    6.5–9 (6.5: about 40% of activity maximum, 9: about 80% of activity maxi-
    mum) [3]

Temperature optimum (°C)
    25 (assay at) [2]

Temperature range (°C)

---

3 ENZYME STRUCTURE

Molecular weight
    More (enzyme exists in more than one form, low MW form: 100000–200000,
    high MW form: 2000000) [2]

Subunits

Glycoprotein/Lipoprotein
    –

---

4 ISOLATION/PREPARATION

Source organism
    Micrococcus luteus [7]; Staphylococcus aureus (Copenhagen) [1, 2, 5, 6,
    9]; Bacillus megaterium (KM) [3]; E. coli (Y-10 [4, 8], K12 [10]) [4, 8, 10]

Source tissue
    Membrane [2, 6, 7]; Envelope [10]

Localization in source

Purification
    E. coli (partial) [10]; Staphylococcus aureus (Copenhagen, solubilization) [2]

Crystallization

–

Cloned

More (amplification of the mraY gene which is very probably the structural gene for EC 2.7.8.13) [8]

Renatured

–

5 STABILITY

pH

Temperature (°C)

Oxidation

Organic solvent

General stability information

Storage

6 CROSSREFERENCES TO STRUCTURE DATABANKS

PIR/MIPS code

PIR2:JC1275 (Bacillus subtilis); PIR2:C47691 (Bacillus subtilis); PIR2:S08395 (Escherichia coli)

Brookhaven code

7 LITERATURE REFERENCES

[1] Struve, W.G., Sinha, R.K., Neuhaus, F.C.: Biochemistry,5,82–93 (1966)
[2] Heydanek, M.G., Neuhaus, F.C.: Biochemistry,8,1474–1481 (1969)
[3] Tanaka, H., Oiwa, R., Matsukura, S., Inokoshi, J., Omura, S.: J. Antibiot.,35, 1216–1221 (1982)
[4] Kimura, K., Miyata, N., Kawanishi, G., Kamio, Y., Izaki, K., Isono, K.: Agric. Biol. Chem., 53,1811–1815 (1989)
[5] Hammes, W.P., Neuhaus, F.C.: J. Biol. Chem.,249,3140–3150 (1974)
[6] Pless, D.D., Neuhaus, F.C.: J. Biol. Chem.,248,1568–1576 (1973)
[7] Umbreit, J.N., Strominger, J.L.: Proc. Natl. Acad. Sci. USA,69,1972–1974 (1972)
[8] Ikeda, M., Wachi, M., Jung, H.K., Ishino, F., Matsuhashi, M.: J. Bacteriol.,173, 1021–1026 (1991)
[9] Weppner, W.A., Neuhaus, F.C.: Biochim. Biophys. Acta,552,418–427 (1979)
[10] Geis, A., Plapp, R.: Biochim. Biophys. Acta,527,414–424 (1978)

1 NOMENCLATURE

EC number
2.7.8.14

Systematic name
CDPribitol:poly(ribitol phosphate) ribitolphosphotransferase

Recommended name
CDPribitol ribitolphosphotransferase

Synonyms
EC 2.4.1.55 (formerly)
Teichoic-acid synthase
Polyribitol phosphate synthetase [1]
Synthetase, teichoate
Poly(ribitol phosphate) synthetase
Polyribitol phosphate polymerase
Teichoate synthase
Teichoic acid synthase

CAS Reg. No.
37277-63-7

---

2 REACTION AND SPECIFICITY

Catalyzed reaction
CDPribitol + (ribitol phosphate)$_n$ $\rightarrow$
$\rightarrow$ CMP + (ribitol phosphate)$_{n+1}$ (mechanism [4])

Reaction type
Substituted phospho group transfer

Natural substrates

Substrate spectrum
1 CDPribitol + (ribitol phosphate)$_n$ (no substrate: D-ribitol 5-phosphate,
   CDP-ribitol treated with pyrophosphatase, acid or alkali) [1]
2 n CDPribitol + lipoteichoic acid (structural requirements of lipoteichoic
   acids for recognition by the enzyme [2], lipoteichoic acid carrier active
   with the Staphylococcus aureus enzyme can be extracted from Bacillus licheniformis, Lactobacillus casei, Lactobacillus plantarum, Streptococcus
   faecalis, Mycobacterium phlei, but not from a variety of other microorganisms [4], single chain mechanism: the enzyme completes a chain of approximately 30 units linked to lipoteichoic acid carrier, before starting a
   new chain [4]) [2–4]

Product spectrum
1 CMP + (ribitol phosphate)$_{n+1}$ [1]
2 Lipoteichoic acid-(P-ribitol)$_n$ + n CMP [2]

Inhibitor(s)
Novobiocin [1]; Gentian violet [1]; Ristocetin [1]; CMP [1, 3];
p-Hydroxymercuribenzoate (inhibition reversed by 2-mercaptoethanol) [3];
CDPglycerol (weak) [3]; More (no inhibition: penicillin G, staphcillin,
alpha-aminobenzyl-penicillin, bacitracin) [1]

Cofactor(s)/prosthetic group(s)/activating agents
Lipoteichoic acid (required as carrier) [3]; Phospholipid (required) [3]

Metal compounds/salts
Co^{2+} (slight stimulation) [1]; Mn^{2+} (Mn^{2+} or Mg^{2+} required, optimal concentration: 10–30 mM) [1]; Mg^{2+} (Mn^{2+} or Mg^{2+} required: 10–30 mM) [1]; More
(no effect: Ni^{2+}, Zn^{2+}, Ca^{2+}) [1]

Turnover number (min^{-1})

Specific activity (U/mg)

K_m-value (mM)
0.1 (CDPribitol, in presence of Mn^{2+}) [1]

pH-optimum
7.8 (Tris-HCl buffer) [1]

pH-range

Temperature optimum (°C)

Temperature range (°C)

3 ENZYME STRUCTURE

Molecular weight

Subunits

Glycoprotein/Lipoprotein
–

4 ISOLATION/PREPARATION

Source organism
Staphylococcus aureus (H [2–4]) [1–4]

Source tissue

Localization in source
  Particle-bound [1]

Purification
  Staphylococcus aureus [3]

Crystallization
  –

Cloned
  –

Renatured
  –

---

5 STABILITY

pH

Temperature (°C)
  25 (24 h, very stable) [3]

Oxidation

Organic solvent

General stability information

Storage

---

6 CROSSREFERENCES TO STRUCTURE DATABANKS

PIR/MIPS code

Brookhaven code

---

7 LITERATURE REFERENCES

[1] Ishimoto, N., Strominger, J.L.: J. Biol. Chem.,241,639–650 (1966)
[2] Fischer, W., Koch, H.U., Rösel, P., Fiedler, F., Schmuck, L.: J. Biol. Chem.,255, 4550–4556 (1980)
[3] Fiedler, F., Glaser, L.: J. Biol. Chem.,249,2684–2689 (1974)
[4] Fiedler, F., Glaser, L.: J. Biol. Chem.,249,2690–2695 (1974)

1 NOMENCLATURE

EC number
2.7.8.15

Systematic name
UDP-N-acetyl-D-glucosamine:dolichyl-phosphate N-acetyl-
D-glucosaminephosphotransferase

Recommended name
UDP-N-acetylglucosamine-dolichyl-phosphate N-acetylglucosaminephos-
photransferase

Synonyms
UDP-D-N-acetylglucosamine N-acetylglucosamine 1-phosphate transferase
[2]
UDP-GlcNAc:dolichyl-phosphate GlcNAc-1-phosphate transferase [4]
UDP-N-acetyl-D-glucosamine:dolichol phosphate N-acetyl-D-glucosa-
mine-1-phosphate transferase [3]
Acetylglucosamine-1-phosphotransferase, uridine diphosphoacetylglucosa-
mine-dolichyl phosphate
Chitobiosylpyrophosphoryldolichol synthase
Dolichol phosphate N-acetylglucosamine-1-phosphotransferase
UDP-acetylglucosamine-dolichol phosphate acetylglucosamine phospho-
transferase
UDP-acetylglucosamine-dolichol phosphate acetylglucosamine-1-phospho-
transferase

CAS Reg. No.
70431-08-2

---

2 REACTION AND SPECIFICITY

Catalyzed reaction
UDP-N-acetyl-D-glucosamine + dolichyl phosphate →
→ UMP + N-acetyl-D-glucosaminyl-diphosphodolichol

Reaction type
Substituted phospho group transfer

Natural substrates
UDP-N-acetyl-D-glucosamine + dolichyl phosphate (first enzyme of dolichol
pathway [1], enzyme initiates the dolichol cycle for the biosynthesis of aspa-
ragine-linked glycoproteins [3]) [1, 3]

Enzyme Handbook © Springer-Verlag Berlin Heidelberg 1997
Duplication, reproduction and storage in data banks are only
allowed with the prior permission of the publishers

Substrate spectrum
    1 UDP-N-acetyl-D-glucosamine + dolichyl phosphate (r [6]) [1–8]

Product spectrum
    1 UMP + N-acetyl-D-glucosaminyl-diphosphodolichol [1, 2, 4]

Inhibitor(s)
    N-Ethylmaleimide [4]; Phosphatidylcholine [7]; Phosphatidylethanolamine
    (stimulates activity of enzyme in endoplasmic membrane vesicles [8]) [7];
    Phosphatidylserine (stimulates [4]) [7]; GDPmannose [7]; UDPglucose [3,
    7]; Amphomycin [7]; Showdomycin [7]; Ca^{2+} (1–2 mM: weak stimulation,
    above 2 mM: inhibition) [7]; Mn^{2+} (1–2 mM: weak stimulation, above 2 mM:
    inhibition) [7]; UDPhexanolamine [3]; UDPxylose [3]; Tunicamycin (both di-
    rections, mechanism of inhibition [6]) [1, 3, 6–8]; UMP [1, 3]; UDP [1, 3]; Di-
    umycin [2, 3, 7]; Hg^{2+} [3]; Dolichyl phosphate (at high concentrations) [3];
    p-Chloromercuribenzoate [3, 4]; Iodoacetamide [3]

Cofactor(s)/prosthetic group(s)/activating agents
    Phospholipid (endogenous microsomal phospholipid is required for reaction
    to proceed normally in rat lung microsomes [5], activity of solubilized en-
    zyme is stimulated by exogenously added phospholipids in the order:
    phosphatidylglycerol > phosphatidylinositol > phosphatidylserine [4]) [4, 5];
    GDPmannose (or other purine sugar nucleotides, stimulates) [4]; Dolichyl-
    phosphoryl-mannose (stimulates) [3, 4]; Phosphatidylglycerol (stimulates)
    [4, 7]; Phosphatidylinositol (stimulates) [4, 7]; Cardiolipin (stimulates activity
    of enzyme in endoplasmic membrane vesicles) [8]; Monogalactosyldiglyce-
    ride (stimulates activity of enzyme in endoplasmic membrane vesicles) [8];
    Phosphatidylethanolamine (stimulates activity of enzyme in endoplasmic
    membrane vesicles [8], inhibits [7]) [8]; More (DEAE-cellulose chromatogra-
    phy separates a heat-stable factor from the enzyme, which when added
    back to the partially purified enzyme stimulates about 5fold) [4]

Metal compounds/salts
    Mg^{2+} (8 mM required for maximal stability [1], supports enzyme activity [2],
    stimulates [3], either Mn^{2+}, 1 mM, or Mg^{2+}, 10 mM, required for optimal ac-
    tivity [4], 10 mM required for optimal activity [7]) [1–4, 7]; Ca^{2+} (1–2 mM:
    weak stimulation, above 2 mM: inhibition) [7]; Mn^{2+} (1–2 mM: weak stimula-
    tion, above 2 mM: inhibition [7], supports enzyme activity [2], stimulates [3],
    either Mn^{2+}, 1 mM, or Mg^{2+}, 10 mM, required for optimal activity [4]) [2–4, 7]

Turnover number (min^{-1})

Specific activity (U/mg)
    More [3]

K_m-value (mM)

0.00042 (UDP-GlcNAc) [4]; 0.0005 (UDP-GlcNAc) [2]; 0.001 (dolichyl phosphate) [1]; 0.0045 (UDP-GlcNAc [3], dolichyl phosphate [2]) [2, 3]; 0.0062 (dolichyl phosphate) [4]; 0.016 (dolichyl phosphate) [3]; 0.180 (dolichyl phosphate) [1]

pH-optimum

7.2–7.6 [7]; 7.4–7.6 [3, 4]; 7.4–7.8 [2]; 7.5 [1]

pH-range

7.0–8.7 (6.0: no activity, 7.0: 85% of activity maximum, 8.7: 64% of activity maximum) [2]

Temperature optimum (°C)

30 [2]

Temperature range (°C)

15–40 (0°C: no activity, 15°C: 11% of activity maximum, 25°C: 73% of activity maximum, 35°C: 96% of activity maximum, 40°C: 57% of activity maximum) [2]

---

3 ENZYME STRUCTURE

Molecular weight

Subunits

More (results indicate that either 70000 MW band in SDS-PAGE is a precursor form of the enzyme or this polypeptide, representing the native enzyme or its subunit, is proteolyzed to smaller, enzymatically active peptides of 50000 and 46000 during purification) [3]

Glycoprotein/Lipoprotein

–

---

4 ISOLATION/PREPARATION

Source organism

Saccharomyces cerevisiae (X2180–1A) [1]; Acanthamoeba castellanii [2]; Glycine max [7]; Bovine [3]; Pig [4, 6]; Rat [5, 8]

Source tissue

Encysting cultures [2]; Aorta [4, 6]; Mammary gland (lactating) [3]; Lung [5]; Cultured cells [7]; Liver [8]

Localization in source

Membrane [1]; Microsomes [3–7]; Endoplasmic reticulum membrane vesicles [8]

Purification
   Pig (partial) [4]

Crystallization
   –

Cloned
   –

Renatured
   –

---

5 STABILITY

pH

Temperature (°C)
   30 (1 h, stable) [2]

Oxidation

Organic solvent

General stability information
   Glycerol, 20%, stabilizes [4]; Phosphatidylglycerol, 0.02 mg/ml protein, stabilizes [4]

Storage
   0–4°C, 20% glycerol, 0.02 mg phosphatidylglycerol/mg of protein, less than 20% loss of activity after 6 days [4]; 4°C, NaN_3, 2 months stable [2]; –20°C or below, indefinitely stable [2]

---

6 CROSSREFERENCES TO STRUCTURE DATABANKS

PIR/MIPS code

Brookhaven code

---

7 LITERATURE REFERENCES

[1] Sharma, C.B., Lehle, L., Tanner, W.: Eur. J. Biochem.,126,319–325 (1982)
[2] Villemez, C.L., Carlo, P.L.: J. Biol. Chem.,255,8174–8178 (1980)
[3] Shailubhai, K., Dong-Yu, B., Saxena, E.S., Vijay, I.K.: J. Biol. Chem.,263,15964–15972 (1988)
[4] Kaushal, G.P., Elbein, A.D.: J. Biol. Chem.,260,16303–16309 (1985)
[5] Plouhar, P.L., Bretthauer, R.K.: J. Biol. Chem.,257,8907–8911 (1982)
[6] Heifetz, A., Keenan, R.W., Elbein, A.D.: Biochemistry,18,2186–2192 (1979)
[7] Kaushal, G.P., Elbein, A.D.: Plant Physiol.,82,748–752 (1986)
[8] Chandra, N.C., Doody, M.B., Bretthauer, R.K.: Arch. Biochem. Biophys.,290,345–354 (1991)

1 NOMENCLATURE

EC number
2.7.8.17

Systematic name
UDP-N-acetyl-D-glucosamine:lysosomal-enzyme N-acetyl-
glucosaminephosphotransferase

Recommended name
UDP-N-acetylglucosamine-lysosomal-enzyme N-acetylglucosaminephos-
photransferase

Synonyms
N-Acetylglucosaminylphosphotransferase
UDP-N-acetylglucosamine:lysosomal enzyme N-acetylglucosamine-1-phos-
photransferase [8]
UDP-GlcNAc:glycoprotein N-acetylglucosamine-1-phosphotransferase [5]
Acetylglucosamine-1-phosphotransferase, uridine diphosphoacetylglucosa-
mine-lysosomal enzyme precursor
Acetylglucosamine-1-phosphotransferase, uridine diphosphoacetylglucosa-
mine-glycoprotein
Lysosomal enzyme precursor acetylglucosamine-1-phosphotransferase
N-Acetylglucosaminyl phosphotransferase
UDP-acetylglucosamine:lysosomal enzyme N-acetylglucosamine-1-phos-
photransferase
UDP-GlcNAc:lysosomal enzyme N-acetylglucosamine-1-phosphotransferase
UDP-N-acetylglucosamine:glycoprotein N-acetylglucosamine-1-phospho-
transferase
UDP-N-acetylglucosamine:glycoprotein N-acetylglucosaminyl-1-phospho-
transferase

CAS Reg. No.
84012-69-1

2 REACTION AND SPECIFICITY

Catalyzed reaction
UDP-N-acetyl-D-glucosamine + lysosomal-enzyme D-mannose →
→ UMP + lysosomal-enzyme N-acetyl-D-glucosaminyl-phospho-D-mannose

Reaction type
Substituted phospho group transfer

Natural substrates

More (participates in the synthesis of the phosphomannosyl recognition marker of lysosomal proteins [6], donates N-acetylglucosamine 1-phosphate to mannose residues of newly synthesized lysosomal enzymes [5], catalyzes the initial determining step by which newly synthesized acid hydrolases are distinguished from other newly synthesized glycoproteins and thus are eventually targeted to lysosomes [3]) [3, 5, 6]

Substrate spectrum

1 UDP-N-acetyl-D-glucosamine + lysosomal-enzyme D-mannose (striking preference for acid hydrolases over other potential glycoprotein acceptors [3], phosphorylates lysosomal enzymes at least 100-fold more efficiently than either other glycoproteins with similar carbohydrate chains or free oligosaccharides [8], alpha-linked N-acetylglucosamine 1-phosphate is transferred en bloc to the 6-hydroxyl of mannose in high mannose oligosaccharides of glycoproteins [5], transfers GlcNAc 1-phosphate to at least 3 different mannose residues in high mannose oligosaccharides in glycoproteins and accepts also oligosaccharides already carrying one phosphorylated mannose residue [1], recognizes the carbohydrate as well as the protein moiety in lysosomal enzymes [1], no substrate: non-lysosomal glycoproteins with high mannose oligosaccharides [1]) [1–8]
2 UDP-N-acetyl-D-glucosamine + high mannose oligosaccharides with 7 or 9 mannose residues (at high concentration) [1]
3 UDP-N-acetyl-D-glucosamine + uteroferrin [7]
4 UDP-N-acetyl-D-glucosamine + cathepsin D [7]
5 UDP-N-acetyl-D-glucosamine + beta-N-acetylhexosaminidase (A or B [3]) [3, 7]
6 UDP-N-acetyl-D-glucosamine + alpha-methylmannoside (poor acceptor [3]) [3, 8]
7 UDP-N-acetyl-D-glucosamine + Man5–8GlcNAc (poor acceptor) [3]
8 UDPglucose + lysosomal enzyme D-mannose (efficiency about 12fold worse than UDP-GlcNAC) [8]

Product spectrum

1 UMP + lysosomal-enzyme N-acetyl-D-glucosaminyl-phospho-D-mannose [1]
2 ?
3 ?
4 ?
5 ?
6 ?
7 ?
8 ?

Inhibitor(s)

UDPglucose (competitive to UDP-N-acetylglucosamine) [8]; Deglycosylated lysosomal enzymes [8]; NaCl [1]; alpha-Methylmannose [1]; p-Nitrophenyl-mannose [1]; High mannose oligosaccharides [1]; Dithiothreitol (20 mM, 21% inhibition) [1]; UDP [2, 8]; UDPglucose [2]; Negatively charged phospholipids (including phosphatidylserine, phosphatidylglycerol, phosphatidic acid) [2]; More (no inhibition by tunicamycin) [5]

Cofactor(s)/prosthetic group(s)/activating agents

Triton X-100 (required) [1]; Tergitol NP-10 (nonionic detergent, required) [2]; More (no stimulation by dolichol phosphate) [5]

Metal compounds/salts

More (highest transfer rate in sodium phosphate buffer [1], with Tris-HCl 50% more active than with sodium cacodylate [8], no effect: Co^{2+} [1, 8], Ca^{2+} [8], Cu^{2+} [8], Cd^{2+} [8], Zn^{2+} [8], Hg^{2+} [8]) [1, 8]; Mg^{2+} (Mn^{2+} or Mg^{2+} required [1, 2], Mn^{2+} more effective than Mg^{2+} [8]) [1, 2, 8]; Mn^{2+} (Mn^{2+} or Mg^{2+} required [1, 2, 8], maximal activity at 10 mM [1], Mn^{2+} more effective than Mg^{2+} [8]) [1, 2, 8]

Turnover number (min^{-1})

Specific activity (U/mg)

0.044 [8]; More (assay) [3, 5, 8]

K_m-value (mM)

0.0056 (human placental beta-hexosaminidase B) [3]; 0.0089 (porcine hepatic alpha-N-acetylglucosaminidase) [3]; 0.009 (UDP-N-acetylglucosamine) [1]; 0.0195 (human placental beta-hexosaminidase A) [3]; 0.024 (UDP-N-acetylglucosamine) [2]; 0.038 (UDP-N-acetylglucosamine) [8]; 0.117 (alpha-methylmannoside) [2]; 0.183 (alpha-methylmannoside) [8]; 0.916 (bovine pancreatic ribonuclease B) [3]; 32 (Man5–8GlcNAc oligosaccharide) [3]; 132 (alpha-methylmannoside) [3]

pH-optimum

6.5 [1]; 6.5–7.5 [8]; 7.2–7.5 [8]

pH-range

Temperature optimum (°C)

37 (assay at) [1, 3, 5, 8]

Temperature range (°C)

3 ENZYME STRUCTURE

Molecular weight
228000 (human placenta, radiation inactivation) [6]
283000 (human skin fibroblasts, radiation inactivation) [6]
1000000 (human) [2]

Subunits

Glycoprotein/Lipoprotein
Glycoprotein [8]

---

4 ISOLATION/PREPARATION

Source organism
Mouse [8]; Chinese hamster [8]; Sheep [8]; Rat [1, 3–5, 8]; Human [2, 6, 7];
Chicken [7]

Source tissue
Murine lymphoma cell line BW5147 [8]; Brain [8]; Spleen [8]; Kidney [8]; Pe-
ripheral blood leukocytes [8]; Chinese hamster ovary cells [8]; Murine mac-
rophage-like cell lines (J774.2 and P388D1) [8]; Fibroblasts (normal, no activity in I-cell fibroblasts [7], cultured skin fibroblasts [6]) [6, 7]; Liver [1, 3–5,
8]; Cultured lymphoblasts [2]; Placenta [6, 8]; Retina [7]

Localization in source
Golgi membranes [1, 6]; Lysosomes [2]; Membrane (integral protein) [8];
Golgi apparatus (activity higher than in ER [4]) [4, 8]; ER (smooth and
rough) [4, 8]

Purification
Rat (partial) [1, 3, 8]; Human (partial) [2]

Crystallization
–

Cloned
–

Renatured
–

---

5 STABILITY

pH

Temperature (°C)
20–22 (16 h, less than 20% loss of activity) [1]; 37 (2 h, less than 20% loss
of activity) [1]; 50 (10 min, complete loss of activity) [1]

Oxidation

Organic solvent

General stability information
Treatment with papain or phospholipase C destroys catalytic acitivity [1]

Storage
0°C, 2 months, less than 10% loss of activity [8]

---

6 CROSSREFERENCES TO STRUCTURE DATABANKS

PIR/MIPS code

Brookhaven code

---

7 LITERATURE REFERENCES

[1] Waheed, A., Hasilik, A., von Figura, K.: J. Biol. Chem.,257,12322–12331 (1982)
[2] Zhao, K.W., Yeh, R., Miller, A.L.: Glycobiology,2,119–125 (1992)
[3] Reitman, M.L., Kornfeld, S.: J. Biol. Chem.,256,11977–11980 (1981)
[4] Waheed, A., Pohlmann, R., Hasilik, A., von Figura, K.: J. Biol. Chem.,256,4150–4152
    (1981)
[5] Reitman, M.L., Kornfeld, S.: J. Biol. Chem.,256,4275–4281 (1981)
[6] Ben-Yoseph, Y., Potier, M., Pack, B.A., Mitchell, D.A., Melancon, S.B., Nadler, H.L.:
    Biochem. J.,235,883–886 (1986)
[7] Hiller, A.M., Koro, L.A., Marchase, R.B.: J. Biol. Chem.,262,4377–4381 (1987)
[8] Reitman, M.L., Lang, L., Kornfeld, S.: Methods Enzymol.,107,163–172 (1984)
    (Review)

1 NOMENCLATURE

EC number
2.7.8.18

Systematic name
UDPgalactose:UDP-N-acetyl-D-glucosamine galactosephosphotransferase

Recommended name
UDPgalactose-UDP-N-acetylglucosamine galactosephosphotransferase

Synonyms
Galactose-1-phosphotransferase, uridine diphosphogalactose-uridine
diphosphoacetylglucosamine
Galactose-1-phosphotransferase [1]
Galactosyl phosphotransferase [1]

CAS Reg. No.
84932-43-4

2 REACTION AND SPECIFICITY

Catalyzed reaction
UDPgalactose + UDP-N-acetyl-D-glucosamine →
→ UMP + UDP-N-acetyl-6-(D-galactose-1-phospho)-D-glucosamine

Reaction type
Substituted phospho group transfer

Natural substrates

Substrate spectrum
1 UDPgalactose + UDP-N-acetyl-D-glucosamine [1]
2 UDPgalactose + UDP-N-acetylgalactosamine (weak activity) [1]
3 UDPgalactose + N-acetylglucosaminyl end groups in glycoproteins (postulated) [1]

Product spectrum
1 UMP + UDP-N-acetyl-6-(D-galactose-1-phospho)-D-glucosamine (i.e.
UDP-N-acetylglucosamine-6-phosphogalactose) [1]
2 ?
3 ?

Inhibitor(s)

Cofactor(s)/prosthetic group(s)/activating agents
Nonidet P-40 ((+ Triton X-100) stimulation) [1]; Nonidet P-40 ((+ Triton 20)
stimulation) [1]

Metal compounds/salts
Mn^{2+} (requirement) [1]

Turnover number (min^{-1})

Specific activity (U/mg)

K_m-value (mM)

pH-optimum
5.4 [1]

pH-range
7.2–8.1 (60% of maximal activity at pH 7.2, 9% of maximal activity at pH 8.1)
[1]

Temperature optimum (°C)
37 (assay at) [1]

Temperature range (°C)

3 ENZYME STRUCTURE

Molecular weight

Subunits

Glycoprotein/Lipoprotein
–

4 ISOLATION/PREPARATION

Source organism
Chicken (hen) [1]

Source tissue
Oviduct (isthmus region) [1]; More (also present in liver, ovary, uterus, magnum of oviduct and kidney of hen) [1]

Localization in source
Microsomes [1]

Purification
Chicken (hen, partial) [1]

2

Crystallization

–

Cloned

–

Renatured

–

5 STABILITY

pH

Temperature (°C)

Oxidation

Organic solvent

General stability information

Storage

6 CROSSREFERENCES TO STRUCTURE DATABANKS

PIR/MIPS code

Brookhaven code

7 LITERATURE REFERENCES

[1] Nakanishi, Y., Otsu, K., Suzuki, S.: FEBS Lett.,151,15–18 (1983)

1 NOMENCLATURE

EC number
2.7.8.19

Systematic name
UDPglucose:glycoprotein-D-mannose glucosephosphotransferase

Recommended name
UDPglucose-glycoprotein glucosephosphotransferase

Synonyms
UDP-glucose:glycoprotein glucose-1-phosphotransferase [1]
GlcPTase [1]
Glc-phosphotransferase [3]
Glucose-1-phosphotransferase, uridine diphosphoglucose-glycoprotein

CAS Reg. No.
84861-40-5

---

2 REACTION AND SPECIFICITY

Catalyzed reaction
UDPglucose + glycoprotein D-mannose →
→ UMP + glycoprotein 6-(D-glucose-1-phospho)-D-mannose

Reaction type
Substituted phospho group transfer

Natural substrates
UDPglucose + parafusin (a protein evidently associated with membrane fusion during exocytosis in Paramecium) [2]
More (may be a controlling enzyme for targeting of certain newly synthesized proteins to the cell surface) [1]

Substrate spectrum
1 UDPglucose + glycoprotein D-mannose (specific for UDPglucose [4], endogenous glycoprotein acceptor [1], acceptor protein in rat liver is cytoplasmic and is glycosylated by the enzyme at a site accessible to cytoplasm [3], endoglycosidase H-sensitive oligosaccharides on acceptor glycoproteins [4], primary acceptor in the chicken retina are peptides having apparent subunit MWs of 66000 and 62000 [4]) [1–4]
2 UDPglucose + parafusin (a protein evidently associated with membrane fusion during exocytosis in Paramecium) [2]

Product spectrum
1 UMP + glycoprotein 6-(D-glucose-1-phospho)-D-mannose [1–4]
2 ?

Inhibitor(s)
Uteroferrin [4]

Cofactor(s)/prosthetic group(s)/activating agents

Metal compounds/salts
Mn^{2+} (activates, maximal activity at 5 mM) [1]

Turnover number (min^{-1})

Specific activity (U/mg)

K_m-value (mM)

pH-optimum
6.3 [1]

pH-range

Temperature optimum (°C)

Temperature range (°C)

---

3 ENZYME STRUCTURE

Molecular weight

Subunits

Glycoprotein/Lipoprotein
–

---

4 ISOLATION/PREPARATION

Source organism
Chicken [1, 4]; Paramecium tetraaurelia [2]; Rat [3]; Human [4]

Source tissue
Retina [1, 4]; Liver [3]; Fibroblasts [4]; I-cells [3]

Localization in source

Purification

Crystallization
–

Cloned

–

Renatured

–

5 STABILITY

pH

Temperature (°C)

Oxidation

Organic solvent

General stability information

Storage

6 CROSSREFERENCES TO STRUCTURE DATABANKS

PIR/MIPS code

Brookhaven code

7 LITERATURE REFERENCES

[1] Koro, L.A., Marchase, R.B.: Cell,31,739–748 (1982)
[2] Satir, B.H., Srisomsap, C., Reichman, M., Marchase, R.B.: J. Cell Biol.,111,901–907 (1990)
[3] Srisomsap, C., Richardson, K.L., Jay, J.C., Marchase, R.B.: J. Biol. Chem.,263, 17792–17797 (1988)
[4] Hiller, A.M., Koro, L.A., Marchase, R.B.: J. Biol. Chem.,262,4377–4381 (1987)

1 NOMENCLATURE

EC number
2.7.8.20

Systematic name
Phosphatidylglycerol:membrane-derived-oligosaccharide-D-glucose glycerophosphotransferase

Recommended name
Phosphatidylglycerol-membrane-oligosaccharide glycerophosphotransferase

Synonyms
Phosphoglycerol transferase
Glycerophosphotransferase, oligosaccharide
Phosphoglycerol transferase I [2, 3]

CAS Reg. No.
80146-86-7

2 REACTION AND SPECIFICITY

Catalyzed reaction
Phosphatidylglycerol + membrane-derived-oligosaccharide D-glucose →
→ 1,2-diacyl-sn-glycerol + membrane-derived-oligosaccharide 6-(glycerophospho)-D-glucose

Reaction type
Substituted phospho group transfer

Natural substrates

Substrate spectrum
1 Phosphatidylglycerol + membrane-derived oligosaccharide D-glucose (1,2-beta- and 1,6-beta-linked glucose residues in membrane polysaccharides and in synthetic glucosides can act as acceptors [1]) [1, 2]
2 Phosphatidylglycerol + p-hydroxyphenyl-beta-D-glucoside (i.e. arbutin) [3]

Product spectrum
1 sn-1,2-Diacylglycerol + beta-glucoside-6-phosphoglycerol [1]
2 Arbutin-6-phosphoglycerol + sn-1,2-diacylglycerol [3]

Inhibitor(s)
EDTA [1]; Arbutin (competitive to membrane-derived oligosaccharides) [1]

Cofactor(s)/prosthetic group(s)/activating agents

Metal compounds/salts

Mn^{2+} (divalent cation required, Mn^{2+} most effective) [1]; Mg^{2+} (divalent cation required, Mg^{2+} is less effective than Mn^{2+}) [1]; Ca^{2+} (divalent cation required, Ca^{2+} is less effective than Mn^{2+}) [1]

Turnover number (min^{-1})

Specific activity (U/mg)

K_m-value (mM)

2–3 (arbutin) [3]

pH-optimum

8.9 [1]

pH-range

7.8–9.7 (50% of activity maximum at pH 7.8 and pH 9.7) [1]

Temperature optimum (°C)

37 (assay at) [1]

Temperature range (°C)

3 ENZYME STRUCTURE

Molecular weight

Subunits

Glycoprotein/Lipoprotein

–

4 ISOLATION/PREPARATION

Source organism

E. coli [1–3]

Source tissue

Localization in source

Inner cytoplasmic membrane (active site is localized on periplasmic face of inner membrane) [1]

Purification

Crystallization

–

Cloned

[2]

Renatured

–

---

5 STABILITY

pH

Temperature (°C)

Oxidation

Organic solvent

General stability information

Storage

---

6 CROSSREFERENCES TO STRUCTURE DATABANKS

PIR/MIPS code

Brookhaven code

---

7 LITERATURE REFERENCES

[1] Jackson, B.J., Kennedy, L.P.: J. Biol. Chem.,258,2394–2398 (1983)
[2] Lanfroy, E., Bohin, J.-P.: J. Bacteriol.,175,5736–5737 (1993)
[3] Bohin, J.-P., Kennedy, E.P.: J. Biol. Chem.,259,8388–8393 (1984)

1 NOMENCLATURE

EC number
    2.7.8.21

Systematic name
    Membrane-derived-oligosaccharide-6-(glycerophospho)-D-glucose:mem-
    brane-derived-oligosaccharide-D-glucose glycerophosphotransferase

Recommended name
    Membrane-oligosaccharide glycerophosphotransferase

Synonyms
    Periplasmic phosphoglycerotransferase
    Phosphoglycerol cyclase

CAS Reg. No.

---

2 REACTION AND SPECIFICITY

Catalyzed reaction
    Transfer of a glycerophospho group from one membrane-derived
    oligosaccharide to another

Reaction type
    Substituted phospho group transfer

Natural substrates

Substrate spectrum
    1 Membrane-derived oligosaccharides (transfer of a glycerophospho group
      from one membrane-derived oligosaccharide to another. beta-Linked glu-
      cose residues in simple glucosides, such as gentiobiose, can act as ac-
      ceptors) [1]

Product spectrum
    1 More (in presence of low concentrations of acceptor, free cyclic
      1,2-phosphoglycerol is formed) [1]

Inhibitor(s)
    EDTA [1]; Gentiobiose (competitive to membrane-derived oligosaccharides)
    [1]

Cofactor(s)/prosthetic group(s)/activating agents

Metal compounds/salts
    Mn^{2+} (absolute requirement for a divalent cation, Mn^{2+} most active) [1]

Turnover number (min^{-1})

Specific activity (U/mg)

K$_m$-value (mM)
    1 (gentiobiose) [1]

pH-optimum
    7.8 [1]

pH-range
    6.5–7.8 (active) [1]

Temperature optimum (°C)
    37 (assay at) [1]

Temperature range (°C)

---

3 ENZYME STRUCTURE

Molecular weight
    56000 (E. coli, gel permeation chromatography) [1]

Subunits

Glycoprotein/Lipoprotein
    –

---

4 ISOLATION/PREPARATION

Source organism
    E. coli (K-12) [1]

Source tissue

Localization in source
    Periplasmic space [1]

Purification
    E. coli (partial) [1]

Crystallization
    –

Cloned
    –

Renatured
    –

5 STABILITY

pH

Temperature (°C)

Oxidation

Organic solvent

General stability information

Storage

6 CROSSREFERENCES TO STRUCTURE DATABANKS

PIR/MIPS code

Brookhaven code

7 LITERATURE REFERENCES

[1] Goldberg, D.E., Rumley, M.K., Kennedy, E.P.: Proc. Natl. Acad. Sci. USA, 78,5513–5517 (1981)

1 NOMENCLATURE

EC number
2.7.8.22

Systematic name
CDPcholine:1-alkenyl-2-acylglycerol cholinephosphotransferase

Recommended name
1-Alkenyl-2-acylglycerol cholinephosphotransferase

Synonyms
CDP-choline-1-alkenyl-2-acyl-glycerol phosphocholinetransferase [1]
Cholinephosphotransferase, 1-alkenyl-2-acylglycerol

CAS Reg. No.
113066-36-7

2 REACTION AND SPECIFICITY

Catalyzed reaction
CDPcholine + 1-alkenyl-2-acylglycerol →
→ CMP + plasmenylcholine

Reaction type
Substituted phospho group transfer

Natural substrates
CDPcholine + 1-alkenyl-2-acylglycerol [1]

Substrate spectrum
1 CDPcholine + 1-alkenyl-2-acylglycerol [1]

Product spectrum
1 CMP + plasmenylcholine [1]

Inhibitor(s)

Cofactor(s)/prosthetic group(s)/activating agents

Metal compounds/salts

Turnover number (min^{-1})

Specific activity (U/mg)

K_m-value (mM)

pH-optimum

pH-range

Temperature optimum (°C)

Temperature range (°C)

---

3 ENZYME STRUCTURE

Molecular weight

Subunits

Glycoprotein/Lipoprotein
  –

---

4 ISOLATION/PREPARATION

Source organism
  Guinea pig [1]

Source tissue
  Heart [1]

Localization in source
  Microsomes [1]

Purification

Crystallization
  –

Cloned
  –

Renatured
  –

---

5 STABILITY

pH

Temperature (°C)

Oxidation

Organic solvent

General stability information

Storage

6 CROSSREFERENCES TO STRUCTURE DATABANKS

PIR/MIPS code

Brookhaven code

7 LITERATURE REFERENCES

[1] Wientzek, M., Man, R.Y.K., Choy, P.C.: Biochem. Cell Biol.,65,860–868 (1987)

1 NOMENCLATURE

EC number
2.7.9.1

Systematic name
ATP:pyruvate,orthophosphate phosphotransferase

Recommended name
Pyruvate,orthophosphate dikinase

Synonyms
Kinase (phosphorylating), pyruvate-phosphate di-
Pyruvate, phosphate dikinase [11]
Pyruvate-inorganic phosphate dikinase
Pyruvate-phosphate dikinase
Pyruvate-phosphate ligase
Pyruvic-phosphate dikinase
Pyruvic-phosphate ligase
Pyruvate, P_i dikinase [10]
PPDK [25]

CAS Reg. No.
9027-40-1

2 REACTION AND SPECIFICITY

Catalyzed reaction
ATP + pyruvate + phosphate →
→ AMP + phosphoenolpyruvate + diphosphate (mechanism [9, 11–13], sequential mechanism for the addition of ATP and phosphate and a ping-pong
mechanims for the addition of pyruvate and release of phosphoenolpyruvate [10], nonclassical three-site tri (uni uni) ping-pong kinetics [19, 21],
two-step bi bi uni uni mechanism [20])

Reaction type
Phospho group transfer

Natural substrates
AMP + phosphoenolpyruvate + diphosphate (the enzyme enables the organism to conserve the energy residing in the diphosphate resulting from
protein and glycogen synthesis) [2]
ATP + pyruvate + phosphate (enzyme functions in regulation of gluconeogenesis and carbohydrate oxidation [8], enzyme in Crassulacean acid me-

tabolism permits the incorporation of pyruvate into carbohydrate in the light [16]) [8, 16]
More (functions in gluconeogenesis pathway in the leaves of tropical grasses and in Acetobacter xylinum. Functions in glycolytic pathway in Bacteroides symbiosus and Entamoeba histolytica [7], involved in C_4 dicarboxylic acid pathway in plant [15]) [7, 15]

Substrate spectrum
1 ATP + pyruvate + phosphate (r [1, 2, 7, 10, 15, 20], HPO_4^{2-} is the substrate [5], in the reverse reaction highly specific for AMP [7], GTP, CTP, ITP or TTP cannot replace ATP [8], dAMP can replace ATP, 20% of the activity [9], activity with UTP, GTP and CTP is 1–3% of the activity with ATP [10]) [1–27]
2 ATP + pyruvate + arsenate (ir [10]) [8, 10]

Product spectrum
1 AMP + phosphoenolpyruvate + diphosphate [1]
2 ?

Inhibitor(s)
Diethyldicarbonate [13]; Bromopyruvate (competitive to PEP) [13]; Iodoacetate (not [9, 14]) [13]; Bromoacetate [13]; 2-Bromopropionate [13]; 3-Bromopropionate [13]; 2-Bromobutyrate [13]; CTP (phosphoenolpyruvate formation [1, 7]) [1, 7, 9]; ITP (phosphoenolpyruvate formation [1, 7]) [1, 7, 9]; $MgHPO_4$ (competitive to HPO_4^{2-}) [5]; Oxalate [7, 22, 24]; Ca^{2+} [7]; AMP (phosphoenolpyruvate formation, competitive to ATP) [8, 10, 19, 27]; ATP (pyruvate formation, competitive to AMP) [8, 19]; p-Hydroxymercuribenzoate [8, 9, 14]; DTNB [9, 14]; NEM [9, 14]; KF (both directions of reaction) [8]; dATP [9]; alpha-beta-Methylene ATP (competitive to ATP) [9]; beta-gamma-Methylene ATP (competitive to ATP) [9]; Methylene diphosphonate (competitive to diphosphate) [9]; Phosphoglycolate (competitive to phosphoenolpyruvate) [9]; ADP (mediates a rapid but reversible inactivation in presence of a thiol) [10]; PCMB [10]; gamma(p-Arsenophenyl)-n-butyrate [10]; Arsenate plus 2,3-dimercaptoethanol [10]; Pyruvate (competitive to PEP) [10, 19]; Phosphoenolpyruvate (competitive to pyruvate) [19]; Phosphate (competitive to diphosphate) [10, 19]; Diphosphate (competitive to phosphate) [19]; Sulfhydryl agents [14]; More (in direction of pyruvate formation no tested 5'-nucleoside monophosphate inhibits the reaction with AMP [7], not: iodoacetamide [9, 14], response of the enzyme to energy charge [27]) [7, 9, 14, 27]

Cofactor(s)/prosthetic group(s)/activating agents
Thiols (required for activity in solution) [10]

Metal compounds/salts

NH_4^+ (activity is almost totally dependent on added monovalent cations in both directions, NH_4^+ most effective [20], in presence of phosphoenolpyruvate NH_4^+ is required for enolization, Mg^{2+} does not significantly increase this rate [21], activates [1, 4], stimulates [15], monovalent cation required [9, 14], NH_4^+ is most effective [9], Tl^+, NH_4^+ or K^+ [14], requirement resides completely in the phosphoenolpyruvate, pyruvate partial reaction [14], marginally increases forward reaction, severalfold increase in reverse reaction, 50 mM NH_4Cl [10]) [1, 4, 9, 10, 14, 15, 20, 21]; K^+ (activity is almost totally dependent on added monovalent cations in both directions, complete activation in forward reaction, partial activation in reverse reaction [20], can replace NH_4^+ in activation [1], stimulates [4], monovalent cation required [9, 14], 70% of the activity with NH_4^+ [9], Tl^+, NH_4^+ or K^+ [14], requirement resides completely in the phosphoenolpyruvate, pyruvate partial reaction [14]) [1, 4, 9, 14, 20]; Tl^+ (monovalent cation required, Tl^+, NH_4^+ or K^+, requirement resides completely in the phosphoenolpyruvate, pyruvate partial reaction) [14]; Mn^{2+} (Mg^{2+}, Mn^{2+} or Co^{2+} required, Mn^{2+} and Co^{2+} each inhibit one of three partial directions at higher concentrations [14], in direction of pyruvate formation divalent metal ion requirement is fulfilled by salts of Ni^{2+}, Mn^{2+}, Mg^{2+} or Co^{2+}, in direction of phosphoenolpyruvate formation only Mg^{2+} salts are effective [1], free Mn^{2+} can replace free Mg^{2+} in activation, MnATP is not a substrate [5], cannot replace Mg^{2+} in either direction [8], absolute requirement for divalent metal [9], 30% of the activity with Mg^{2+} in direction of pyruvate formation [9]) [1, 5, 8, 9, 14]; Ni^{2+} (in direction of pyruvate formation divalent metal ion requirement is fulfilled by salts of Ni^{2+}, Mn^{2+}, Mg^{2+} or Co^{2+}, in direction of phosphoenolpyruvate formation only Mg^{2+} salts are effective) [1]; Mg^{2+} (required [19], Mg^{2+}, Mn^{2+} or Co^{2+} required, Mg^{2+} most effective [14], in direction of pyruvate formation divalent metal ion requirement is fulfilled by salts of Ni^{2+}, Mn^{2+}, Mg^{2+} or Co^{2+}, in direction of phosphoenolpyruvate formation only Mg^{2+} salts are effective [1], free Mg^{2+} activates [5], MgATP is the substrate [5], optimum concentration: 7.5 mM, higher concentrations inhibit [5], required [8, 10], K_m: 2.2 mM (phosphoenolpyruvate formation), 0.76 mM (pyruvate formation) [8], 1.7 mM (phosphoenolpyruvate formation) [19], 0.87 mM (pyruvate formation) [19], absolute requirement for a divalent metal ion, Mg^{2+} most effective [9]) [1, 5, 8–10, 14, 19]; Co^{2+} (in direction of pyruvate formation divalent metal ion requirement is fulfilled by salts of Ni^{2+}, Mn^{2+}, Mg^{2+} or Co^{2+}, in direction of phosphoenolpyruvate formation only Mg^{2+} salts are effective [1], absolute requirement for divalent metal [9], Mg^{2+}, Mn^{2+} or Co^{2+} required, Mn^{2+} and Co^{2+} each inhibit one of three partial directions at higher concentrations [14], 45% of the activity with Mg^{2+} in direction of pyruvate formation [9]) [1, 9, 14]

Turnover number (min^{-1})

Specific activity (U/mg)
More [1, 2, 12, 13, 15, 26]; 0.88 [8]; 12 [7]; 35 [9]

K_m-value (mM)
0.0014 (inorganic phosphate) [4]; 0.0016 (AMP) [8]; 0.0035 (AMP) [1, 7]; 0.005 (AMP) [4]; 0.024 (phosphoenolpyruvate) [4]; 0.029 (inorganic diphosphate) [4]; 0.03 (ATP) [4]; 0.04 (diphosphate) [10]; 0.06 (diphosphate) [8]; 0.07 (pyruvate) [4]; 0.08 (pyruvate) [1, 7]; 0.09 (ATP, pH 8.3) [10]; 0.1 (ATP [1, 7], diphosphate [7], HPO_4^{2-} [5], phosphoenolpyruvate [8]) [1, 5, 7, 8]; 0.11 (pyruvate, phosphoenolpyruvate, pH 8.3) [10]; 0.2 (pyruvate) [8]; 0.4 (ATP) [8]; 0.5 (phosphate, pH 8.3) [10]; 0.6 (phosphate) [1]; 0.8 (phosphate) [8]; More [9, 11, 17, 19, 20, 24, 27]

pH-optimum
6.4 (pyruvate formation [1, 7, 11], Bacteroides symbiosus [11]) [1, 7, 11]; 6.4–6.7 (pyruvate formation) [9]; 6.5 (pyruvate formation [8, 11], Acetobacter xylinum [11]) [8, 11]; 6.5–7.0 (Propionibacterium shermanii, both directions of reaction [11]) [4, 11]; 6.9 (pyruvate formation) [20]; 7.0–7.8 (phosphoenolpyruvate formation) [9]; 7.1 (pyruvate formation) [19]; 7.2–7.8 (phosphoenolpyruvate formation [1, 7], Bacteroides symbiosus, PEP formation [11]) [1, 7, 11]; 8.0 (phosphoenolpyruvate formation) [19]; 8.2 (phosphoenolpyruvate formation [8, 11, 20], Acetobacter xylinum [11]) [8, 11, 20]; 8.3 (sugarcane, phosphoenolpyruvate formation) [10]

pH-range

Temperature optimum (°C)
22 (assay at) [15]; 25 (assay at) [9]; 30 (assay at) [7, 8, 19]

Temperature range (°C)

---

3 ENZYME STRUCTURE

Molecular weight
105800 (Rhodospirillum rubrum, sucrose density gradient sedimentation) [26]
150000–160000 (Propionibacterium shermanii, equilibrium sedimentation, meniscus depletion method, gel filtration) [9]
170000 (Bacteroides symbiosus, gel filtration) [12]
330000 (Acetobaceter aceti, gel filtration) [19]
370000 (Zea mays, gel filtration, dark-treated inactive enzyme form) [23]
372000 (Giardia intestinalis) [4]
387000 (Zea mays, sedimentation analysis) [10, 15]

Subunits

? (x × 94000, Bacteroides symbiosus, SDS-PAGE [7], x × 58000, Ace-
tobacter aceti, SDS-PAGE [19], x × 103900, Flaveria trinervia [25]) [7, 19, 25]
Dimer (2 × 75000, Propionibacterium shermanii, SDS-PAGE [9], 2 × 94000,
Bacteroides symbiosus, SDS-PAGE [12], 2 × 92700, Rhodospirillum rubrum,
SDS-PAGE [26]) [9, 12, 26]
Tetramer (4 × 94000, Zea mays, SDS-PAGE) [10, 15]

Glycoprotein/Lipoprotein

–

4 ISOLATION/PREPARATION

Source organism

Bacteroides symbiosus (ATCC 14940 [1, 7]) [1, 7, 11–14, 21, 22]; Entamoeba histolytica [2, 3, 11]; Giardia intestinalis [4]; Zea mays (L., var. Golden
Cross Bantam T51 [18]) [5, 6, 10, 15, 18, 20, 23, 27]; Acetobacter xylinum
[8, 11]; Propionibacterium shermanii [9]; Sugarcane [10, 11]; Bryophyllum
tubiflorum [16]; Sedum prealtum [16]; Hordeum vulgare [17]; Triticum aestivum (wheat) [17, 24]; Oryza sativa [17]; Hordeum murinum [17]; Secale
cereale [17]; Secale italica [17]; Avena fatua [17]; Avena sativa [17]; Acetobacter aceti [19]; Flaveria trinervia [25]; Rhodospirillum rubrum [26]; More
(the only plants in which the enzyme has been detected are those utilizing
C_4-pathway photosynthesis) [10]

Source tissue

Cells [1]; Trophozoites [2]; Leaves (dark-treated [23]) [6, 10, 15, 18, 20, 23];
Grains (developing [17], green immature [24], aleurone layer [17]) [17, 24]

Localization in source

Purification

Bacteroides symbiosus [1, 7, 12, 13]; Entamoeba histolytica (partial) [2]; Giardia intestinalis [4]; Acetobacter xylinum [8]; Propionibacterium shermanii
[9]; Zea mays (inactive enzyme form from dark-treated leaves [23]) [10, 15,
23, 27]; Bryophyllum tubiflorum (partial) [16]; Sedum prealtum (partial) [16];
Triticum aestivum (partial) [17, 24]; Avena sativa (partial) [17]; Hordeum vulgare (partial) [17]; Rhodospirillum rubrum (partial) [26]; Secale cereale (partial) [17]; Acetobacter aceti [19]

Crystallization

[13]

Cloned

[3, 25]

Enzyme Handbook © Springer-Verlag Berlin Heidelberg 1997
Duplication, reproduction and storage in data banks are only
allowed with the prior permission of the publishers

Renatured

–

---

5 STABILITY

pH
5.5–7.0 (stability optimum) [9]

Temperature (°C)
0 (rapid inactivation, activity is recovered by rewarming to 20–30°C [10], 20 min, about 70% loss of activity [18]) [10, 18]; 10 (30 min, about 40% loss of activity) [18]; 20–40 (30 min, stable) [18]; More (cold labile, reactivation by several min incubation at 22°C) [24]

Oxidation

Organic solvent

General stability information
Does not require thiol compounds to maintain stability during storage or assay [1]; Mg^{2+} stabilizes the oligomeric structure of the enzyme [5]; Glycerol in vitro protects the active day-form and the inactive night-form [6]; Sensitive to dilution, particularly at concentrations below 0.3 mg/ml [9]; Freezing and thawing inactivates [9]

Storage
0°C, as a precipitate in a 66% saturated solution of $(NH_4)_2SO_4$ [10]; At room temperature 15% loss of activity after two weeks [8]

---

6 CROSSREFERENCES TO STRUCTURE DATABANKS

PIR/MIPS code
PIR1:KIQAPO (Clostridium symbiosum); PIR2:A53505 (Clostridium symbiosum (fragments)); PIR2:S55478 (common ice plant); PIR2:S49497 (common ice plant); PIR2:S36601 (Entamoeba histolytica); PIR2:S53297 (Flaveria pringlei); PIR2:PQ0190 (1 cytosolic maize (fragment)); PIR2:PQ0191 (2 cytosolic maize (fragment)); PIR1:KIZMPO (precursor maize); PIR2:S12894 (precursor chloroplast Flaveria trinervia); PIR3:S61410 (cytosolic Flaveria trinervia (fragment))

Brookhaven code

7 LITERATURE REFERENCES

[1] Reeves, R.E.: Biochem. J.,125,531–539 (1971)
[2] Reeves, R.E.: J. Biol. Chem.,243,3202–3204 (1968)
[3] Saavedra Lira, E., Robinson, O., Perez Montfort, R.: Arch. Med. Res.,23,39–40 (1992)
[4] Hrdy, I., Mertens, E., Nohynkova, E.: Exp. Parasitol.,76,438–441 (1993)
[5] Nakamoto, H., Edwards, G.E.: Plant Physiol. Biochem.,28,553–559 (1990)
[6] Salahas, G., Manetas, Y., Gavalas, N.A.: Photosynth. Res.,26,9–17 (1990)
[7] South, D.J., Reeves, R.E.: Methods Enzymol.,42C,187–191 (1975) (Review)
[8] Benziman, M.: Methods Enzymol.,42C,192–199 (1975) (Review)
[9] Milner, Y., Michaels, G., Wood, H.G.: Methods Enzymol.,42C,199–212 (1975) (Review)
[10] Hatch, M.D., Slack, C.R.: Methods Enzymol.,42C,212–219 (1975) (Review)
[11] Cooper, R.A., Kornberg, H.L. in "Enzymes",3rd Ed. (Boyer, P.D., ed.) 10,631–649 (1974) (Review)
[12] Goss, N.H., Wood, H.G.: Methods Enzymol.,87,51–66 (1982) (Review)
[13] Yoshida, H., Wood, H.G.: J. Biol. Chem.,253,7650–7655 (1978)
[14] Michaels, G., Milner, Y., Moskovitz, B.R., Wood, H.G.: J. Biol. Chem.,253,7656–7661 (1978)
[15] Sugiyama, T.: Biochemistry,12,2862–2868 (1973)
[16] Kluge, M., Osmond, C.B.: Naturwissenschaften,58,414–415 (1971)
[17] Meyer, A.O., Kelly, G.J., Latzko, E.: Plant Physiol.,69,7–10 (1982)
[18] Shirahashi, K., Hayakawa, S., Sugiyama, T.: Plant Physiol.,62,826–830 (1978)
[19] Schwitzguebel, J.-P., Ettlinger, L.: Arch. Microbiol.,122,103–108 (1979)
[20] Jenkins, C.L.D., Hatch, M.D.: Arch. Biochem. Biophys.,239,53–62 (1985)
[21] Moskovitz, B.R., Wood, H.G.: J. Biol. Chem.,253,884–888 (1978)
[22] Michaels, G., Milner, Y., Reed, G.H.: Biochemistry,14,3213–3219 (1975)
[23] Sugiyama, T., Iwaki, H.: Agric. Biol. Chem.,41,1239–1244 (1977)
[24] Meyer, A.O., Kelly, G.J., Latzko, E.: Plant Sci. Lett.,12,35–40 (1978)
[25] Rosche, E., Westhoff, P.: FEBS Lett.,273,116–121 (1990)
[26] Ernst, S.M., Budde, R.J.A., Chollet, R.: J. Bacteriol.,165,483–488 (1986)
[27] Nakamoto, H., Edwards, G.E.: Biochim. Biophys. Acta,924,360–368 (1987)

1 NOMENCLATURE

EC number

2.7.9.2

Systematic name

ATP:pyruvate,water phosphotransferase

Recommended name

Pyruvate,water dikinase

Synonyms

Kinase (phosphorylating), pyruvate–water di-
PEP synthetase
Phosphoenolpyruvate synthase
Phoephoenolpyruvate synthetase
Phosphoenolpyruvic synthase
Phosphopyruvate synthetase, phosphopyruvate
Synthetase, phosphopyruvate

CAS Reg. No.

9013-09-6

2 REACTION AND SPECIFICITY

Catalyzed reaction

ATP + pyruvate + H_2O →
→ AMP + phosphoenolpyruvate + phosphate (reaction sequence [2, 5],
mechanism [5], identification of phosphohistidine in phosphoenzyme intermediate [8])

Reaction type

Phospho group transfer

Natural substrates

ATP + pyruvate + H_2O (enzyme is essential for gluconeogenesis in E. coli
and Salmonella typhimurium during the growth on pyruvate, lactate, alanine
or serine, in certain circumstances the enzyme may also provide
phosphoenolpyruvate under glycolytic conditions [5], essential step in gluconeogenesis if pyruvate or lactate are used as carbon source [7]) [5, 7]

Substrate spectrum

1 ATP + pyruvate + H_2O (r [1, 3, 5], equilibrium lies far to the side of phosphoenolpyruvate formation [3], phosphorylated enzyme as an intermediate [4], highly specific [5], in the reverse reaction dAMP yields 1% of the rate obtained with AMP, 3'-AMP gives no reaction [5]) [1-5]

Product spectrum

1 AMP + phosphoenolpyruvate + phosphate [1]

Inhibitor(s)

Sulfhydryl reagents [1]; Ca^{2+} (inhibits Mn^{2+}-activated enzyme) [1]; Iodoacetate [1]; PCMB [1]; ATP (excess of ATP inhibits at high concentrations of $MgCl_2$ or $MnCl_2$) [1]; 5'-Adenylylmethylene diphosphonate (competitive to ATP) [2]; F^- [3]; AMP [6]; ADP [5]; Oxalacetate [5]; 2-Oxoglutarate [5]; Malate [6]; Phosphoenolpyruvate [6]; ADPglucose [6]; 3-Phosphoglyceraldehyde [5]; Mg^{2+} (divalent metal ion Mg^{2+} or Mn^{2+} required for forward reaction, inhibition at high concentrations of Mg^{2+} or Mn^{2+}) [1]; Mn^{2+} (divalent metal ion Mg^{2+} or Mn^{2+} required for forward reaction, inhibition at high concentrations of Mg^{2+} or Mn^{2+}) [1]; More (not: arsenate) [3]

Cofactor(s)/prosthetic group(s)/activating agents

More (enzyme contains sulfhydryl groups essential for activity) [5]

Metal compounds/salts

Mn^{2+} (divalent metal ion Mg^{2+} or Mn^{2+} required for forward reaction [1, 5], Mg^{2+} is more effective [5], inhibition at high concentrations of Mg^{2+} or Mn^{2+} [1], 3 to 4 mol of Mn^{2+} bound per mol of enzyme [1], 4.2 to 5.6 equivalent binding sites for Mn^{2+} per mol of enzyme [5]) [1, 5]; Mg^{2+} (divalent metal ion Mg^{2+} or Mn^{2+} required for forward reaction [1, 5], Mg^{2+} is more effective [5], inhibition at high concentrations of Mg^{2+} or Mn^{2+} [1]) [1, 5]

Turnover number (min^{-1})

More [5]

Specific activity (U/mg)

8.9 [1]; More [8]

K_m-value (mM)

0.028 (ATP) [1]; 0.083 (pyruvate) [1]; 10.5 (phosphate) [1]; More [5]

pH-optimum

6.8 (pyruvate formation) [5]; 8.4 (phosphoenolpyruvate formation) [5]

pH-range

Temperature optimum (°C)

30 (assay at) [1]

Temperature range (°C)

3 ENZYME STRUCTURE

Molecular weight
150000 (E. coli, sedimentation equilibrium studies) [8]
180000 (E. coli, gel filtration) [4]
250000 (E. coli, gel filtration) [8]

Subunits
? (x × 87430, E. coli, calculation from nucleotide sequence) [7]
Dimer (2 × 77000, E. coli, enzyme tends to dissociate to monomers at low protein concentration) [8]

Glycoprotein/Lipoprotein
–

---

4 ISOLATION/PREPARATION

Source organism
E. coli (strain Bm, a mutant of strain B devoid of phosphoenolpyruvate carboxylase [3], B [5, 6]) [1–8]

Source tissue

Localization in source

Purification
E. coli [1, 5]

Crystallization
–

Cloned
(sequence homology with other phosphohistidine-containing enzymes, including pyruvate,phosphate dikinase from plants and Bacteroides symbiosus and Enzyme I of the bacterial PEP:carbohydrate phosphotransferase system) [7]

Renatured
–

---

5 STABILITY

pH
5.5–6.8 (most stable at, rapid loss of activity above pH 6.8 and below pH 5.5) [1, 5]

Temperature (°C)
22 (retains full activity for several days if stored at room temperature in the presence of EDTA and Mg^{2+}) [3]; More (not extremely cold-labile) [8]

Oxidation

Organic solvent

General stability information

Sucrose, 1.0 M, stabilizes against inactivation by heat and during storage [1, 5]

Storage

4°C, 10 mM Tris-HCl buffer, pH 6.8, containing 1 M sucrose, 0.2 mM EDTA, 0.2 mM dithioerythritol, no loss of activity after 1 year [1]; Unstable if stored in ice, but retains full activity for several days if stored at room temperature in the presence of EDTA and Mg^{2+} [3]; 4°C, dephosphorylated form of enzyme, 50 mM Tris/HCl, pH 6.8, 0.2 mM EDTA, 0.2 mM DTT, 1 M sucrose, stable over a period of 12 months [8]

---

6 CROSSREFERENCES TO STRUCTURE DATABANKS

PIR/MIPS code

PIR2:F64367 ((intein-containing) Methanococcus jannaschii); PIR2:S20554 (Escherichia coli); PIR2:JC4176 (Pyrococcus furiosus)

Brookhaven code

---

7 LITERATURE REFERENCES

[1] Berman, K.M., Cohn, M.: J. Biol. Chem.,245,5309–5318 (1970)
[2] Berman, K.M., Cohn, M.: J. Biol. Chem.,245,5319–5325 (1970)
[3] Cooper, R.A., Kornberg, H.L.: Biochim. Biophys. Acta,104,618–620 (1965)
[4] Cooper, R.A., Kornberg, H.L.: Biochem. J.,105,49c-50c (1967)
[5] Cooper, R.A., Kornberg, H.L. in "Enzymes",3rd Ed. (Boyer, P.D., ed.) 10,631–649, Academic Press, New York (1974) (Review)
[6] Chulavatnatol, M., Atkinson, D.E.: J. Biol. Chem.,248,2712–2715 (1973)
[7] Niersbach, M., Kreuzaler, F., Geerse, R.H., Postma, P.W., Hirsch, H.J.: Mol. Gen. Genet.,231,332–336 (1992)
[8] Narindrasorasak, S., Bridger, W.A.: J. Biol. Chem.,252,3121–3127 (1977)

1 NOMENCLATURE

EC number
2.8.1.1

Systematic name
Thiosulfate:cyanide sulfurtransferase

Recommended name
Thiosulfate sulfurtransferase

Synonyms
Thiosulfate cyanide transsulfurase
Thiosulfate thiotransferase
Rhodanese
Rhodanase
Sulfurtransferase, thiosulfate
More (rhodanese activity is found to be a minor function of erythrocytic beta-mercaptopyruvate) [1]

CAS Reg. No.
9026-04-4

2 REACTION AND SPECIFICITY

Catalyzed reaction
Thiosulfate + cyanide →
→ sulfite + thiocyanate (double displacement mechanism [2, 9, 17, 24, 25] involving a covalent sulfur-enzyme intermediate [9], mechanism [8])

Reaction type
Sulfur atom transfer

Natural substrates
More (role in aerobic energy metabolism [24], cyanide-detoxifying function [8], overview: function of the enzyme in certain bacterial species [8]) [8, 24]

Substrate spectrum
1 Thiosulfate + cyanide (r [8]) [1–25]
2 Thiosulfate + dihydrolipoate (more efficient than cyanide [24]) [2, 9, 13, 19, 24]
3 Thiosulfate + sulfite (r [8]) [8, 9]
4 Thiosulfate + borohydride [8, 9]
5 Thiosulfate + dithionite [8]
6 Thiosulfate + thiosulfinate [9]

Enzyme Handbook © Springer-Verlag Berlin Heidelberg 1997
Duplication, reproduction and storage in data banks are only
allowed with the prior permission of the publishers

 7 Thiosulfate + monothiol [9]
 8 Thiosulfate + benzene thiosulfonate [19]
 9 Thiosulfate + a thiol (single displacement mechanism) [25]
10 Thiosulfate + N-acetyl-L-cysteine (5% of the activity with cyanide) [25]
11 Thiosulfate + L-cysteine (12% of the activity with cyanide) [25]
12 Thiosulfate + D,L-homocysteine (19% of the activity with cyanide) [25]
13 Thiosulfate + glutathione (20% of the activity with cyanide) [25]
14 Thiosulfate + 2-mercaptoethanol (20% of the activity with cyanide) [25]
15 Thiosulfate + dithiothreitol [25]
16 Thiosulfonate + cyanide [8, 9]
17 Alkyl sulfinate + cyanide (r) [8]
18 Aryl sulfinate + cyanide [8]
19 Persulfide + cyanide [9]
20 4-(Dimethylamino)-4'-azobenzene sulfinate-SO_2^- + sulfite [11]
21 4-(Dimethylamino)-4'-azobenzene sulfinate-$S(O_2)S^-$ + cyanide [11]
22 4-(Dimethylamino)-4'-azobenzene sulfinate-$S(O_2)S^-$ + GS^- [11]
23 5-Dimethyl-1-naphthalene sulfinate + thiosulfate [11]
24 5-Dimethyl-1-naphthalene sulfinate + cyanide [11]
25 5-Dimethyl-1-naphthalene sulfinate + GS^- [11]
26 More (no substrates: dithiols which oxidize to cyclic disulfides having
     more than 5 ring members, i.e. larger than the dithiolane ring of oxidized
     lipoate) [8]

Product spectrum

 1 Sulfite + thiocyanate [1–25]
 2 ?
 3 ?
 4 ?
 5 ?
 6 ?
 7 ?
 8 ?
 9 ?
10 ?
11 ?
12 ?
13 ?
14 ?
15 ?
16 ?
17 ?
18 ?
19 ?
20 4-(Dimethylamino)-4'-azobenzene sulfinate-$S(O_2^-)S^-$ + sulfite [11]
21 4-(Dimethylamino)-4'-azobenzene sulfinate-SO_2^- + thiocyanide [11]

2

22 4-(Dimethylamino)-4'-azobenzene sulfinate-SO_2^- + GSS⁻ [11]
23 5-Dimethyl-1-naphthalene sulfinate-$S(O_2)S^-$ + sulfite [11]
24 5-Dimethyl-1-naphthalene sulfinate-SO_2 + thiocyanate [11]
25 5-Dimethyl-1-naphthalene sulfinate + GSS⁻ [11]
26 ?

Inhibitor(s)

Triton X-100 (slight) [21]; SDS [21]; Dithiothreitol [8]; Dithioerythritol (maximal activation at 0.005 mM, inhibition above [23]) [8, 23]; $CaCl_2$ (0.02 M, 35% inhibition) [8]; CN⁻(substrate inhibition at high concentrations [19]) [8, 19]; 2-Naphthalene sulfonate [9]; Dinitrobenzene [9]; Phenylglyoxal [9]; Sulfhydryl reagents [9]; Iodoacetate [19, 21, 24]; 2-Oxoglutarate (incubation of mitochondria with 2-oxoglutarate causes a significant decrease in activity [17]) [12, 17]; DL-Dihydrolipoate (inactivation of rhodanese, no inactivation of sulfur-free form of the enzyme) [13]; Lipoate (no inactivation of rhodanese, inactivation of sulfur-free form of the enzyme [13]) [13, 23]; DL-Isocitrate [17]; Citrate [17]; Malate [17]; Pyruvate [17]; Cysteine (0.01 mM enhances activity maximally, 0.2 mM: 29% inhibition [23], pH 8.0 [19]) [19, 23]; Oxalacetate [17]; Fumarate [17]; Succinate [17]; NEM [21]; Sulfite [8, 21, 23]; 2-Mercaptoethanol (weak [22], pH 8.0 [19]) [18, 19, 22]; PCMB (at pH 9.5, not at pH 8.0 [19], 10 mM, 5% inhibition [21]) [19, 21]; Sodium borohydride (pH 8.0) [19]; Thiosulfate (substrate inhibition at high concentrations) [19]; Dinitrofluorobenzene [24]; Sodium arsenite [19]; Sulfate (incubation of mitochondria with sulfate causes a significant decrease in activity) [17]; N-Bromosuccinimide [23]; Anions (inhibited by most anions at rather high concentration, the most active inhibitors are aromatic anions [9], incubation of mitochondria with sulfate and 2-oxoglutarate causes a significant decrease in activity [17]) [9, 17]

Cofactor(s)/prosthetic group(s)/activating agents

Dithioerythritol (maximal activation at 0.005 mM, inhibition above) [23]; Cysteine (activates in direction of SCN-formation at low CN⁻ concentrations [6], L- [6], enhances activity [21], 0.01 mM enhances activity maximally, 0.2 mM: 29% inhibition [23]) [6, 21, 23]; 2-Mercaptoethanol (activates in direction of SCN-formation at low CN⁻ concentrations [6], enhances activity [21]) [6, 21]; Reduced glutathione (enhances activity) [21]

Metal compounds/salts

Turnover number (min⁻¹)

15600 (thiosulfate + cyanide) [24]

Specific activity (U/mg)

More (assay [8]) [3, 8, 9, 14, 17–19, 21, 23–25]

K_m-value (mM)
  4 (thiosulfate) [9]; 5.7 (CN⁻, rat) [6]; 14.2 (CN⁻, guinea pig) [6]; 22.7 (thiosulfate, rat) [6]; 26.9 (thiosulfate, mitochondrial enzyme) [4]; 35.7 (thiosulfate,
  guinea pig) [6]; 127.8 (thiosulfate, cytosolic) [6]; More [14, 19, 21, 22, 24]

pH-optimum
  6.5 (mitochondria) [4]; 7.0–8.5 [22]; 7.5 (cytosol [4]) [4, 6]; 7.5–8.5 [19];
  8–8.5 [21]; 11 [23]

pH-range
  6.0–11.0 (6.0: about 75% of activity maximum, 11.0: about 70% of activity
  maximum) [19]

Temperature optimum (°C)
  20 [6]; 30 (mitochondria) [4]; 35 (cytosol) [4]; 50 [21]; 60 [23]

Temperature range (°C)

3 ENZYME STRUCTURE

Molecular weight
  9000 (Rana temporaria, mitochondria, thin layer gel filtration) [4]
  14000 (E. coli, gel filtration, autooxidation to a polymeric form which is probably an inert dimer) [24]
  16500 (Rhodopseudomonas palustris) [3]
  20300 (Rana temporaria, cytosol, thin layer gel filtration) [4]
  29000 (Azotobacter vinelandii) [2]
  32000 (rat) [6]
  33000 (Manihot esculenta, sedimentation data) [18]
  35000 (Acinetobacter calcoaceticus, gel filtration) [21]
  37000 (Cercopithecus aethiops [5], bovine [7], guinea pig [6]) [5–7]
  37500 (bovine, sedimentation velocity-diffusion method) [8]
  51700 (Trametes sanguinea, gel filtration) [19]
  78000 (Thiobacillus novellus, gel filtration) [22]
  More (rat, purification yields two active fractions, of MW 17500 and 12600
  by gel filtration [17], primary, secondary, tertiary and quarternary structure
  [8]) [8, 17]

Subunits
  Monomer (1 × 17000, Acinetobacter calcoaceticus lwoffi, denaturing PAGE
  [25], 1 × 29000, Azotobacter vinelandii [2], 1 × 16500, Rhodopseudomonas
  palustris [3], 1 × 35000, Acinetobacter calcoaceticus, SDS-PAGE [21]) [2, 3,
  21, 25]

Dimer (2 × 38000, Thiobacillus novellus, SDS-PAGE [22], 2 × 19000, bovine,
SDS-PAGE [8], 2 × 18500, bovine, crystallographic data [10], 1 × 16000 +
1 × 17000, Manihot esculenta, SDS-PAGE [18]) [8, 10, 18, 22]
? (x × 33000, bovine, bovine liver and recombinant E. coli BL21(DE3) enzyme, SDS-PAGE) [14]
More (sedimentation equilibrium studies have shown that crystalline rhodanese preparation of full specific activity may contain both a nondissocia-
ble species of 33000 and a 37000 MW species undergoing slow dissocia-
tion to species of near 19000 MW) [9]

Glycoprotein/Lipoprotein

–

---

4 ISOLATION/PREPARATION

Source organism

Blowfly [8]; Thiobacillus denitrificans [8]; Thiobacillus novellus [8, 22]; Ferro-
bacillus ferrooxidans [8]; Pseudomonas aeruginosa [8]; Desulfotomaculum
nigrificans [8]; Manihot utilissima [8]; Spinach [8]; Parsley [8]; Cabbage [8];
Turnips [8]; Manihot esculenta [18]; Trametes sanguinea [19]; Acinetobacter
calcoaceticus (lwoffi [25]) [21, 25]; Methanosarcina frisia [23]; Bovine (bovine liver and recombinant E. coli BL21(DE3) enzyme [14]) [7–16, 20];
E. coli [8, 24]; Cat [8]; Dog [8]; Rabbit [8]; Rat [6, 17]; Cercopithecus aethiops (vervet monkey) [5]; Rana temporaria [4]; Human [1, 8, 20]; Azotobacter vinelandii [2]; Rhodopseudomonas palustris [3]; Guinea pig [6]; Chicken
[20]

Source tissue

Adrenals [8, 15]; Larvae [8]; Pupae [8]; Hepatomas (only about 20% of the
activity of normal liver) [8]; Leaf [18]; Culture filtrate [19]; Kidney [7];
Erythrocytes [1, 8]; Liver [4–6, 8–10, 12–14, 16, 17]; More (in all mammalian
tissues except blood and muscle) [8]

Localization in source

Soluble [21]; Mitochondria (mammalian liver enzyme occurs exclusively in
mitochondrial matrix [9]) [4, 8, 9]; Cytosol [4]; Chloroplast [8]

Purification

Thiobacillus novellus [22]; Manihot esculenta [18]; Trametes sanguinea [19];
Acinetobacter calcoaceticus (lwoffi [25]) [21, 25]; Methanosarcina frisia
[23]; E. coli [24]; Guinea pig (partial) [6]; Rat (partial [6]) [6, 17]; Human [1];
Azotobacter vinelandii [2]; Rana temporaria (partial) [4]; Cercopithecus
aethiops [5]; Bovine (bovine liver and recombinant E. coli BL21(DE3) enzyme [14]) [7, 9, 14]

Crystallization

[7, 9, 10, 16]

Cloned

[14, 15, 20]

Renatured

(refolds from 8 M urea to enzymatically active species) [14]

---

5 STABILITY

pH

4–8 (30°C, 44 h, stable) [19]; 9–13 (stable) [23]

Temperature (°C)

50 (rapid denaturation above [21], 20 min, stable up to, without stabilizer [23]) [21, 23]; 60 (20 min, stable up to, thiosulfate as stabilizer) [23]; 70 (10 min, pH 5.0, stable up to) [19]

Oxidation

Organic solvent

General stability information

At 2 M ammonium sulfate, crystals of the enzyme are stable when substrates are added [16]; At 1.4 M ammonium sulfate, crystals rapidly dissolve in 1 mM CN⁻ but are relatively stable in 1 mM $S_2O_3^{2-}$ [16]

Storage

–20°C, several months [21]

---

6 CROSSREFERENCES TO STRUCTURE DATABANKS

PIR/MIPS code

PIR1:ROBO (bovine); PIR1:ROHU (human); PIR2:JC4398 (mouse); PIR2:S15081 (rat (fragment)); PIR2:A37209 (hepatic chicken)

Brookhaven code

1RHD (Bovine (Bos Taurus) liver)

7 LITERATURE REFERENCES

[1] Scott, E.M., Wright, R.C.: Biochem. Biophys. Res. Commun.,97,1334–1338 (1980)
[2] Pagani, S., Sessa, G., Sessa, F., Colnaghi, R.: Biochem. Mol. Biol. Int.,29,595–604 (1993)
[3] Vazquez, E.S., Buzaleh, A.M., Wider, E.A., Battle, A.M.Del C.: Int. J. Biochem.,19, 1193–1197 (1987)
[4] Wrobel, M., Frendo, J.: Bull. Pol. Acad. Sci., Biol.,32,303–313 (1984)
[5] Van Rensburg, L.J., Schabort, J.C.: Int. J. Biochem.,16,539–546 (1984)
[6] Anosike, E.O., Jack, A.S.: Indian J. Biochem. Biophys.,19,13–16 (1982)
[7] Cannella, C., Pecci, L., Federici, G.: Ital. J. Biochem.,21,1–7 (1972)
[8] Westley, J.: Adv. Enzymol. Relat. Areas Mol. Biol.,39,327–368 (1973) (Review)
[9] Westley, J.: Methods Enzymol.,77,285–291 (1981) (Review)
[10] Drenth, J., Smit, J.D.G.: Biochem. Biophys. Res. Commun.,45,1320–1322 (1971)
[11] Burrous, M.R., Lane, J., Westley, A., Westley, J.: Methods Enzymol.,143,235–239 (1987) (Review)
[12] Oi, S.: J. Biochem.,76,455–458 (1974)
[13] Pagani, S., Bonomi, F., Cerletti, P.: Biochim. Biophys. Acta,742,116–121 (1983)
[14] Miller, D.M., Kurzban, G.P., Mendoza, J.A., Chirgwin, J.M., Hardies, S.C., Horowitz, P.M.: Biochim. Biophys. Acta,1121,286–292 (1992)
[15] Miller, D.M., Delgado, R., Chirgwin, J.M.: J. Biol. Chem.,266,4686–4691 (1991)
[16] Horowitz, P.M., Patel, K.: Biochem. Biophys. Res. Commun.,94,419–423 (1980)
[17] Oi, S.: J. Biochem.,78,825–834 (1975)
[18] Boey, C.G., Yeoh, H.H., Chew, M.Y.: Phytochemistry,15,1343–1344 (1976)
[19] Oi, S.: Agric. Biol. Chem.,37,629–635 (1973)
[20] Pallini, R., Guazzi, G.C., Cannella, C., Cacace, M.G.: Biochem. Biophys. Res. Commun.,180,887–893 (1991)
[21] Vandenbergh, P.A., Berk, R.S.: Can. J. Microbiol.,26,281–286 (1980)
[22] Fukumori, Y., Hoshiko, K., Yamanaka, T.: FEMS Microbiol. Lett.,65,159–164 (1989)
[23] Turkowsky, A., Blotevogel, K.-H., Fischer, U.: FEMS Microbiol. Lett.,81,251–256 (1991)
[24] Alexander, K., Volini, M.: J. Biol. Chem.,262,6595–6604 (1987)
[25] Aird, B.A., Heinrikson, R.L., Westley, J.: J. Biol. Chem.,262,17327–17335 (1987)

1 NOMENCLATURE

EC number
2.8.1.2

Systematic name
3-Mercaptopyruvate:cyanide sulfurtransferase

Recommended name
3-Mercaptopyruvate sulfurtransferase

Synonyms
Sulfurtransferase, 3-mercaptopyruvate
beta-Mercaptopyruvate sulfurtransferase
More (rhodanese activity, EC 2.8.1.1 is a minor function of human erythrocytic beta-mercaptopyruvate sulfurtransferase) [9]

CAS Reg. No.
9026-05-5

2 REACTION AND SPECIFICITY

Catalyzed reaction
3-Mercaptopyruvate + cyanide →
→ pyruvate + thiocyanate (sequential formal mechanism [6, 10], when
2-mercaptoethanol is sulfur acceptor addition of the substrate is random [6],
rapid equilibrium-ordered mechanism [6, 7], with 3-mercaptopyruvate as the
first substrate [6])

Reaction type
Sulfur atom transfer

Natural substrates
More (the enzyme plays a role in iron-sulfur chromophore formation in adrenal cortex) [11]

Substrate spectrum
1 3-Mercaptopyruvate + cyanide [1–11]
2 2-Mercaptoethanol + cyanide [5, 10]

Product spectrum
1 Pyruvate + thiocyanate [1, 2, 6, 7]
2 Ethanol + thiocyanate

Enzyme Handbook © Springer-Verlag Berlin Heidelberg 1997
Duplication, reproduction and storage in data banks are only
allowed with the prior permission of the publishers

Inhibitor(s)

Cyanide (inhibits at short-time intervals, slight enhancement at longer periods) [2]; Mercaptoethanol (high concentrations) [2]; Cysteamine (slight) [2]; Thioglycolic acid (slight) [2]; Mercaptosuccinamic acid (slight) [2]; Mercaptopropionic acid (slight) [2]; Cysteine [2]; Glutathione [2]; Pyruvate (10 mM: 17% inhibition, 20 mM: 45% inhibition [2], product inhibition [6, 10]) [2, 6, 10]

Cofactor(s)/prosthetic group(s)/activating agents

Metal compounds/salts

KCl (0.02 M, 70% activation) [2]; K_2SO_4 (0.02 M, 70% activation) [2]; Na_2SO_4 (0.02 M, 70% activation) [2]; Zinc (E. coli enzyme contains 0.1 mol of zinc [6], zinc protein, 1 atom/mol [2], no indication of a function in the mechanism of catalysis [2]) [2, 6]; Copper (no copper protein [1], contains 0.5 mol copper per mol of protein [2], E. coli enzyme contains 0.5 mol of copper [6]) [2, 6]; More (no effect: 0.02 M $CdCl_2$, 0.5 mM arsenite, 0.01 mM copper acetate) [2]

Turnover number (min^{-1})

750 (3-mercaptopyruvate + cyanide) [2]

Specific activity (U/mg)

540 [2]; 1240 [6, 10]; More [9]

K_m-value (mM)

7.3 (3-mercaptopyruvate, cytosol) [3]; 7.6 (3-mercaptopyruvate, mitochondria) [3]; 8.34 (mercaptopyruvate, pyruvate determined) [2]; 12.5 (mercaptopyruvate, thiocyanate determined) [2]

pH-optimum

9.3–9.6 [2]

pH-range

More [2]

Temperature optimum (°C)

30 (assay at) [1, 2, 6]; 45–50 [2]

Temperature range (°C)

45–60 (45–50°C: temperature optimum, 60°C: no activity) [2]

3 ENZYME STRUCTURE

Molecular weight
10900 (Rana temporaria, mitochondria, gel filtration) [3]
23800 (E. coli, sedimentation equilibrium ultracentrifugation) [2]
30200 (Rana temporaria, cytosol, gel filtration) [3]
33000 (bovine, gel filtration) [6]
33000–34000 (rat, gel filtration) [5]
36000 (rat, gel filtration) [4]

Subunits
Monomer (1 × 36000, rat, SDS-PAGE) [4]

Glycoprotein/Lipoprotein
Sialoprotein (219 amino acids and 38 carbohydrate residues) [4]

---

4 ISOLATION/PREPARATION

Source organism
Rat [1, 4, 5, 8]; E. coli [2]; Rana temporaria [3]; Bovine [6, 7, 10, 11]; Human [9]

Source tissue
Kidney [6, 7, 10, 11]; Liver [1, 3, 5, 8, 11]; Erythrocytes [1, 4, 9]; Adrenal gland [10]; Heart [11]

Localization in source
Mitochondria (highest activity in the matrix, followed by intramembrane space, low activity in inner and outer membrane [8]) [3, 5, 8]; Cytosol [3, 5]

Purification
Bovine [6, 10, 11]; Human [9]; Rat (partial) [1, 4, 5]; E. coli [2]; Rana temporaria (partial) [3]

Crystallization
–

Cloned
–

Renatured
–

5 STABILITY

pH

Temperature (°C)

Oxidation

Organic solvent

General stability information

Markedly stabilized during purification and storage by the presence of monovalent cations, maximal stability is obtained if purification and storage are
carried out at pH 6.5–7.5 in presence of 0.8 M KCl and 2 mM mercaptoethanol [2]; Unstable to thawing and refreezing [1]; Dialysis against urea inactivates, effect is reversed by dialysis, dilution or electrophoresis [2]; Stabilized in 0.8 M KCl [2]; Very unstable, spontaneous inactivation can be partly
prevented by glycerol [4]

Storage

4°C, no loss of activity after 10 days [2]; 4°C, several days [1]; –35°C, 50
mM potassium phosphate buffer, pH 7.4, 0.5 mg/ml bovine serum albumin,
50% glycerol, slow decrease of activity [10]

6 CROSSREFERENCES TO STRUCTURE DATABANKS

PIR/MIPS code
PIR2:A57483 (rat (fragment))

Brookhaven code

7 LITERATURE REFERENCES

[1] Van den Hamer, C.J.A., Morell, A.G., Scheinberg, I.H.: J. Biol. Chem.,242,
    2514–2516 (1967)
[2] Vachek, H., Wood, J.L.: Biochim. Biophys. Acta,258,133–146 (1972)
[3] Wrobel, M., Frendo, J.: Bull. Pol. Acad. Sci., Biol.,32,303–313 (1984)
[4] Wlodek, L., Ostrowski, W.S.: Acta Biochim. Pol.,29,121–133 (1982)
[5] Kasperczyk, H., Koj, A., Wasylewski, Z.: Bull. Acad. Pol. Sci., Ser. Sci. Biol.,25,7–13
    (1977)
[6] Jarabak, R.: Methods Enzymol.,77,291–297 (1981) (Rewiew)
[7] Jarabak, R., Westley, J.: Biochemistry,19,900–904 (1980)
[8] Koj, A., Frendo, J., Wojtczak, L.: FEBS Lett.,57,42–46 (1975)
[9] Scott, E.M., Wright, R.C.: Biochem. Biophys. Res. Commun.,97,1334–1338 (1980)
[10] Jarabak, R., Westley, J.: Arch. Biochem. Biophys.,185,458–465 (1978)
[11] Taniguchi, T., Kimura, T.: Biochim. Biophys. Acta,364,284–295 (1974)

1 NOMENCLATURE

EC number
2.8.1.3

Systematic name
Thiosulfate:thiol sulfurtransferase

Recommended name
Thiosulfate-thiol sulfurtransferase

Synonyms
Glutathione-dependent thiosulfate reductase
Sulfane reductase
Sulfurtransferase, sulfane
Sulfane sulfurtransferase

CAS Reg. No.
111070-24-6

2 REACTION AND SPECIFICITY

Catalyzed reaction
Thiosulfate + 2 glutathione →
→ sulfite + oxidized glutathione + sulfide (substrates add in a random fashion [1])

Reaction type
Sulfur atom transfer

Natural substrates

Substrate spectrum
1 Thiosulfate + glutathione [1–3]
2 Benzenethiosulfonate + glutathione (rapid equilibrium-ordered mechanism with glutathione as leading substrate) [3]
3 L-Cysteine + glutathione [1]
4 More (the enzyme has two distinct closely situated substrate binding sites, one for compounds with an RSO_3-structure and one for the sulfhydryl substrate) [1]

Product spectrum
1 Sulfite + oxidized glutathione + sulfide (the primary product is glutathione hydrodisulfide, which reacts with glutathione to give oxidized glutathione and sulfide) [1–3]

Enzyme Handbook © Springer-Verlag Berlin Heidelberg 1997
Duplication, reproduction and storage in data banks are only
allowed with the prior permission of the publishers

2 More (glutathione persulfide as an immediate product) [3]
3 ?
4 ?

Inhibitor(s)
Thiosulfonates [1]; Sulfite (product inhibition) [1]; Sulfide (product inhibition) [1]; Oxidized glutathione (product inhibition) [1]; More (alkylation of cysteine residues with iodoacetate or iodoacetamide does not inactivate) [2]

Cofactor(s)/prosthetic group(s)/activating agents

Metal compounds/salts

Turnover number (min^{-1})

Specific activity (U/mg)
More [1, 2]

K_m-value (mM)
0.24 (benzenethiosulfonate) [3]; 0.89 (glutathione (+ benzenethiosulfonate)) [3]; 2.9 (thiosulfate) [3]; 4.0 (glutathione (+ thiosulfate)) [3]

pH-optimum

pH-range

Temperature optimum (°C)

Temperature range (°C)

3 ENZYME STRUCTURE

Molecular weight
17000 (Saccharomyces cerevisiae, SDS-PAGE, gel filtration) [1]

Subunits
Monomer (1 × 17000, Saccharomyces cerevisiae, SDS-PAGE) [1]

Glycoprotein/Lipoprotein
–

4 ISOLATION/PREPARATION

Source organism
Saccharomyces cerevisiae (Red Star brand) [1–3]

Source tissue

Localization in source

Purification
    Saccharomyces cerevisiae [1, 2]

Crystallization
    –

Cloned
    –

Renatured
    –

---

5 STABILITY

pH

Temperature (°C)

Oxidation

Organic solvent

General stability information
    Freezing and thawing during the later stages of purification destabilize [1];
    Extremely labile during all ion-exchange steps [2]

Storage
    –35°C, concentrated enzyme in solution of 50 mM Tris, pH 8.0, 0.5 mM
    $Na_2S_2O_3$, 50% glycerol, stable for at least 4 months [1]; –40°C, 50 mM
    Tris-acetate, pH 8.0, 0.5 mM $Na_2S_2O_3$, 50% glycerol, stable for at least 9
    months [2]

---

6 CROSSREFERENCES TO STRUCTURE DATABANKS

PIR/MIPS code

Brookhaven code

---

7 LITERATURE REFERENCES

[1] Uhteg, L., Westley, J.: Arch. Biochem. Biophys.,195,211–222 (1979)
[2] Chauncey, T.R., Westley, J.: Biochim. Biophys. Acta,744,304–311 (1983)
[3]Chauncey, T.R., Westley, J.: J. Biol. Chem.,258,15037–15045 (1983)

1 NOMENCLATURE

EC number
2.8.1.4

Systematic name
L-Cysteine:tRNA sulfurtransferase

Recommended name
tRNA sulfurtransferase

Synonyms
Sulfurtransferase, transfer ribonucleate
RNA sulfurtransferase
Sulfurtransferase, ribonucleate
Transfer RNA sulfurtransferase
TransferRNA thiolase

CAS Reg. No.
9055-57-6

2 REACTION AND SPECIFICITY

Catalyzed reaction
L-Cysteine + activated tRNA →
→ L-serine + tRNA containing a thionucleotide

Reaction type
Sulfur atom transfer

Natural substrates
More (the enzyme is part of the sulfur-transferase system which forms
4-thiouridylate in tRNA) [1]

Substrate spectrum
1 Cysteine + activated tRNA (E. coli B tRNA [1], yeast tRNA [1, 2], denatured deoxyribonucleic acid prepared from various sources [2], synthetic
  ribohomopolymers [2, 5], tRNA of higher organsims (Pseudomonas aeruginosa enzyme) [3], rat liver tRNA is the poorest tRNA substrate for the
  rat liver enzyme [4], sulfur-deficient tRNA accepts nearly 4times more sulfur in vitro than normal tRNA, sulfur incorporation is inversely proportional
  to the 4-thiouridine content [6], not: other polyribo- or polydeoxyribonucle-
  otide sulfur acceptors [1]) [1–6]
2 Cysteine + rRNA [3–5]
3 Cysteine + DNA (native and denatured [3]) [3, 4]
4 3-Mercaptopyruvate + activated tRNA (sulfurtransferase system [1]) [1, 2,
  4, 5, 7]

Product spectrum
1 tRNA containing a thionucleotide (major thionucleotide formed during in vitro thiolation is thiocytidine, major product in vivo is 4-thiouridine [6], product of rat brain enzyme is not identical with 4-thiouridine [5]) [5, 6]
2 ?
3 ?
4 More (the product is sensitive to deacylation) [4]

Inhibitor(s)
2-Mercaptoethanol (stimulates at low concentration, inhibits at higher concentration) [5]; 3-Mercaptopyruvate [2]; Cysteine (inhibits reaction with 3-mercaptopyruvate) [4]; More (presence of an inhibitor in Morris hepatomas: dialyzable heat-stable with a MW below 5000) [7]

Cofactor(s)/prosthetic group(s)/activating agents
Supernatant protein (required) [4]; 2-Mercaptoethanol (stimulates at low concentration, inhibits at higher concentration) [5]; ATP (required) [4–7]

Metal compounds/salts
Mg^{2+} (required [3–7], $MgCl_2$ required [3]) [3–7]; More (sulfurtransferase system requires a divalent metal) [2]

Turnover number (min⁻¹)

Specific activity (U/mg)
More [1, 5]; 0.00813 [2]

K_m-value (mM)
0.0012–0.0016 (cysteine) [1]; 0.0027 (sulfur-deficient tRNA) [6]; 0.26 (cysteine) [6]; 1 (cysteine) [2]

pH-optimum
7.5–8.0 [4]; 8.0 [3]

pH-range
7.0–9.0 (7.0: about 75% of activity maximum, 9.0: about 55% of activity maximum) [4]

Temperature optimum (°C)
37 (assay at) [1, 2, 5]

Temperature range (°C)

3 ENZYME STRUCTURE

Molecular weight

Subunits

Glycoprotein/Lipoprotein
–

4 ISOLATION/PREPARATION

Source organism
E. coli (B [1], HfrC, RCrel, met-, cys-, lambda [6]) [1, 6]; Bacillus subtilis (W168) [2]; Pseudomonas aeruginosa [3]; Rat (Buffalo rat [7], tRNA sulfurtransferase, tRNA methyltransferase and aminoacyl-tRNA synthetase activity are associated in a complex [8]) [4, 5, 7, 8]

Source tissue
Heart [4]; Adrenals [4]; Brain (cerebral hemispheres [5]) [4, 5]; Morris hepatomas (9618A2, 7777, 5123TC, 7800, 5123B, 7787) [7]; Testes [4]; Liver [4, 7, 8]; Muscle (highest activity of rat tissues) [4]; Lung [4]; Kidney [4]

Localization in source
Cytoplasm (soluble) [5]

Purification
E. coli (enzyme system forming 4-thiouridylate in tRNA) [1]; Bacillus subtilis [2]; Pseudomonas aeruginosa [3]; Rat (partial) [5]

Crystallization
–

Cloned
–

Renatured
–

---

5 STABILITY

pH

Temperature (°C)

Oxidation

Organic solvent

General stability information

Storage
–75°C, 40–70% ammonium sulfate, stable for several months [5]; 0–4°C, 30–60% loss of activity after 1 week, even in presence of 50% glycerol [8]

---

6 CROSSREFERENCES TO STRUCTURE DATABANKS

PIR/MIPS code

Brookhaven code

7 LITERATURE REFERENCES

[1] Abrell, J.W., Kaufman, E.E., Lipsett, M.N.: J. Biol. Chem.,246,294–301 (1971)
[2] Wong, T.-W., Weiss, S.B., Eliceiri, G.L., Bryant, J.: Biochemistry,9,2376–2386 (1970)
[3] Thimmappaya, B., Cherayil, J.D.: Indian J. Biochem. Biophys.,12,405–407 (1975)
[4] Harris, C.L.: Nucleic Acids Res.,5,599–613 (1978)
[5] Wong, T.-W., Harris, M.A., Jankowicz, C.A.: Biochemistry,13,2805–2812 (1974)
[6] Harris, C.L., Titchener, E.B.: Biochemistry,10,4207–4212 (1971)
[7] Wong, T.-W., Harris, M.A., Morris, H.P.: Biochem. Biophys. Res. Commun.,65, 1137–1145 (1975)
[8] Harris, C.L., Marin, K., Stewart, D.: Biochem. Biophys. Res. Commun.,79,657–662 (1977)

1 NOMENCLATURE

EC number
2.8.1.5

Systematic name
Thiosulfate:dithioerythritol sulfurtransferase

Recommended name
Thiosulfate-dithiol sulfurtransferase

Synonyms
Thiosulfate reductase
TSR [1]
Reductase, thiosulfate
More (may be identical with EC 2.8.1.1)

CAS Reg. No.
9059-49-8

2 REACTION AND SPECIFICITY

Catalyzed reaction
Thiosulfate + dithioerythritol →
→ sulfite + dithioerythritol disulfide + sulfide

Reaction type
Sulfur atom transfer

Natural substrates

Substrate spectrum
1 Thiosulfate + dithioerythritol [1]
2 Cyanate + dithioerythritol (activity only in presence of thiols) [1]
3 More (little activity with: glutathione, L-cysteine, beta-mercaptoethanol) [1]

Product spectrum
1 Sulfite + dithioerythritol disulfide + sulfide [1]
2 ?
3 ?

Inhibitor(s)
Thiosulfate (above 0.5 mM) [1]

Cofactor(s)/prosthetic group(s)/activating agents
Thioredoxin (stimulates thiosulfate reductase IIIa and IIIb) [1]

Metal compounds/salts

Enzyme Handbook © Springer-Verlag Berlin Heidelberg 1997
Duplication, reproduction and storage in data banks are only
allowed with the prior permission of the publishers

Turnover number (min^{-1})

Specific activity (U/mg)

K$_m$-value (mM)

0.156 (thiosulfate (+ dithioerythritol), isoenzyme TSR II) [1]; 0.164 (thiosulfate (+ dithioerythritol), isoenzyme TSR I) [1]; 1.54 (dithioerythritol (+ thiosulfate), isoenzyme TSR II) [1]; 3.1 (dithioerythritol (+ thiosulfate), isoenzyme TSR I, dithioerythritol (+ KCN)) [1]; 20 (KCN (+ dithioerythritol)) [1]

pH-optimum

8.5 (isoenzyme TSR IIIa) [1]; 9.0 (isoenzyme TSR I and TSR II) [1]; 9.5 (isoenzyme TSR IIIb) [1]

pH-range

Temperature optimum (°C)

37 (assay at) [1]

Temperature range (°C)

---

3 ENZYME STRUCTURE

Molecular weight

24000 (Chlorella fusca, isoenzyme IIIb, gel filtration) [1]
26500 (Chlorella fusca, isoenzyme II, gel filtration) [1]
28000 (Chlorella fusca, isoenzyme I, gel filtration) [1]
55000 (Chlorella fusca, isoenzyme IIIa, gel filtration) [1]

Subunits

Glycoprotein/Lipoprotein

–

---

4 ISOLATION/PREPARATION

Source organism

Chlorella fusca (strain 211–8b) [1]

Source tissue

Localization in source

Purification

Chlorella fusca (isoenzymes: TSR I, TSR II, TSR IIIa, TSR IIIb) [1]

Crystallization

–

2

Cloned

–

Renatured

–

5 STABILITY

pH

Temperature (°C)

Oxidation

Organic solvent

General stability information

Storage

6 CROSSREFERENCES TO STRUCTURE DATABANKS

PIR/MIPS code

PIR2:A57143 (phsA chain Salmonella typhimurium); PIR2:B57143 (phsB chain Salmonella typhimurium); PIR2:C57143 (phsC chain Salmonella typhi-murium)

Brookhaven code

7 LITERATURE REFERENCES

[1] Schmidt, A., Erdle, I., Gamon, B.: Planta,162,243–249 (1984)

1 NOMENCLATURE

EC number
2.8.2.1

Systematic name
3'-Phosphoadenylylsulfate:phenol sulfotransferase

Recommended name
Aryl sulfotransferase

Synonyms
Phenol sulfotransferase
Sulfokinase
Sulfotransferase, aryl
1-Naphthol phenol sulfotransferase
2-Naphtholsulfotransferase
4-Nitrocatechol sulfokinase
Arylsulfotransferase
Dopamine sulfotransferase
p-Nitrophenol sulfotransferase
Phenol sulfokinase
Ritodrine sulfotransferase
PST [3]
More (cf. EC 2.8.2.9)

CAS Reg. No.
9026-09-9

---

2 REACTION AND SPECIFICITY

Catalyzed reaction
3'-Phosphoadenylylsulfate + a phenol →
→ adenosine 3',5'-bisphosphate + an aryl sulfate (isozyme M-PST: ordered bisubstrate reaction mechanism [18], isozyme P-PST: sequential ordered bisubstrate reaction mechanism [16])

Reaction type
Sulfate group transfer

Natural substrates

3'-Phosphoadenylylsulfate + a phenol (enzyme may be considered as detoxification enzyme which catalyzes the conjugation of xenobiotics containing a phenol group or of phenolic compounds generated by endogenous oxidation) [2, 3]

3'-Phosphoadenylylsulfate + 3-hydroxyindole (in vivo function may include the production of the normal tryptophan metabolite indican) [2]

3'-Phosphoadenylylsulfate + dopamine (accounts for approximately 10% of the enzymic activity directed towards catabolism of dopamine) [11]

Substrate spectrum

1 3'-Phosphoadenylylsulfate + dopamine (low activity [14]) [1, 2, 4, 9, 11, 14, 15, 18]

2 3'-Phosphoadenylylsulfate + nitrophenol (4-nitrophenol [1, 2, 4, 8, 15, 17], 3-nitrophenol [2]) [1, 2, 4, 8, 15–17]

3 3'-3'-Phosphoadenylylsulfate + 2-hydroxyestrone [1]

4 3'-Phosphoadenylylsulfate + 2-hydroxyestradiol [1]

5 3'-Phosphoadenylylsulfate + noradrenaline [1]

6 3'-Phosphoadenylylsulfate + adrenaline [1]

7 3'-Phosphoadenylylsulfate + naphthol (1-naphthol [5], 2-naphthol [2, 7, 12], highest activity among simple phenols tested [5]) [2, 5, 7, 12]

8 3'-Phosphoadenylylsulfate + 3-hydroxyindole [2]

9 3'-Phosphoadenylylsulfate + 5-hydroxyindole [2]

10 3'-Phosphoadenylylsulfate + 5-hydroxytryptamine [2]

11 3'-Phosphoadenylylsulfate + phenol [2, 9, 12, 14]

12 3'-Phosphoadenylylsulfate + chlorophenol (2-chlorophenol, 3-chlorophenol- [2], 4-chlorophenol [2, 12]) [2, 12]

13 3'-Phosphoadenylylsulfate + methylphenol (3-methylphenol, 4-methylphenol) [2]

14 3'-Phosphoadenylylsulfate + methoxyphenol (4-methoxyphenol [12]) [2, 12]

15 3'-Phosphoadenylylsulfate + hydroquinone [2]

16 3'-Phosphoadenylylsulfate + hydroxybiphenyl [2]

17 3'-Phosphoadenylylsulfate + 4-acetamidophenol [2]

18 3'-Phosphoadenylylsulfate + 2,6-dichloro-4-nitrophenol [3]

19 3'-Phosphoadenylylsulfate + 1-phenylethanol ((+)-(R)- and (-)-(S)-) [6]

20 3'-Phosphoadenylylsulfate + (-)-(1R,2S)-ephedrine (absolute stereospecificity) [6]

21 3'-Phosphoadenylylsulfate + (-)-(1R,2R)-pseudoephedrine (absolute stereospecificity) [6]

22 3'-Phosphoadenylylsulfate + (-)-(S)-2-methyl-1-phenyl-1-propanol (absolute stereospecificity) [6]

23 3'-Phosphoadenylylsulfate + 1,2,3,4-tetrahydro-1-naphthol (only the (-)-(R)-enantiomer is active as substrate) [6]

24 3'-Phosphoadenylylsulfate + catecholamine metabolites (deaminated and (or) O-methylated) [8]

25 3'-Phosphoadenylylsulfate + salicylamide [8]

26 3'-Phosphoadenylylsulfate + vanillin [8]

27 3'-Phosphoadenylylsulfate + 4-methylumbelliferone [10]

28 3'-Phosphoadenylylsulfate + 3-methoxy-4-hydroxyphenylethyleneglycol [10]

29 3'-Phosphoadenylylsulfate + tyramine [18]

30 3'-Phosphoadenylylsulfate + 3-methoxy-4-hydroxyphenylglycol [14]

31 3'-Phosphoadenylylsulfate + 3,4-dihydroxyphenylglycol [14]

32 3'-Phosphoadenylylsulfate + homovanillic acid [14]

33 3'-Phosphoadenylylsulfate + vanillylmandelic acid [14]

34 3'-Phosphoadenylylsulfate + dihydroxyphenylacetic acid [14]

35 3'-Phosphoadenylylsulfate + 4-methoxytyramine [18]

36 More (isoenzyme B catalyzes the sulfurylation of a wider range of substrates than A which is preferentially active with dopamine [1], biogenic amines: the absence of a meta substituent on the phenolic ring, or the presence of a beta-OH group on the aliphatic amine side chain greatly reduces their binding affinities [11], activity with various substrates at different pH values [12], 2 forms, TL: thermolabile, sulfate conjugation of dopamine and other phenolic monoamines, TS: thermostable, sulfate conjugation of simple phenols, e.g. p-nitrophenol [13], no activity: serotonin [13], tyrosine and its derivatives [5]) [1, 5, 11–13]

Product spectrum

1 Adenosine 3',5'-bisphosphate + ?

2 Adenosine 3',5'-bisphosphate + nitrophenyl sulfate

3 Adenosine 3',5'-bisphosphate + ?

4 Adenosine 3',5'-bisphosphate + ?

5 Adenosine 3',5'-bisphosphate + ?

6 Adenosine 3',5'-bisphosphate + ?

7 Adenosine 3',5'-bisphosphate + naphthyl sulfate

8 Adenosine 3',5'-bisphosphate + 3-indoxyl sulfate [2]

9 Adenosine 3',5'-bisphosphate + 5-indoxyl sulfate

10 Adenosine 3',5'-bisphosphate + 3-(2-aminoethyl)-5-indoxyl sulfate

11 Adenosine 3',5'-bisphosphate + phenyl sulfate

12 Adenosine 3',5'-bisphosphate + chlorophenyl sulfate

13 Adenosine 3',5'-bisphosphate + methylphenyl sulfate

14 Adenosine 3',5'-bisphosphate + methoxyphenyl sulfate

15 Adenosine 3',5'-bisphosphate + ?

16 Adenosine 3',5'-bisphosphate + phenylphenyl sulfate

17 Adenosine 3',5'-bisphosphate + 4-acetamidophenyl sulfate

18 Adenosine 3',5'-bisphosphate + 2,6-dichloro-4-nitrophenyl sulfate [4]

19 Adenosine 3',5'-bisphosphate + 1-phenylethyl sulfate

20 Adenosine 3',5'-bisphosphate + 2-(methylamino)-1-phenylpropyl sulfate

21 Adenosine 3',5'-bisphosphate + 2-(methylamino)-1-phenylpropyl sulfate
22 Adenosine 3',5'-bisphosphate + 2-methyl-1-phenylpropyl 1-sulfate
23 Adenosine 3',5'-bisphosphate + 1,2,3,4-tetrahydronaphthyl 1-sulfate
24 Adenosine 3',5'-bisphosphate + ?
25 Adenosine 3',5'-bisphosphate + ?
26 Adenosine 3',5'-bisphosphate + 4-methoxybenzaldehyde 4-sulfate
27 Adenosine 3',5'-bisphosphate + ?
28 Adenosine 3',5'-bisphosphate + ?
29 Adenosine 3',5'-bisphosphate + 4-(2-aminoethyl)phenyl sulfate
30 Adenosine 3',5'-bisphosphate + ?
31 Adenosine 3',5'-bisphosphate + ?
32 Adenosine 3',5'-bisphosphate + ?
33 Adenosine 3',5'-bisphosphate + ?
34 Adenosine 3',5'-bisphosphate + ?
35 Adenosine 3',5'-bisphosphate + ?
36 ?

Inhibitor(s)

Mg^{2+} (10 mM) [1]; Ca^{2+} (10 mM) [1]; Zn^{2+} (10 mM) [1]; Cu^{2+} (10 mM) [1]; KCl (50 mM, 60% inhibition of isoenzyme A, slight activation of isoenzyme B) [1]; Adenosine 3',5'-bisphosphate [2, 18]; ADP (isoenzyme I) [2]; ATP (isoenzyme I [2]) [2, 13, 18]; Cibacron Blue (transferase I) [2]; Pyridoxal 5'-phosphate (excess of 2-naphthol protects) [3]; 2,6-Dichloro-4-nitrophenol (sulfation of 4-nitrophenol (weak, non-competitive [4]) [4, 8] or dopamine [4], 2 forms, TL: relatively resistant to inhibition by 2,6-dichloro-4-nitrophenol, TS: relatively sensitive to 2,6-dichloro-4-nitrophenol [13]) [4, 5, 8, 12, 13, 15]; 4-Nitrophenol (sulfation of 2,6-dichloro-4-nitrophenol or dopamine, competitive [4], at high concentration [8]) [4, 8]; Chlorpromazine [5]; (+)-(R)-2-Methyl-1-phenyl-1-propanol (competitive to 1-naphthalenemethanol) [6]; (+)-(S)-1,2,3,4-Tetrahydro-1-naphthol (competitive to 1-naphthalenemethanol) [6]; 3,4-Dihydroxyphenylethyleneglycol [10]; 3,4-Dihydroxymandelic acid [10]; 3,4-Dihydroxyphenylacetic acid [10]; 3-Methoxy-4-hydroxyphenylethylene-glycol [10]; 3-Methoxy-4-hydroxyphenethanol [10]; 3-Methoxy-4-hydroxy-phenylacetic acid [10]; Pentachlorophenol [12]; NaCl (50% inhibition at: 325 mM [18], 100 mM [16]) [16, 18]; NEM [17]; Phenylglyoxal [17]; More (not: adrenaline [10], normetanephrine, metanephrine, dopamine, the presence of an amino group on the side chain which is positively charged at pH 7.4 drastically decreases inhibitory power, substrate inhibition for most compounds at concentrations exceeding approximately 5times the K_m [13]) [10, 13]

Cofactor(s)/prosthetic group(s)/activating agents

EDTA (10 mM, stimulates) [1]; Reduced glutathione (increases activity) [2]

4

Metal compounds/salts

KCl (50 mM, 60% inhibition of isoenzyme A, slight activation of isoenzyme B [1], aryl sulfotransferase I and II are activated appreciably in the presence of 0.5 M KCl or NaCl [12]) [1, 12]; NaCl (aryl sulfotransferase I and II are activated appreciably in the presence of 0.5 M NaCl or KCl) [12]; More (no effect: Mg^{2+} [13]) [12, 13]

Turnover number (min^{-1})

0.6 (5-hydroxytryptamine, isoenzyme II) [2]; 1.5 (5-hydroxytryptamine, isoenzyme I) [2]; 6.8 (phenol, isoenzyme I) [2, 12]; 7.3 (phenol, isoenzyme II) [2, 12]; 36 (5-hydroxyindole, isoenzyme II) [2]; 39 (5-hydroxyindole, isoenzyme I, 3-hydroxyindole, isoenzyme II) [2]; 40 (4-chlorophenol, isoenzyme II) [12]; 44 (3-hydroxyindole, 2-chlorophenol, isoenzyme I) [2]; 47 (2-naphthol, isoenzyme I) [2, 12]; 48 (4-chlorophenol, isoenzyme I) [12]; 54 (2-naphthol, isoenzyme II) [2, 12]; 93 (4-methoxyphenol, isoenzyme I) [2, 12]; 120 (4-methoxyphenol, isoenzyme II) [2, 12]; More [2, 12]

Specific activity (U/mg)

0.00363 [1]; 0.282 [2]; 0.25 (arylsulfotransferase I) [12]; 0.282 (arylsulfotransferase II) [12]; 0.34 [17]; More (rapid, simple sensitive radioassay [13]) [13, 16, 18]

K$_m$-value (mM)

0.0017 (3'-phosphoadenylylsulfate (+ dopamine), isoenzyme A) [1]; 0.0062 (dopamine, isoenzyme B) [1]; 0.0065 (3'-phosphoadenylylsulfate (+ 2-naphthol), isoenzyme I) [2]; 0.01 (dihydroxyphenylacetic acid) [14]; 0.012 (3'-phosphoadenylylsulfate (+ 2-naphthol), isoenzyme II) [2]; 0.014 (homovanillic acid) [14]; 0.017 (3,4-dihydroxyphenylglycol) [14]; 0.0177 (dopamine, isoenzyme A) [1]; 0.018 (3-methoxy-4-hydroxyphenylglycol) [14]; 0.02 (dopamine) [14]; 0.026 (3'-phosphoadenylylsulfate (+ dopamine), isoenzyme B) [1]; 0.04 (phenol) [14]; 0.06 (2-naphthol, isoenzyme I) [2]; 0.07 (3-hydroxyindole, isoenzyme II) [2]; 0.08 (3-hydroxyindole, isoenzyme I) [2]; 0.09 (2-naphthol, isoenzyme II) [2]; 0.1 (vanillylmandelic acid, norepinephrine) [14]; 0.12 (3-chlorophenol, isoenzyme I) [2]; 0.15 (2-chlorophenol, isoenzyme I) [2]; 0.16 (3-chlorophenol, isoenzyme II) [2]; 0.18 (p-hydroxybiphenyl, isoenzyme I) [2]; 0.19 (o-chlorophenol, isoenzyme II) [2]; 0.25 (p-hydroxybiphenyl, isoenzyme II) [2]; 0.29 (5-hydroxyindole, isoenzyme I) [2]; 0.3 (5-hydroxyindole, isoenzyme II) [2]; 0.44 (3-nitrophenol, isoenzyme II) [2]; 0.53 (5-hydroxytryptamine, isoenzyme II) [2]; 0.6 (hydroquinone, isoenzyme II) [2]; 0.833 (tyrosylglycine) [16]; 1.0 (3-nitrophenol, isoenzyme I) [2]; 1.2 (4-chlorophenol, isoenzyme II) [2]; 1.4 (3-methylphenol, isoenzyme II) [2]; 1.5 (4-chlorophenol, isoenzyme I) [2]; 1.6 (4-nitrophenol, 5-hydroxytryptamine, isoenzyme I, 4-acetamidophenol, isoenzyme II) [2]; 1.8 (hydroquinone, phenol, isoenzyme I) [2]; 2.1 (4-acetamidophenol, isoenzyme I) [2]; 2.2 (4-methylphenol, isoenzyme II) [2]; 2.5 (4-nitrophenol, isoenzyme II) [2];

5

2.6 (phenol, isoenzyme II) [2]; 2.9 (4-methylphenol, isoenzyme I) [2]; 4.2 (4-methoxyphenol, isoenzyme I) [2]; 6.5 (4-methoxyphenol, isoenzyme II) [2]; More [7, 8, 11, 13, 15, 16, 18]

pH-optimum

5.5 (4-nitrophenol, isoenzyme B [1], 2-naphthol, aryl sulfotransferase I and II [12]) [1, 12]; 5.6 [8]; 5.7 [5]; 5.8 (phenol, homovanillic acid, dihydroxy-phenylacetic acid) [14]; 6 (dopamine, isoenzyme A [1], 3,4-dihydroxy-phenylglycol [14]) [1, 14]; 6.2 (vanillylmandelic acid) [14]; 6.4 (3-methoxy-4-hydroxyphenylglycol) [14]; 6.5 (4-nitrophenol [1, 2], 2-naphthol (a second optimum at pH 9.5) [2], isoenzyme A [1], 2-naphthol, aryl sulfotransferase II [12]) [1, 2, 12]; 7.0 (dopamine) [18]; 8.5 [16]; 9.0 (dopamine, norepine-phrine) [14]; 9.5 (dopamine [1], 2-naphthol (a second optimum at pH 6.5) [2], isoenzyme B [1]) [1, 2]

pH-range

More [1, 2, 16]

Temperature optimum (°C)

37 (assay at) [1, 2, 12]

Temperature range (°C)

More (incubation at 37°C causes a selective decrease in activity towards dopamine compared with phenol) [9]

---

3 ENZYME STRUCTURE

Molecular weight

60000 (dog, gel filtration) [1]
61000 (rat, arylsulfotransferase III, gel filtration) [12]
64000 (rat, arylsulfotransferase I or II, gel filtration) [12]
64000–65000 (rat, gel filtration) [2]
68000 (rat, gel filtration [8], human, isozyme P-PST, gel filtration [17]) [8, 17]
250000 (human, isozyme M-PST, gel filtration) [18]

Subunits

Dimer (2 × 35000, rat, SDS-PAGE [2], 2 × 33500, rat, arylsulfotransferase III, SDS-PAGE [12], 2 × 64000, rat, arylsulfotransferase I or II [12], 2 × 32000, human, SDS-PAGE [17]) [2, 12, 17]

Glycoprotein/Lipoprotein

–

4 ISOLATION/PREPARATION

Source organism
Dog [1]; Rat [2, 4, 6, 8, 10, 12, 14]; Bovine [3, 5]; Human (2 forms [9], 2 forms: TL, TS [13], 2 forms: P-PST, M-PST, expressed on COS-7 cells as HAST1 and HAST3 [15]) [7, 9, 11, 13, 15–18]

Source tissue
Liver [1, 2, 4–6, 12, 15, 17]; Lung [3]; Ileum [7]; Colon mucosa [7]; Brain [8–11, 13–16, 18]; Platelets [9, 13]; Jejunum [9]; Adrenal [9]

Localization in source
Cytosol [4, 7, 15, 17]; Microsomes (membrane) [5]

Purification
Dog (isoenzyme A and B [1]) [1]; Rat (isoenzyme I and II [2], arylsulfotransferase I, II, III (IV, see EC 2.8.2.9) [12], partial [14]) [2, 8, 12, 14]; Human (partial [13], phenol-sulfating form (P-PST [17]) [16, 17], monoamine-sulfating form, M-PST [18]) [11, 13, 16–18]

Crystallization
–

Cloned
[15]

Renatured
–

---

5 STABILITY

pH
6.0 (4°C, 82 h, 60% loss of activity) [14]; 8.0 (4°C, 82 h, 80% loss of activity) [14]

Temperature (°C)
More (human: one thermolabile form (TL) and one thermostable form (TS) [13], HAST1 is considerably more thermostable than HAST3 [15]) [13, 15]

Oxidation

Organic solvent

General stability information
Sucrose and 2-mercaptoethanol stabilize during purification [2]; Glycerol reduces activity [2]; Transferase II is unstable to isoelectric focusing [12]

Storage

4°C, stable for at least 2 months [2]; –80°C, storage over night, 10%, trans-ferase I, and 35%, transferase II, loss of activity [2]; 4°C, sodium phosphate, pH 7.0, 0.25 M sucrose, 5 mM mercaptoethanol, 3 mM NaN_3, protein con-centration 0.5–1 mg/ml, transferase I, II lose about 5% of their activity per week [12]; –80°C, transferase III may deteriorate at any stage of purification [12]; –20°C, 7 days, 20% loss of activity of partially purified enzyme [14]

6 CROSSREFERENCES TO STRUCTURE DATABANKS

PIR/MIPS code

PIR2:S52399 (human); PIR2:S52791 (human); PIR2:S52794 (human); PIR2:JC$_2$523 (brain isoform human); PIR2:JN0714 (HAST2 human); PIR2:A55451 (HAST3 /estrogen sulfotransferase EST human); PIR2:S10329 (IV rat); PIR2:S28183 (p1 mouse)

Brookhaven code

7 LITERATURE REFERENCES

[1] Romain, Y., Demassieux, S., Carriere, S.: Biochem. Biophys. Res. Commun., 106,999–1005 (1982)
[2] Sekura, R.D., Jakoby, W.B.: J. Biol. Chem.,254,5658–5663 (1979)
[3] Bartzatt, R., Beckmann, J.D.: Biochem. Pharmacol.,47,2087–2095 (1994)
[4] Seah, V.M.Y., Wong, K.P.: Biochem. Pharmacol.,47,1743–1749 (1994)
[5] Fernando, P.H.P., Sakakibara, Y., Nakatsu, S., Suiko, M., Han, J.R., Liu, M.C.: Biochem. Mol. Biol. Int.,30,433–441 (1993)
[6] Rao, S.I., Duffel, M.W.: Chirality,3,104–111 (1991)
[7] Pacifici, G.M., Franchi, M., Giuliani, L.: Pharmacology,38,146–150 (1989)
[8] Baranczyk-Kuzma, A., Borchardt, R.T., Schasteen, C.S., Pinnick, C.L. in "Phenol-sulfotransferase, Ment. Health Res." (Sandler, M., Usdin, E. Eds.) ,55–73, Macmil-lan, London, UK (1981)
[9] Rein, G., Glover, V., Sandler, M.: Biochem. Pharmacol.,31,1893–1897 (1982)
[10] Pennings, E.J.M., Vrielink, R., Wolters, W.L., van Kempen, G.M.J.: J. Neurochem., 27,915–920 (1976)
[11] Roth, J.A., Rivett, J., Renskers, K. in "Phenolsulfotransferase, Ment. Health Res." (Sandler, M., Usdin, E. Eds.) ,74–85, Macmillan, London, UK (1981)
[12] Sekura, R.D., Duffel, M.W., Jakoby, W.B.: Methods Enzymol.,77,197–206 (1981) (Review)
[13] Weinshilboum, R.M.: Fed. Proc.,45,2223–2228 (1986) (Review)
[14] Foldes, A., Meek, J.L.: Biochim. Biophys. Acta,327,365–374 (1973)
[15] Veronese, M.E., Burgess, W., Zhu, X., McManus, M.E.: Biochem. J.,302,497–502 (1994) (Review)
[16] Whittemore, R.M., Pearce, L.B., Roth, J.A.: Arch. Biochem. Biophys.,249,464–471 (1986)
[17] Falany, C.N., Vazquez, M.E., Heroux, J.A., Roth, J.A.: Arch. Biochem. Biophys., 278,312–318 (1990)
[18] Whittemore, R.M., Pearce, L.B., Roth, J.A.: Biochemistry,24,2477–2482 (1985)

1 NOMENCLATURE

EC number
2.8.2.2

Systematic name
3'-Phosphoadenylylsulfate:alcohol sulfotransferase

Recommended name
Alcohol sulfotransferase

Synonyms
Hydroxysteroid sulfotransferase
Sulfotransferase, 3beta-hydroxy steroid
DELTA^5–3beta-Hydroxysteroid sulfokinase
3beta-Hydroxy steroid sulfotransferase
3-Hydroxysteroid sulfotransferase
HST [3]
5alpha-Androstenol sulfotransferase
Cholesterol sulfotransferase
Dehydroepiandrosterone sulfotransferase
Estrogen sulfokinase
Estrogen sulfotransferase
Steroid alcohol sulfotransferase
Steroid sulfokinase
Steroid sulfotransferase
Sterol sulfokinase
Sterol sulfotransferase
Alcohol/hydroxysteroid sulfotransferase [4]
3beta-Hydroxysteroid sulfotransferase [4]

CAS Reg. No.
9032-76-2

2 REACTION AND SPECIFICITY

Catalyzed reaction
3'-Phosphoadenylylsulfate + an alcohol →
→ adenosine 3',5'-bisphosphate + an alkyl sulfate

Reaction type
Sulfate group transfer

Natural substrates

Substrate spectrum

1 3'-Phosphoadenylylsulfate + dehydroepiandrosterone (i.e. 3beta-hy-droxyandrost-5-en-17-one) [1–7]
2 3'-Phosphoadenylylsulfate + testosterone (i.e. 17beta-hydroxyandrost-4-en-3-one) [1, 2, 5]
3 3'-Phosphoadenylylsulfate + beta-estradiol [1, 2]
4 3'-Phosphoadenylylsulfate + cortisol [1, 2]
5 3'-Phosphoadenylylsulfate + corticosterone [1, 5]
6 3'-Phosphoadenylylsulfate + hydrocortisone [1]
7 3'-Phosphoadenylylsulfate + aldosterone [1, 2]
8 3'-Phosphoadenylylsulfate + 11-deoxycorticosterone [1, 2, 5]
9 3'-Phosphoadenylylsulfate + androst-5-en-3beta,17alpha-diol [1]
10 3'-Phosphoadenylylsulfate + androst-5-en-3beta,17beta-diol [5]
11 3'-Phosphoadenylylsulfate + ascorbic acid [1]
12 3'-Phosphoadenylylsulfate + chlorephedrine [1]
13 3'-Phosphoadenylylsulfate + retinol [1]
14 3'-Phosphoadenylylsulfate + oubain [1]
15 3'-Phosphoadenylylsulfate + 2-propanol [1, 2]
16 3'-Phosphoadenylylsulfate + L-propanol (also active on D-isomer) [1]
17 3'-Phosphoadenylylsulfate + cortisone [2]
18 3'-Phosphoadenylylsulfate + allopregnanolone (i.e. 3alpha-hydroxy-5alpha-pregnan-20-one) [3]
19 3'-Phosphoadenylylsulfate + 17-hydroxypregnenolone (i.e. 3beta,17alpha-dihydroxypregn-4-en-3,20-dione) [3]
20 3'-Phosphoadenylylsulfate + methanol [2]
21 3'-Phosphoadenylylsulfate + ethanol [2]
22 3'-Phosphoadenylylsulfate + 1-propanol [2]
23 3'-Phosphoadenylylsulfate + 1-butanol [2, 5]
24 3'-Phosphoadenylylsulfate + 1-pentanol [2]
25 3'-Phosphoadenylylsulfate + 1-hexanol [2]
26 3'-Phosphoadenylylsulfate + 3-methyl-1-butanol [2]
27 3'-Phosphoadenylylsulfate + epiandrosterone (ie. 3beta-hydroxy-5alpha-androstan-17-one) [5]
28 3'-Phosphoadenylylsulfate + 5alpha-pregnane-3beta,20alpha-diol [5]
29 3'-Phosphoadenylylsulfate + 5alpha-androstane-3beta,17beta-diol [5]
30 3'-Phosphoadenylylsulfate + pregnenolone (i.e. 3beta-hydroxy-5-pregnen-20-one) [3, 5]
31 3'-Phosphoadenylylsulfate + androsterone (i.e. 3alpha-hydroxy-5alpha-androstan-17-one) [3, 5]
32 3'-Phosphoadenylylsulfate + 3beta-hydroxy-5beta-androstan-17-one [5]
33 3'-Phosphoadenylylsulfate + 5alpha-androstane-3alpha,17beta-diol [5]
34 3'-Phosphoadenylylsulfate + 5beta-androstane-3alpha,17beta-diol [5]
35 3'-Phosphoadenylylsulfate + 4-nitrophenol [5]
36 3'-Phosphoadenylylsulfate + 1-naphthol [5]

37 3'-Phosphoadenylylsulfate + 3alpha-hydroxy-5beta-androstan-17-one [5]
38 More (no sulfate acceptors: cholesterol [1, 5], 2-naphthylamine [1, 5], 2-naphthol [1, 5], taurolithocholic acid [1], estrone [1, 5], progesterone [1], N-hydroxy-2-acetylaminofluorene [1], estradiol [5], 11beta-hydroxy-pregn-4-ene-3,20-dione [5], bilirubin [5], diethylstilbestrol [5]) [1, 5]

Product spectrum

1 Adenosine 3',5'-bisphosphate + dehydroepiandrosterone 3-sulfate [2, 7]
2 ?
3 ?
4 ?
5 ?
6 ?
7 ?
8 ?
9 ?
10 ?
11 ?
12 ?
13 ?
14 ?
15 Adenosine 3',5'-bisphosphate + 2-propyl sulfate
16 ?
17 ?
18 ?
19 ?
20 Adenosine 3',5'-bisphosphate + methyl sulfate
21 Adenosine 3',5'-bisphosphate + ethyl sulfate
22 Adenosine 3',5'-bisphosphate + 1-propyl sulfate
23 Adenosine 3',5'-bisphosphate + 1-butyl sulfate
24 Adenosine 3',5'-bisphosphate + 1-pentyl sulfate
25 Adenosine 3',5'-bisphosphate + 1-hexyl sulfate
26 Adenosine 3',5'-bisphosphate + 3-methyl-1-butyl sulfate
27 Adenosine 3',5'-bisphosphate + 5alpha-androstan-17-one 3-sulfate
28 ?
29 ?
30 Adenosine 3',5'-bisphosphate + 5-pregnen-20-one 3-sulfate
31 Adenosine 3',5'-bisphosphate + 5alpha-androstan-17-one 3-sulfate
32 Adenosine 3',5'-bisphosphate + 5beta-androstan-17-one 3-sulfate
33 ?
34 ?
35 Adenosine 3',5'-bisphosphate + 4-nitrophenyl sulfate
36 Adenosine 3',5'-bisphosphate + 1-naphthyl sulfate
37 Adenosine 3',5'-bisphosphate + 5beta-androstan-17-one 3-sulfate
38 ?

Inhibitor(s)

Adenosine 3',5'-bisphosphate (competitive) [1, 2, 5]; ATP (less than 10% inhibition at 10 mM) [2]; ADP (less than 10% inhibition at 10 mM) [2]; AMP (less than 10% inhibition at 10 mM) [2]; 2-Mercaptoethanol [2]; $HgCl_2$ [5]; 2,5-Dichloro-4-nitrophenol [7]; More (substrate inhibition) [5]

Cofactor(s)/prosthetic group(s)/activating agents

Glutathione (activation) [5]

Metal compounds/salts

Fe^{2+} (activation) [5]; Co^{2+} (activation) [5]; Mn^{2+} (activation) [5]

Turnover number (min^{-1})

120 (dehydroepiandrosterone) [2]; 35 (beta-estradiol) [2]; 28 (testosterone) [2]; 27 (1-pentanol) [2]; 25 (3-methyl-1-butanol, 1-hexanol) [2]; 17 (1-butanol) [2]; 13 (2-propanol) [2]; 11 (cortisol, 11-deoxycorticosterone, 1-propanol) [2]; 10 (estriol) [2]; 9 (cortisone, ethanol) [2]; 8 (d-aldosterone) [2]; 5 (methanol) [2]

Specific activity (U/mg)

0.51 (sulfotransferase 1) [1]; 0.18 (sulfotransferase 2) [1]; 0.75 (sulfotransferase 3) [1]; 0.110 (substrate estradiol) [2]; 0.15 (substrate butanol) [2]

K_m-value (mM)

0.00014 (3'-phosphoadenylylsulfate) [7]; 0.000175 (androsterone) [3]; 0.000257 (pregnenolone) [3]; 0.00083 (allopregnanolone) [3]; 0.000897 (17-hydroxypregnenolone) [3]; 0.001 (1-hexanol) [2]; 0.0012 (3-methyl-1-butanol) [1]; 0.0017 (1-pentanol) [2]; 0.002 (dehydroepiandrosterone) [7]; 0.003 (1-butanol) [2]; 0.006 (dehydroepiandrosterone) [5]; 0.012–0.013 (dehydroepiandrosterone [2], 3'-phosphoadenylylsulfate [1, 2, 5], sulfotransferase 1 [1, 2]) [1, 2, 5]; 0.017 (2-propanol) [2]; 0.020 (3'-phosphoadenylylsulfate, sulfotransferase 3) [1]; 0.024 (1-propanol) [2]; 0.035 (beta-estradiol) [2]; 0.042 (ethanol) [2]; 0.047 (3'-phosphoadenylylsulfate, sulfotransferase 2 [1], methanol [2]) [1, 2]; 0.07 (testosterone) [2]; 0.17 (d-aldosterone) [2]; 0.29 (cortisol) [2]; 0.35 (estriol) [2]; 0.44 (cortisone) [2]; 0.52 (11-deoxycorticosterone) [2]

pH-optimum

5.0 (substrate dehydroepiandrosterone) [5]; 5.5 (substrate dehydroepiandrosterone, sulfotransferases 2 and 3) [1]; 5.5–6.0 (steroid substrates) [2]; 6.0 (substrate dehydroepiandrosterone, sulfotransferase 1) [1]; 7.5 (substrate 1-butanol [1, 2], sulfotransferases 1, 2 and 3 [1]) [1, 2]

pH-range

Temperature optimum (°C)

Temperature range (°C)

3 ENZYME STRUCTURE

Molecular weight
120000 (rat, sulfotransferase 3) [1]
180000 (rat, sulfotransferase 1, gel filtration, calculation from Stokes radius, partial specific volume, sedimentation coefficient) [2]
290000 (rat, sulfotransferase 2) [1]

Subunits
? (x × 33760–33765, human, calculation from sequence of cDNA [4, 6], x × 32000, guinea pig, allopregnenolone-specific, i.e. 3beta-specific enzyme, x × 33000, guinea pig, pregnenolone-specific, i.e. 3alpha-specific enzyme, SDS-PAGE [3], x × 28000, rat, sulfotransferase 1, x × 32000, rat, sulfotransferase 2, x × 60000, rat, sulfotransferase 3 [1]) [1, 3, 4, 6]

Glycoprotein/Lipoprotein
–

---

4 ISOLATION/PREPARATION

Source organism
Rat (female [2]) [1, 2, 5]; Guinea pig [3]; Human [4, 5–7]

Source tissue
Liver [1, 2, 4–7]; Adrenal gland [3]

Localization in source
Cytosol [3]; Soluble [2]

Purification
Rat (sulfotransferases 1, 2 and 3 [1], sulfotransferase 1 [2]) [1, 2]; Guinea pig [3]

Crystallization
–

Cloned
[4, 6]

Renatured
–

---

5 STABILITY

pH
6–9 (30 min at 37°C stable) [5]

Temperature (°C)
50 (gradual loss of activity) [2]

Oxidation

Organic solvent

General stability information

Storage
    −80°C, sulfotransferases 1 and 3 stable, sulfotransferase 2 loses 10–20% of
    activity per week [1]; −20°C, several months stable [5]; 4°C, 4 weeks, 60%
    loss of activity [5]

---

6 CROSSREFERENCES TO STRUCTURE DATABANKS

PIR/MIPS code
    PIR2:I60190 (black rat); PIR2:A54026 (guinea pig); PIR2:JC1223 (human);
    PIR2:A33569 (rat); PIR2:I52849 (rat); PIR2:I65760 (rat); PIR2:JC4531 (2
    guinea pig); PIR2:A34822 (a rat)

Brookhaven code

---

7 LITERATURE REFERENCES

[1] Lyon, E.S., Marcus, C.J., Wang, J.-L., Jakoby, W.B.: Methods Enzymol.,77,206–213
    (1981)
[2] Lyon, E.S., Jakoby, W.B.: Arch. Biochem. Biophys.,202,474–481 (1980)
[3] Driscoll, W.J., Martin, B.M., Chen, H.-C., Strott, C.A.: J. Biol. Chem.,268,23496–23503
    (1993)
[4] Kong, A.-N.T., Yang, L., Ma, M., Tao, D., Bjornsson, T.D.: Biochem. Biophys. Res.
    Commun.,187,448–454 (1992)
[5] Ryan, R.A., Carroll, J.: Biochim. Biophys. Acta,429,391–401 (1976)
[6] Comer, K.A., Falany, J.L., Falany, C.N.: Biochem. J.,289,233–240 (1993)
[7] Hernandez, J.S., Watson, R.W.G., Wood, T.C., Weinshilboum, R.M.: Drug Metab.
    Dispos.,20,413–422 (1992)

1 NOMENCLATURE

EC number
2.8.2.3

Systematic name
3'-Phosphoadenylylsulfate:amine N-sulfotransferase

Recommended name
Amine sulfotransferase

Synonyms
Arylamine sulfotransferase
Sulfotransferase, aryl amine
Amine N-sulfotransferase [1]

CAS Reg. No.
9026-08-8

2 REACTION AND SPECIFICITY

Catalyzed reaction
3'-Phosphoadenylylsulfate + an amine →
→ adenosine 3',5'-bisphosphate + a sulfamate

Reaction type
Sulfate group transfer

Natural substrates

Substrate spectrum
1 3'-Phosphoadenylylsulfate + aniline [1]
2 3'-Phosphoadenylylsulfate + 2-naphthylamine (ir [1]) [1, 2]
3 3'-Phosphoadenylylsulfate + cyclohexylamine [1]
4 3'-Phosphoadenylylsulfate + octylamine [1]
5 3'-Phosphoadenylylsulfate + 1,2,3,4-tetrahydroquinoline [1]
6 3'-Phosphoadenylylsulfate + 1,2,3,4-tetrahydroisoquinoline [1]
7 3'-Phosphoadenylylsulfate + desmethylimipramine [1]
8 More (acceptor: primary amines, secondary amines, purified preparation also has O-sulfotransferase activities, suggesting that transfer to oxygen could represent an intrinsic function of N-sulfotransferase (substrates for O-sulfation)) [1]

Product spectrum
1 ?
2 Adenosine 3',5'-bisphosphate + 2-naphthylsulfamate [1]
3 Adenosine 3',5'-bisphosphate + cyclamate [1]
4 ?
5 ?
6 ?
7 ?
8 ?

Inhibitor(s)
2,6-Dichloro-4-nitrophenol [2]

Cofactor(s)/prosthetic group(s)/activating agents
More (activity is dependent on the presence of unprotonated amino groups) [1]

Metal compounds/salts
Mg^{2+} (required [1], optimal stimulation: 20 mM $MgCl_2$) [1]

Turnover number (min^{-1})

Specific activity (U/mg)
0.05 [1]

K_m-value (mM)
0.00013 (3'-phosphoadenylylsulfate, biphasic kinetics, K_m: 0.00013 mM and 0.0022 mM) [2]; 0.0022 (3'-phosphoadenylylsulfate, biphasic kinetics, K_m: 0.00013 mM and 0.0022 mM) [2]; 0.030 (3'-phosphoadenylylsulfate (+ 2-naphthylamine)) [1]; 0.322 (2-naphthylamine) [2]; 14 (1,2,3,4-tetrahydroisoquinoline) [1]; 38 (1,2,3,4-tetrahydroquinoline) [1]; 3400 (2-naphthylamine) [1]

pH-optimum
8.0 (assay at) [1]

pH-range

Temperature optimum (°C)
37 (assay at) [1]

Temperature range (°C)

3 ENZYME STRUCTURE

Molecular weight
60000 (guinea pig, gel filtration) [1]

Subunits
? (x × 33000 + x × 34000, guinea pig, SDS-PAGE) [1]

Glycoprotein/Lipoprotein
–

4 ISOLATION/PREPARATION

Source organism
Guinea pig [1]; Rat [1]; Rabbit [1]; Human [2]

Source tissue
Liver [1, 2]; Intestinal mucosa (rat or rabbit, not guinea pig) [1]; Kidney (rat, not guinea pig or rabbit) [1]

Localization in source
Cytosol [2]

Purification
Guinea pig [1]

Crystallization
–

Cloned
–

Renatured
–

5 STABILITY

pH

Temperature (°C)
More [2]

Oxidation

Organic solvent

General stability information

Storage

6 CROSSREFERENCES TO STRUCTURE DATABANKS

PIR/MIPS code

Brookhaven code

7 LITERATURE REFERENCES

[1] Ramaswamy, S.G., Jakoby, W.B.: J. Biol. Chem.,262,10039–10043 (1987)
[2] Hernandez, J.S., Powers, S.P., Weinshilboum, R.M.: Drug Metab. Dispos.,19,1071–1079 (1991)

1 NOMENCLATURE

EC number
   2.8.2.4

Systematic name
   3'-Phosphoadenylylsulfate:estrone 3-sulfotransferase

Recommended name
   Estrone sulfotransferase

Synonyms
   Sulfotransferase, estrone
   3'-Phosphoadenylyl sulfate-estrone 3-sulfotransferase
   Estrogen sulfotransferase [6–8, 11, 14–20]
   Estrogen sulphotransferase [1–5, 9]
   Oestrogen sulphotransferase [10, 12, 13]
   3'-Phosphoadenylylsulfate:oestrone sulfotransferase [12]

CAS Reg. No.
   9026-06-6

---

2 REACTION AND SPECIFICITY

Catalyzed reaction
   3'-Phosphoadenylylsulfate + estrone →
   → adenosine 3',5'-bisphosphate + estrone 3-sulfate

Reaction type
   Sulfate group transfer

Natural substrates

Substrate spectrum
   1 3'-Phosphoadenylylsulfate + estrone [1, 3–8, 10–13, 17, 19–22]
   2 3'-Phosphoadenylylsulfate + 17beta-estradiol [1–3, 9, 11, 13, 19, 20]
   3 3'-Phosphoadenylylsulfate + 17-deoxyestrone [3]
   4 3'-Phosphoadenylylsulfate + 16-epiestriol (i.e. 1,3,5(10)estra-
     trien-3,16beta,17beta-triol) [3, 7]
   5 3'-Phosphoadenylylsulfate + estriol [3, 13, 19, 20]
   6 3'-Phosphoadenylylsulfate + 3-hydroxy-1,3,5(10)-estratrien-16-one [7]
   7 3'-Phosphoadenylylsulfate + 3,16alpha-dihydroxy-1,3,5(10)-estra-
     trien-17-one [7]
   8 3'-Phosphoadenylylsulfate + 1,3,5(10)-estratrien-3,16alpha-diol [7]
   9 3'-Phosphoadenylylsulfate + 1,3,5(10)-estratrien-3,16alpha,17beta-triol [7]
   10 3'-Phosphoadenylylsulfate + 1,3,5(10)-estratrien-3-ol [7]

11 3'-Phosphoadenylylsulfate + 1,3,5(10)-estratrien-3,17alpha-diol [7]
12 3'-Phosphoadenylylsulfate + 1,3,5(10)-estratrien-3,17beta-diol-17-mono-
    acetate [7]
13 3'-Phosphoadenylylsulfate + 17alpha-ethynyl-1,3,5(10)-estratrien-3,17beta-
    diol [7]
14 3'-Phosphoadenylylsulfate + 1,3,5(10)-estratrien-3,17beta-diol-17-mono-
    valerinate [7]
15 3'-Phosphoadenylylsulfate + 1,3,5(10)-estratrien-3,17beta-diol-17-mono-
    glucosiduronate (very low sulfation rate) [7]
16 3'-Phosphoadenylylsulfate + 3,17beta-dihydroxy-1,3,5(10),9-estratetra-
    en-12-one [7]
17 3'-Phosphoadenylylsulfate + 3-hydroxy-1,3,5(10),9-estratetraen-12,17-
    dione [7]
18 3'-Phosphoadenylylsulfate + 11-hydroxy-1,3,5(10)-estratrien-17-one [7]
19 3'-Phosphoadenylylsulfate + 3,17beta-dihydroxy-1,3,5(10)-estra-
    trien-6-one (high sulfation rate) [7]
20 3'-Phosphoadenylylsulfate + 3-hydroxy-1,3,5(10),7-estratetraen-17-one [7]
21 3'-Phosphoadenylylsulfate + 7alpha-methyl-3-hydroxy-1,3,5(10)-estra-
    trien-17-one [7]
22 3'-Phosphoadenylylsulfate + 7alpha-methyl-1,3,5(10)-estratrien-3,17beta-
    diol [7]
23 3'-Phosphoadenylylsulfate + 3-hydroxy-1,3,5(10),6-estratetraen-17-one [7]
24 3'-Phosphoadenylylsulfate + 3-hydroxy-1,3,5(10),6,8-estra-
    pentaen-17-one [7]
25 3'-Phosphoadenylylsulfate + 4-nitro-1,3,5(10)estratrien-3,17beta-diol
    (overview: similar substrates with functional groups in ring A of estra-
    trienes) [7]
26 More (almost no sulfation of 4-hydroxybenzoic acid esters [7], analogs
    of 3'-adenylylsulfate in sulfation of estrone [6], no sulfation of dehydro-
    epiandrosterone [1, 13], etiocholanolone [13], 2-naphthylamine [1, 13],
    4-nitrophenol [1, 13, 20], 11-deoxycorticosterone [13], 17beta-estradiol
    3-methylether [1, 13], testosterone [13], pregnenolone [13], phenol [1],
    1-naphthol [1], 2-naphthol [1], neutral steroids [20]) [1, 6, 7, 13, 20]

Product spectrum

1 Adenosine 3',5'-bisphosphate + estrone 3-sulfate [11]
2 Adenosine 3',5'-bisphosphate + estradiol 3-sulfate [11]
3 ?
4 ?
5 ?
6 ?
7 ?
8 ?
9 ?
10 ?

11 ?
12 ?
13 ?
14 ?
15 ?
16 ?
17 ?
18 ?
19 ?
20 ?
21 ?
22 ?
23 ?
24 ?
25 ?
26 ?

Inhibitor(s)

DTT (activation up to 0.005 mM, inhibition above) [13]; p-Chloro-mercuribenzoate (inhibits at 1 mM and above) [1]; o-Iodosobenzoate (inhibits at 10 mM and above) [1]; Zn^{2+} [1]; Co^{2+} [1]; Ni^{2+} [1]; ADP [1, 12]; ATP [12]; Retinoic acid [4]; Unsaturated fatty acid of chain length C_{11}-C_{14} [4]; SDS [4]; Cetyltriammonium bromide [4]; Analogs of 3'-phosphoadenylylsulfate [6]; Analogs of estrogen [8]; Analogs of estrogen sulfate [8]; p-Hydroxy-mercuribenzoate [12]; N-Ethylmaleimide [18]; Iodoacetamide [18]; Iodo-acetate [18]

Cofactor(s)/prosthetic group(s)/activating agents

Cysteine (activation) [1, 12]; 2-Mercaptoethanol (activation) [12, 13]; DTT (activation up to 0.005 mM, inhibition above) [13]; Monothioglycerol (stimulation) [20]

Metal compounds/salts

Mg^{2+} (activation) [1, 11, 20]; Ca^{2+} (activation) [1, 11, 20]; Mn^{2+} (activation) [1, 20]; Zn^{2+} (activation) [11]

Turnover number (min^{-1})

Specific activity (U/mg)

0.0028 [1]; 0.01 [13]; More [2, 11]

K_m-value (mM)

0.0027 (estrone) [4]; 0.005 (estriol) [3]; 0.008 (17-epiestriol) [3]; 0.014–0.015 (17beta-estradiol [1, 3], estrone [3, 12], 17-deoxyestrone [3], 16-epiestriol [3]) [1, 3, 12]; 0.037–0.044 (3'-phosphoadenylylsulfate) [6, 12]; 0.07 (3'-phosphoadenylylsulfate (+ 0.1 mM 17beta-estradiol)) [1]; More (values for analogs of 3'-phosphoadenylylsulfate [6], kinetics [12]) [6, 12]

pH-optimum
    6.2 [13]; 8.0 [1, 10]

pH-range

Temperature optimum (°C)

Temperature range (°C)

---

3 ENZYME STRUCTURE

Molecular weight
    50000 (guinea pig, chorion, FPLC) [11]
    52300 (guinea pig, liver, FPLC) [11]
    67000 (bovine, enzyme form A, gel chromatography) [2]
    70000–76000 (bovine, PAGE [9, 13], sedimentation equilibrium centrifuga-
    tion in presence and absence of 6 M guanidine-HCl, sucrose density gradi-
    ent centrifugation at pH 5.8–9.0 [12]) [9, 12, 13]
    191000 (bovine, enzyme form B, gel chromatography) [2]

Subunits
    Monomer (1 × 74000, bovine, SDS-PAGE) [12]
    Dimer (2 × 35000, bovine, SDS-PAGE) [9]
    ? (x × 35161, guinea pig, calculated from sequence of cDNA [16],
    x × 34600, bovine, calculated from sequence of cDNA [22], x × 33000, hu-
    man liver, gel electrophoresis, immunoblotting [21], x × 36000, fetal human
    liver, SDS-PAGE [17], x × 68000, human placenta, SDS-PAGE [19]) [16, 17,
    19, 21, 22]

Glycoprotein/Lipoprotein
    –

---

4 ISOLATION/PREPARATION

Source organism
    Bovine [1–9, 12, 13, 22]; Guinea pig [10, 11, 14, 16, 18, 20, 22]; Human [15,
    17, 19, 21]

Source tissue
    Adrenal gland [1–8, 10, 12, 14, 18]; Placenta [9, 13, 15, 19, 22]; Liver [11,
    17, 21, 22]; Chorion [11, 20, 22]; Uterus [20]

Localization in source
    Cytosol [10, 11, 17, 19]

Purification

Bovine (enzyme form A [1], enzyme forms A and B [2]) [1, 2, 12, 13]; Guinea pig (separation from hydroxysteroid sulfotransferase [18], 2 enzyme forms [22]) [14, 18, 22]; Human [17, 19]

Crystallization

–

Cloned

[15, 16, 22]

Renatured

–

5 STABILITY

pH

Temperature (°C)

Oxidation

Organic solvent

General stability information

Not stabilized by 20% glycerol or 25 mM NaCl [11]; Purified enzyme unstable [19]

Storage

–20°C, presence of 25 mM monothioglycerol, 20% loss of activity in 1 week [11]; 0°C or –20°C, 0.1 M Tris-HCl buffer, pH 7.5 [12]

6 CROSSREFERENCES TO STRUCTURE DATABANKS

PIR/MIPS code

PIR2:S29045 (bovine); PIR2:A41930 (rat)

Brookhaven code

7 LITERATURE REFERENCES

[1] Adams, J.B., Poulos, A.: Biochim. Biophys. Acta,146,493–508 (1967)
[2] Adams, J.B., Chulavatnatol, M.: Biochim. Biophys. Acta,146,509–521 (1967)
[3] Adams, J.B.: Biochim. Biophys. Acta,146,522–528 (1967)
[4] Adams, J.B., Ellyard, R.K.: Biochim. Biophys. Acta,260,724–730 (1972))
[5] Horwitz, J.P., Misra, R.S., Rozhin, J., Neenan, J.P., Huo, A., Godefroi, V.E., Philips, K.D., Chung, H.L., Butke, G., Brooks, S.C.: Biochim. Biophys. Acta,525,364–372 (1978)

[6] Horwitz, J.P., Misra, R.S., Rozhin, J., Helmer, S., Bhuta, A., Brooks, S.C.: Biochim. Biophys. Acta,613,85–94 (1980)

[7] Rozhin, J., Soderstrom, R.L., Brooks, S.C.: J. Biol. Chem.,249,2079–2087 (1974)

[8] Rozhin, J., Huo, A., Zemlicka, J., Brooks, S.C.: J. Biol. Chem.,252,7214–7220 (1977)

[9] Adams, J.B.: Biochim. Biophys. Acta,1076,282–288 (1991)

[10] Hobkirk, R., Glasier, M.A., Brown, L.Y.: Biochem. J.,268,759–764 (1990)

[11] Dick, C.M., Hobkirk, R.: Biochim. Biophys. Acta,925,362–370 (1987)

[12] Adams, J.B., Ellyard, R.K., Low, J.: Biochim. Biophys. Acta,370,160–188 (1974)

[13] Adams, J.B., Low, J.: Biochim. Biophys. Acta,370,189–196 (1974)

[14] Lee, Y.C., Komatsu, K., Driscoll, W.J., Strott, C.: Mol. Endocrinol.,8,1627–1635 (1994)

[15] Bernier, F., Lopez, S.I., Labrie, F., van Luu, T.: Mol. Cell. Endocrinol.,99, R11-R15 (1994)

[16] Oeda, T., Lee, Y.C., Driscoll, W.J., Chen, H.C., Strott, C.A.: Mol. Endocrinol.,6, 1216–1226 (1992)

[17] Hondoh, T., Suzuki, T., Hirato, K., Saitoh, H., Kadofuku, T., Sato, T., Yanahara, T.: Biomed. Res.,14,129–136 (1993)

[18] Glasier, M.A., Glutek, S.M., Hobkirk, R.: Steroids,57,295–300 (1992)

[19] Tseng, L., Lee, L.Y., Mazella, J.: J. Steroid Biochem.,22,611–615 (1985)

[20] Freeman, D.J., Saidi, F., Hobkirk, R.: J. Steroid Biochem.,18,23–27 (1983)

[21] Nash, A.R., Glenn, W.K., Morre, S.S., Kerr, J. Thompson, A.R., Thompson, E.O.P.: Aust. J. Biol. Sci.,41,507–516 (1988)

[22] Hobkirk, R.: J. Steroid Biochem.,29,87–91 (1988)

1 NOMENCLATURE

EC number
2.8.2.5

Systematic name
3'-Phosphoadenylylsulfate:chondroitin 4'-sulfotransferase

Recommended name
Chondroitin 4-sulfotransferase

Synonyms
Sulfotransferase, chondroitin
Chondroitin sulfotransferase
Sulfotransferase, chondroitin 4-
More (not identical with EC 2.8.2.17)

CAS Reg. No.
9026-07-7; 83589-04-2

2 REACTION AND SPECIFICITY

Catalyzed reaction
3'-Phosphoadenylylsulfate + chondroitin →
→ adenosine 3',5'-bisphosphate + chondroitin 4'-sulfate

Reaction type
Sulfate group transfer

Natural substrates
More (involved in biosynthesis of chondroitin sulfate) [1]

Substrate spectrum
1 3'-Phosphoadenylylsulfate + chondroitin [1–6]
2 3'-Phosphoadenylylsulfate + chondroitin-derived oligosaccharides [4, 6]

Product spectrum
1 Adenosine 3',5'-bisphosphate + chondroitin 4'-sulfate [1–6]
2 Adenosine 3',5'-bisphosphate + mixture of 4'-monosulfated oligosaccharides [6]

Inhibitor(s)
ATP [2]; ADP [2]; Heparin [2]; Oversulfated glucosaminoglycan [2]; Detergents (with the exception of Triton X-100 inhibit sulfation in the mast cell system) [3]; 2-(N-Morpholino)ethanesulfonic acid buffer (pH above 6.0) [6]

Cofactor(s)/prosthetic group(s)/activating agents
Protamine (stimulates [1, 2], optimum concentration: 0.075 mg/ml [1], molar ratio of protamine to repeating disaccharide unit of chondroitin is 1:100 [2]) [1, 2]; Histone (stimulates, optimum concentration: 0.5 mg/ml) [1]; Lysozyme (stimulates, optimum concentration: 4.0 mg/ml) [1]; Spermine (stimulates [1, 2], optimum concentration: 0.6 mM [1]) [1, 2]; Spermidine (stimulates, optimum concentration: 4.0 mM) [1]; Dithiothreitol (stimulates, optimum concentration: 2.0 mM) [1]; Glutathione (stimulates, optimum concentration: 5.0 mM) [1]; 2-Mercaptoethanol (stimulates, optimum concentration: 10 mM) [1]; Triton X-100 (increases activity in mast cell system) [4]; Polyamines (stimulate) [1]; Basic proteins (stimulate) [1]; More (stimulation by basic substances is much higher than that by Mn^{2+}, however increasing Mn^{2+} concentration immediately reduces the stimulation by basic substances) [2]

Metal compounds/salts
Mn^{2+} (stimulates [1, 2], optimum concentration: 1–10 mM [1]) [1, 2]

Turnover number (min^{-1})

Specific activity (U/mg)

K_m-value (mM)
More [2, 6]; 0.3 (3'-phosphoadenylylsulfate, presence of protamine) [1]; 0.8 (3'-phosphoadenylylsulfate, addition of spermine) [1]; 1.4 (3'-phosphoadenylylsulfate, addition of Mn^{2+}) [1]

pH-optimum
5.5–7.5 [6]; 6.2 [5]; 6.4 [1]

pH-range
More [1, 5]

Temperature optimum (°C)
37 (assay at) [6]

Temperature range (°C)

---

3 ENZYME STRUCTURE

Molecular weight

Subunits

Glycoprotein/Lipoprotein
–

---

4 ISOLATION/PREPARATION

Source organism
Chicken [1, 2, 5, 6]; Mouse [3, 4]

Source tissue

Chondrocytes (embryo) [5, 6]; Cartilage (embryo [1, 2], epiphyseal [1]) [1, 2]; Eye cornea [2]; Mastocytoma cells [3, 4]; Serum (in hepatitis serum 4-fold more active than in normal) [1]; Synovial fluid [1]

Localization in source

Microsomes [4, 6]; Golgi apparatus (may be secreted to the extracellular space in a soluble form under the culture conditions) [5]

Purification

Crystallization

–

Cloned

–

Renatured

–

---

5 STABILITY

pH

Temperature (°C)

Oxidation

Organic solvent

General stability information

Storage

---

6 CROSSREFERENCES TO STRUCTURE DATABANKS

PIR/MIPS code

Brookhaven code

---

7 LITERATURE REFERENCES

[1] Habuchi, O., Miyashita, N.: Biochim. Biophys. Acta,717,414–421 (1982)
[2] Habuchi, O., Miyata, K.: Biochim. Biophys. Acta,616,208–217 (1980)
[3] Sugumaran, G., Silbert, J.E.: J. Biol. Chem.,263,4673–4678 (1988)
[4] Sugumaran, G., Cogburn, J.N., Silbert, J.E.: J. Biol. Chem.,261,12659–12664 (1986)
[5] Habuchi, O., Tsuzuki, M., Takeuchi, I., Hara, M., Matsui, Y., Ashikari, S.: Biochim. Biophys. Acta,1133,9–16 (1991)
[6] Delfert, D.M., Conrad, H.E.: J. Biol. Chem.,260,14446–14451 (1985)

1 NOMENCLATURE

EC number
2.8.2.6

Systematic name
3'-Phosphoadenylylsulfate:choline sulfotransferase

Recommended name
Choline sulfotransferase

Synonyms
Sulfotransferase, choline
Choline sulphokinase [1]

CAS Reg. No.
9047-23-8

---

2 REACTION AND SPECIFICITY

Catalyzed reaction
3'-Phosphoadenylylsulfate + choline →
→ adenosine 3',5'-bisphosphate + choline sulfate (rapid equilibrium random binding sequence [2])

Reaction type
Sulfate group transfer

Natural substrates

Substrate spectrum
1 3'-Phosphoadenylylsulfate + choline (ir (in vitro) [1], equilibrium lies far towards the direction of choline O-sulfate formation, the equilibrium constant for choline O-sulfate synthesis at pH 7.8 and 26°C is at least 10000 [2]) [1–3]
2 3'-Phosphoadenylylsulfate + N,N-dimethylaminoethanol (35% of the activity with choline [2]) [1, 2]
3 3'-Phosphoadenylylsulfate + dimethylethanolamine [2]
4 3'-Phosphoadenylylsulfate + dimethylethylaminoethanol [1]
5 3'-Phosphoadenylylsulfate + trimethylaminoethanol (2% of the activity with choline) [2]

Product spectrum
    1 Adenosine 3',5'-bisphosphate + choline sulfate [1–3]
    2 ?
    3 ?
    4 ?
    5 ?

Inhibitor(s)
    Trimethylammonium [2]; Neurine [2]; Chlorocholine [2]; Choline O-phos-
    phate [2]; Mercaptoethanol (50 mM, irreversible denaturation) [2]; Pyri-
    dine-2-carbinol [1]; 4-Nitrophenol [1]; 2,4-Dinitrophenol [1]; Prostigmine [1];
    1-Naphthylamine [1]; Thiocholine [1, 2]; Dimethylaminoethanethiol [1];
    n-Propanol (activates, maximal activation at 2.5 mM, inhibition above 25
    mM) [1]; Choline analogues (overview) [1]; Carnitine [1]; Dimethylaminopro-
    pen-1-ol [1]; Diethanolaminopropan-1-ol [1]; Diethanolamine [1]; Triethanol-
    amine [1]; Acetylcholine [1]; Neurine bromide [1]; Hexadecyltrimethylammo-
    nium bromide [1]; Phenol [1]; 3-Aminophenol [1]; 4-Dimethylaminophenol
    [1]; 2'-AMP (weak) [2]; 2',5'-ADP [2]; Adenosine 3',5'-diphosphate (PAP) [2];
    Neurine [2]; Chlorocholine [2]; Mn^{2+} [1, 3]; Ni^{2+} [1]; Fe^{2+} [1]; Zn^{2+} [3]; Ca^{2+}
    [3]; Co^{2+} [1]; Fe^{3+} [1]; PCMB [1]; NEM [1]; Iodoacetate [1]; Cyanide (weak)
    [1]; SO_4^{2-} (weak) [1]; 3'-AMP [2]; Tetramethylammonium [2]; More (if Mg^{2+} is
    replaced by other divalent cations activity is inhibited, inhibition increasing
    in the order $Fe^{2+} < Ca^{2+} < Mn^{2+} < Zn^{2+}$) [3]

Cofactor(s)/prosthetic group(s)/activating agents
    Ethanol (activates) [1]; n-Propanol (activates, maximal activation, at 2.5 mM,
    inhibition above 25 mM) [1]; Glycol (activates) [1]

Metal compounds/salts
    Mg^{2+} (partial requirement [1], activates [3]) [1, 3]

Turnover number (min^{-1})

Specific activity (U/mg)

K_m-value (mM)
    0.0055 (3'-phosphoadenylylsulfate) [3]; 0.012 (3'-phosphoadenylylsulfate)
    [2]; 0.017 (choline) [2]; 0.0222 (3'-phosphoadenylylsulfate) [1]; 0.025 (cho-
    line) [3]; 12 (choline) [1]; 20 (N,N-dimethylethylaminoethanol) [1]; 25
    (N,N-dimethylaminoethanol) [1]

pH-optimum
    7.1–7.2 [2]; 7.3 (assay at) [1]; 7.8 [1]; 9.0 [3]

pH-range
    5.8–10.1 (about 50% of activity maximum at pH 5.8 and 10.1) [1]; 7.6–9.9
    (7.6: about 50% of activity maximum, 9.9: about 70% of activity maximum)
    [3]

Temperature optimum (°C)
20–30 [1]; 37 (assay at) [1, 3]

Temperature range (°C)
20–45 (20–30°C: activity maximum, 45°C: about 25% of activity maximum)
[1]

---

3 ENZYME STRUCTURE

Molecular weight
90000 (Penicillium chrysogenum, gel filtration) [2]

Subunits

Glycoprotein/Lipoprotein
–

---

4 ISOLATION/PREPARATION

Source organism
Limonium sinuatum [3]; Limonium perezii (constitutive enzyme in roots and
leaves of Limonium perezii, the activity is increased at least 4-fold by
salinization with 40% v/v artificial sea water) [3]; Limonium latifolium [3]; Limonium ramosissimum [3]; Limonium nashii [3]; Aspergillus nidulans [1];
Pseudomonas sp. C12B [4]; Penicillium chrysogenum [2]; More (enzyme
activity is very low or absent from species which do not accumulate choline
O-sulfate) [3]

Source tissue
Mycelium [2]; Roots [3]; Leaf [3]; Cell culture [3]

Localization in source

Purification
Aspergillus nidulans (partial) [1]; Penicillium chrysogenum (partial) [2]

Crystallization
–

Cloned
–

Renatured
–

3

5 STABILITY

pH

5.0 (46°C, 10 min, 34% loss of activity, 58.5°C, 10 min, complete loss of activity) [1]; 7–8.8 (stable for at least 15 min) [2]; 7.3 (58.5°C, 10 min, complete loss of activity, 46°C, 10 min, 30% loss of activity) [1]; 8–11 (room temperature, 15 min, stable) [1]; 9.2 (46°C, 10 min, no loss of activity, 58.5°C, 10 min, 15% loss of activity) [1]; 12 (room temperature, 15 min, 80% loss of activity) [1]

Temperature (°C)

22 (room temperature, 15 min, pH 12: 80% loss of activity, pH 8–11, stable) [1]; 46 (10 min, pH 9.2: no loss of activity, pH 7.3: 30% loss of activity, 5.0: 34% loss of activity) [1]; 58.5 (10 min, pH 5.0 or pH 7.3: complete loss of activity, pH 9.2: 15% loss of activity) [1]

Oxidation

Organic solvent

General stability information

Sucrose stabilizes [2]

Storage

Frozen in presence of 25% sucrose, purified enzyme stable [2]

---

6 CROSSREFERENCES TO STRUCTURE DATABANKS

PIR/MIPS code

Brookhaven code

---

7 LITERATURE REFERENCES

[1] Orsi, B.A., Spencer, B.: J. Biochem.,56,81–91 (1964)
[2] Renosto, F., Segel, I.H.: Arch. Biochem. Biophys.,180,416–428 (1977)
[3] Rivoal, J., Hanson, A.D.: Plant Physiol.,106,1187–1193 (1994)
[4] Fitzgerald, J.W., Luschinski, P.C.: Can. J. Microbiol.,23,483–490 (1977)

1 NOMENCLATURE

EC number
2.8.2.7

Systematic name
3'-Phosphoadenylylsulfate:UDP-N-acetyl-D-galactosamine-4-sulfate 6-sulfo-transferase

Recommended name
UDP-N-acetylgalactosamine-4-sulfate sulfotransferase

Synonyms
Sulfotransferase, uridine diphosphoacetylgalactosamine 4-sulfate
Uridine diphospho-N-acetylgalactosamine 4-sulfate sulfotransferase
Uridine diphosphoacetylgalactosamine 4-sulfate sulfotransferase

CAS Reg. No.
37278-32-3

---

2 REACTION AND SPECIFICITY

Catalyzed reaction
3'-Phosphoadenylylsulfate + UDP-N-acetyl-D-galactosamine 4-sulfate →
→ adenosine 3',5'-bisphosphate + UDP-N-acetylgalactosamine 4,6-bissul-fate

Reaction type
Sulfate group transfer

Natural substrates
3'-Phosphoadenylylsulfate + UDP-N-acetylgalactosamine 4-sulfate (compo-nent of microsomal multienzyme system involved in UDP-N-acetylgalactosa-mine 6-sulfate biosynthesis) [2]

Substrate spectrum
1 3'-Phosphoadenylylsulfate + UDP-N-acetylgalactosamine 4-sulfate (i.e. 3'-phosphoadenosine 5'-phosphosulfate or PAPS, catalyzes the transfer of sulfate to position 6 of N-acetylgalactosamine moiety of UDP-N-acetyl-galactosamine sulfate [1]) [1, 2]
2 3'-Phosphoadenylylsulfate + Delta4,5-glucuronido-N-acetylgalactosamine 4-sulfate (poor substrate) [1]
3 3'-Phosphoadenylylsulfate + chondroitin (poor substrate) [1]
4 3'-Phosphoadenylylsulfate + N-acetylgalactosamine 4-sulfate [1]
5 3'-Phosphoadenylylsulfate + N-acetylgalactosamine 1-phosphate 4-sulfate [1]

6 More (no substrates are N-acetylgalactosamine, UDP-N-galactosamine, chondrosin, Delta4,5-glucuronido-N-acetylgalactosamine, N-acetyl-galactosamine 6-sulfate, Delta4,5-glucuronido-N-acetylgalactosamine 6-sulfate, chondroitin sulfate A, B or C, oversulfated chondroitin sulfate from shark cartilage, hyaluronic acid, kerato sulfate or heparin) [1]

Product spectrum
1 Adenosine 3',5'-bisphosphate + UDP-N-acetylgalactosamine-4,6-bissul-fate [1, 2]
2 ?
3 ?
4 ?
5 ?
6 ?

Inhibitor(s)

Cofactor(s)/prosthetic group(s)/activating agents

Metal compounds/salts

Turnover number (min^{-1})

Specific activity (U/mg)
0.0000069 [1]

K_m-value (mM)
0.05 (UDP-N-acetylgalactosamine 4-sulfate) [1]; 0.13 (N-acetylgalactosa-mine 1-phosphate 4-sulfate) [1]; 1.4 (N-acetylgalactosamine 4-sulfate) [1]; 2 (Delta4,5-glucuronido-N-acetylgalactosamine 4-sulfate) [1]

pH-optimum
4.8 [1]

pH-range

Temperature optimum (°C)
37 (assay at) [2]; 38 (assay at) [1]

Temperature range (°C)

3 ENZYME STRUCTURE

Molecular weight

Subunits

Glycoprotein/Lipoprotein
–

4 ISOLATION/PREPARATION

Source organism
Chicken (White Leghorn hen) [1]; Coturnix coturnix (quail) [2]

Source tissue
Oviduct (magnum, i.e. albumen-secreting region, predominantly in tubular glands [2] and isthmus [1], distribution in magnum [2]) [1, 2]

Localization in source
Soluble [1]; Microsomes (predominantly) [2]; More (subcellular distribution) [2]

Purification
Chicken (partial) [1]; Coturnix coturnix (partial) [2]

Crystallization
–

Cloned
–

Renatured
–

5 STABILITY

pH

Temperature (°C)

Oxidation

Organic solvent

General stability information

Storage
–18°C, 1 month [1]

6 CROSSREFERENCES TO STRUCTURE DATABANKS

PIR/MIPS code

Brookhaven code

7 LITERATURE REFERENCES

[1] Harada, T., Shimizu, S., Nakanishi, Y., Suzuki, S.: J. Biol. Chem.,242,2288–2294 (1967)
[2] Otsu, K., Inoue, H., Nakanishi, Y., Kato, S., Tsuji, M., Suzuki, S.: J. Biol. Chem., 259,6403–6410 (1984)

1 NOMENCLATURE

EC number
2.8.2.8

Systematic name
3'-Phosphoadenylylsulfate:N-desulfoheparin N-sulfotransferase

Recommended name
Desulfoheparin sulfotransferase

Synonyms
Sulfotransferase, desulfoheparin
Heparin N-sulfotransferase
3'-Phosphoadenylylsulfate:N-desulfoheparin sulfotransferase [10]
PAPS:N-desulfoheparin sulfotransferase [10]
PAPS:DSH sulfotransferase [10]
More (may be identical with EC 2.8.2.12)

CAS Reg. No.
9026-75-9

---

2 REACTION AND SPECIFICITY

Catalyzed reaction
3'-Phosphoadenylylsulfate + N-desulfoheparin →
→ adenosine 3',5'-bisphosphate + heparin

Reaction type
Sulfate group transfer

Natural substrates
More (N-sulfated residues of heparan sulfate participate in the binding of
this polymer to proteins as basic fibroblast growth factor [4], the sulfation of
the nitrogen of glucosamine in heparan sulfate is an obligatory step for subsequent epimerization of D-glucuronic to L-iduronic acid and of O-sulfation
of the sugar chains [7], may be involved in biosynthesis of heparin and not
of heparan sulfate [8]) [4, 7, 8]

Substrate spectrum
1 3'-Phosphoadenylylsulfate + heparitin [1–3, 7]
2 3'-Phosphoadenylylsulfate + heparan sulfate [2, 4, 6–10]
3 3'-Phosphoadenylylsulfate + N-desulfated heparan sulfate (chemically
de-N-sulfated heparan sulfate is a better substrate than heparan sulfate
[12]) [9, 12]

4 3'-Phosphoadenylylsulfate + N,O-desulfated heparan sulfate [2, 3]

5 3'-Phosphoadenylylsulfate + oligosaccharides derived from
N-desulfoheparan sulfate [3]

6 3'-Phosphoadenylylsulfate + N,O-desulfoheparan sulfate tetrasaccharides with the nonreducing terminus occupied by glucuronic acid (not
iduronic acid) [3]

7 3'-Phosphoadenylylsulfate + N-desulfated heparin (best substrate [7],
much poorer substrate than N-desulfated heparan sulfate [9]) [7, 9–11]

8 3'-Phosphoadenylylsulfate + N-deacetylated K5-polysaccharide (derived
from E. coli K5-derived capsular polysaccharide) [8]

9 3'-Phosphoadenylylsulfate + dermatan sulfate (weak activity [10]) [9, 10]

10 3'-Phosphoadenylylsulfate + chondroitin 4-sulfate (weak activity) [10]

11 3'-Phosphoadenylylsulfate + N-acetylated heparan sulfate [11]

12 More (poor acceptors: N-desulfo-N-acetylheparan [3], heparin [3],
N-desulfoheparin [3], no acceptors: N-acetylated heparan sulfate [7],
N-acetylated heparin [7], chondroitin [7], chondroitin sulfate [7, 9], tyrosine-containing tripeptides [7], heparin [9, 10], hyaluronic acid [9], p-nitrophenol [10]) [3, 7, 9, 10]

Product spectrum

1 Adenosine 3',5'-bisphosphate + N-sulfoheparitin [1]

2 ?

3 Adenosine 3',5'-bisphosphate + heparan sulfate [9, 12]

4 ?

5 ?

6 More (the enzyme transfers sulfate to the 2-amino groups and to the
6-hydroxy groups of glucosamine units of the acceptor substrate, the ratio of the N/O-sulfation ranges between 3:1 and 2:1) [3]

7 ?

8 ?

9 ?

10 ?

11 ?

12 ?

Inhibitor(s)

More (NEM: no effect) [4]; NaCl (above 200 mM [4], 0.125 M [11]) [4, 11];
3',5'-ADP [4]; EDTA [7]; PCMB [10]; Phenylmercuric acetate [10]; Cu^{2+} [10];
Zn^{2+} [10]

Cofactor(s)/prosthetic group(s)/activating agents

Estrogen (enhances activity with N,O-desulfated heparan sulfate as acceptor, progesterone suppresses the effect of estrogen) [2]

Metal compounds/salts

Mn^{2+} (divalent cation required [3, 11], Mn^{2+} most effective [3, 11], maximal activation at: 5 mM [3], 10 mM (4- to 5-fold activation) [11], 62% of the activity with Mg^{2+} [7]) [3, 7, 11]; Mg^{2+} (activates at 10 mM [10], 52% of the activity with Mn^{2+} [3], metal ion required, maximal activity with 5 mM [7], little effect [11]) [3, 7, 10, 11]; Ca^{2+} (41% [3], 22% [7] of the activity with Mn^{2+} [3, 7], can partially replace Mn^{2+} in activation [11]) [3, 7, 11]; More (greatest activity in presence of sodium phosphate buffer of ionic strength of 0.075, in imidazole-HCl reaction rate is lower, maximal activity at ionic strength of 0.125) [10]

Turnover number (min^{-1})

Specific activity (U/mg)

1.95 (heparan sulfate) [7]; More [3]

K_m-**value** (mM)

0.0009 (N-deacetylase K5-polysaccharide, rat liver) [8]; 0.005 (3'-phosphoadenylylsulfate) [7]; 0.01 (3'-phosphoadenylylsulfate); 0.02 (3'-phosphoadenylylsulfate) [11]; 0.0224 (N-deacetylase K5-polysaccharide, MST cells) [8]; 0.0407 (3'-phosphoadenylylsulfate (+ N-deacetylase K5-polysaccharide), MST cells) [8]; 0.108 (3'-phosphoadenylylsulfate (+ N-deacetylase K5-polysaccharide), rat liver) [8]; 1.89 (N-desulfoheparan sulfate, calculated from disaccharide units) [3]; 2.5 (N,O-desulfoheparan sulfate tetrasaccharide, calculated from disaccharide units) [3]

pH-optimum

6.2 [3]; 6.7–7.2 [10]; 7.2 [7]; 7.5 [11]

pH-range

5.7–8.1 (5.7: 33% of activity maximum, 8.1: 54% of activity maximum) [7]

Temperature optimum (°C)

30 (assay at) [3]; 37 (assay at) [10]

Temperature range (°C)

3 ENZYME STRUCTURE

Molecular weight

92000 (rat, radiation inactivation analysis) [6]
97000 (rat, gel filtration) [7]
More (bovine, peaks with enzyme activity: 200000 and 110000 MW) [3]

Subunits

Monomer (1 × 94000, rat, SDS-PAGE) [7]

Glycoprotein/Lipoprotein

Glycoprotein [3, 7]

4 ISOLATION/PREPARATION

Source organism
Mouse [8, 10, 11]; Chicken (hen) [1, 9]; Rabbit [2]; Bovine (calf [3], ox [12]) [3, 12]; Rat (overexpressed in CHO cells [4, 5], a single protein possesses both N-deacetylase and N-sulfotransferase activity [4, 6]) [4–7]

Source tissue
Oviduct [1]; Uterus (endometrium [2]) [2, 9]; Mastocytoma cells (MST cells [8]) [8, 11]; Arterial tissue [3]; Liver [4–7]; Lung [12]; Furth mouse mast cell tumor [10]

Localization in source
Golgi vesicles (lumen [4], membrane [7]) [4, 6, 7]; Membranes [7]; Microsomes [11]; More (associated with postmicrosomal fraction) [10]

Purification
Rat [7]; Bovine (N-desulfo-N-acetylheparan sulfate deacetylase activity co-purifies) [3]; Mouse [10]

Crystallization
–

Cloned
[4–6, 8]

Renatured
–

5 STABILITY

pH

Temperature (°C)
40 (2 min, no effect) [10]; 50 (1 min, 15% loss of activity) [11]; 55 (2 min, 85% loss of activity) [10]; 70 (2 min, 87% loss of activity) [10]; 85 (2 min, complete loss of activity) [10]

Oxidation

Organic solvent

General stability information

Storage
–18°C, 20 h, stable [1]; –18°C, stable for at least 4 months [10]; –70°C, stable for at least 6 months in presence of 20% glycerol [7]

6 CROSSREFERENCES TO STRUCTURE DATABANKS

PIR/MIPS code

Brookhaven code

7 LITERATURE REFERENCES

[1] Suzuki, S., Trenn, R.H., Strominger, J.L.: Biochim. Biophys. Acta,50,169–170 (1961)

[2] Hiroshi, M., Isemura, M., Yosizawa, Z.: Int. J. Biochem.,17,1077–1083 (1985)

[3] Göhler, D., Niemann, R., Buddecke, E.: Eur. J. Biochem.,138,301–308 (1984)

[4] Wei, Z., Swiedler, S.J., Ishihara, M., Orellana, A., Hirschberg, C.B.: Proc. Natl. Acad. Sci. USA,90,3885–3888 (1993)

[5] Hashimoto, Y., Orellana, A., Gil, G., Hirschberg, C.B.: J. Biol. Chem.,267, 15744–15750 (1992)

[6] Mandon, E., Kempner, E.S., Ishihara, M., Hirschberg,.C.B.: J. Biol. Chem.,269, 11729–11733 (1994)

[7] Brandan, E., Hirschberg, C.B.: J. Biol. Chem.,263,2417–2422 (1988)

[8] Orellana, A., Hirschberg, C.B., Wei, Z., Swiedler, S.J., Ishihara, M.: J. Biol. Chem., 269,2270–2276 (1994)

[9] Johnson, A.H., Baker, J.R.: Biochim. Biophys. Acta,320,341–351 (1973)

[10] Eisenman, R.A., Balasubramanian, A.S., Marx, W.: Arch. Biochem. Biophys.,119, 387–397 (1967)

[11] Jansson, L., Höök, M., Wasteson, A., Lindahl, U.: Biochem. J.,149,49–55 (1975)

[12] Foley, T., Baker, J.R.: Biochem. J.,124,25P-26P (1971)

1 NOMENCLATURE

EC number
2.8.2.9

Systematic name
3'-Phosphoadenylylsulfate:L-tyrosine-methyl-ester sulfotransferase

Recommended name
Tyrosine-ester sulfotransferase

Synonyms
Sulfotransferase, tyrosine ester
Aryl sulfotransferase IV
L-Tyrosine methyl ester sulfotransferase
Tyrosine ester sulfotransferase
More (cf. EC 2.8.2.1)

CAS Reg. No.
9055-56-5

---

2 REACTION AND SPECIFICITY

Catalyzed reaction
3'-Phosphoadenylylsulfate + L-tyrosine methyl ester →
→ adenosine 3',5'-bisphosphate + L-tyrosine methyl ester 4-sulfate (random rapid equilibrium bi bi kinetic mechanism [2, 4])

Reaction type
Sulfate group transfer

Natural substrates
More (it appears unlikely that the enzyme is involved in the biosynthesis of proteins containing L-tyrosine O-sulfate residues, it is probably important in sulfation of physiologically active amines) [6]

Substrate spectrum
1 3'-Phosphoadenylylsulfate + L-tyrosine methyl ester [1–4, 6, 8, 9]
2 3'-Phosphoadenylylsulfate + 2-naphthol [1, 9]
3 3'-Phosphoadenylylsulfate + phenol [1, 9]
4 3'-Phosphoadenylylsulfate + chlorophenol (3-chlorophenol [1, 8], 4-chlorophenol [1, 8, 9], 2-chlorophenol [8]) [1, 8, 9]
5 3'-Phosphoadenylylsulfate + methylphenol (3-methylphenol or 4-methyl-phenol) [1]

---

6 3'-Phosphoadenylylsulfate + nitrophenol (3-nitrophenol or 4-nitrophenol)
   [1, 8]
7 3'-Phosphoadenylylsulfate + 4-methoxyphenol [1, 9]
8 3'-Phosphoadenylylsulfate + 3-methoxy-4-hydroxyphenylglycol [1]
9 3'-Phosphoadenylylsulfate + N-acetylserotonin [1]
10 3'-Phosphoadenylylsulfate + 5-hydroxytryptophol [1]
11 3'-Phosphoadenylylsulfate + 2-cyanoethyl-N-hydroxythioacetamide [1]
12 3'-Phosphoadenylylsulfate + epinephrine [1, 9]
13 3'-Phosphoadenylylsulfate + tyramine [1, 3, 4, 6, 9]
14 3'-Phosphoadenylylsulfate + dopamine [1]
15 3'-Phosphoadenylylsulfate + 2-chloro-4-nitrophenol [2]
16 3'-Phosphoadenylylsulfate + N-acetyl-L-tyrosine ethyl ester [4]
17 3'-Phosphoadenylylsulfate + L-tyrosine amide [6, 8]
18 3'-Phosphoadenylylsulfate + 5-hydroxytryptamine [6]
19 3'-Phosphoadenylylsulfate + tert-butoxycarbonyl-Asp-Tyr [7]
20 3'-Phosphoadenylylsulfate + tert-butoxycarbonyl-Asp-Tyr-Met [7]
21 3'-Phosphoadenylylsulfate + tert-butoxycarbonyl-Asp-Tyr-Met-Gly [7]
22 3'-Phosphoadenylylsulfate + tert-butoxycarbonyl-Asp-Tyr-Met-Gly-Trp [7]
23 3'-Phosphoadenylylsulfate + tert-butoxycarbonyl-Asp-Tyr-Met-Gly-Trp-Met
   [7]
24 3'-Phosphoadenylylsulfate + tert-butoxycarbonyl-
   Asp-Tyr-Met-Gly-Trp-Met-Asp-Phe-NH$_2$ (i.e. Boc-CCK-8, ns) [7]
25 3'-Phosphoadenylylsulfate + cholecystokinin (nonsulfated) [7]
26 3'-Phosphoadenylylsulfate + tert-butoxycarbonyl-
   Asp-Arg-Asp-Tyr-Met-Gly [7]
27 3'-Phosphoadenylylsulfate + Asp-Arg-Asp-Tyr-Met-Gly [7]
28 3'-Phosphoadenylylsulfate + caerulein (nonsulfated) [7]
29 3'-Phosphoadenylylsulfate + L-tyrosine [8]
30 3'-Phosphoadenylylsulfate + L-tyrosine ethyl ester [8]
31 3'-Phosphoadenylylsulfate + L-tyrosine allyl ester [8]
32 3'-Phosphoadenylylsulfate + L-tyrosine tert-butyl ester [8]
33 3'-Phosphoadenylylsulfate + L-tyrosine benzyl ester [8]
34 3'-Phosphoadenylylsulfate + L-p-hydroxyphenylglycine [8]
35 3'-Phosphoadenylylsulfate + glycyltyrosine [8]
36 3'-Phosphoadenylylsulfate + tyrosylglycine [8]
37 3'-Phosphoadenylylsulfate + alanyltyrosine [8]
38 3'-Phosphoadenylylsulfate + 2-cyanoethyl-N-hydroxythioacetimidate [9]
39 More (enzyme is specific for substrate molecules with a free and unprotonated amino group and an unionized hydroxyl group [4], not: adenosine 5'-phosphosulfate [8]) [4, 8]

Product spectrum

1 Adenosine 3',5'-bisphosphate + L-tyrosine methyl ester 4-sulfate
2 Adenosine 3',5'-bisphosphate + 2-naphthyl sulfate
3 Adenosine 3',5'-bisphosphate + phenyl sulfate
4 Adenosine 3',5'-bisphosphate + chlorophenyl sulfate
5 Adenosine 3',5'-bisphosphate + methylphenyl sulfate
6 Adenosine 3',5'-bisphosphate + nitrophenyl sulfate
7 Adenosine 3',5'-bisphosphate + 4-methoxyphenyl sulfate
8 Adenosine 3',5'-bisphosphate + ?
9 Adenosine 3',5'-bisphosphate + ?
10 Adenosine 3',5'-bisphosphate + ?
11 Adenosine 3',5'-bisphosphate + ?
12 Adenosine 3',5'-bisphosphate + ?
13 Adenosine 3',5'-bisphosphate + ?
14 Adenosine 3',5'-bisphosphate + ?
15 Adenosine 3',5'-bisphosphate + 2-chloro-4-nitrophenyl sulfate [2]
16 Adenosine 3',5'-bisphosphate + ?
17 Adenosine 3',5'-bisphosphate + ?
18 Adenosine 3',5'-bisphosphate + ?
19 Adenosine 3',5'-bisphosphate + ?
20 Adenosine 3',5'-bisphosphate + ?
21 Adenosine 3',5'-bisphosphate + ?
22 Adenosine 3',5'-bisphosphate + ?
23 Adenosine 3',5'-bisphosphate + ?
24 Adenosine 3',5'-bisphosphate + ?
25 Adenosine 3',5'-bisphosphate + ?
26 Adenosine 3',5'-bisphosphate + ?
27 Adenosine 3',5'-bisphosphate + ?
28 Adenosine 3',5'-bisphosphate + ?
29 Adenosine 3',5'-bisphosphate + tyrosine O^4-sulfate [8]
30 Adenosine 3',5'-bisphosphate + tyrosine ethyl ester 4-sulfate
31 Adenosine 3',5'-bisphosphate + tyrosine allyl ester 4-sulfate
32 Adenosine 3',5'-bisphosphate + tyrosine tert-butyl ester 4-sulfte
33 Adenosine 3',5'-bisphosphate + tyrosine benzyl ester 4-sulfate
34 Adenosine 3',5'-bisphosphate + ?
35 Adenosine 3',5'-bisphosphate + ?
36 Adenosine 3',5'-bisphosphate + ?
37 Adenosine 3',5'-bisphosphate + ?
38 Adenosine 3',5'-bisphosphate + ?
39 ?

Inhibitor(s)

Adenosine 3',5'-bisphosphate (PAP, competitive to 3'-phosphoadenylylsulfate [8]) [8, 9]; 3'-Phosphoadenylylsulfate [2]; Adenosine 5'-triphosphate [7]; 2-Chloro-4-nitrophenyl sulfate (product inhibition) [2]; 2-Chloro-4-nitrophenol [2]; o-Iodosobenzoate [3]; PCMB [3, 7]; NEM [3]; Iodoacetate [3]; Iodoacetamide (pH 7.5, 30°C, inhibition prevented by presence of 3'-phosphate 5'-sulfatophosphate but not by L-tyrosine methyl ester) [4]; 2'-O-[(R)-Formyl(adenin-9-yl)]-(S)-glyceraldehyde 3'-triphosphate (inhibition prevented by either adenosine 3'5'-diphosphate or 3'-phosphoadenylylsulfate, binds to Lys65 and Cys66) [5]; 2,6-Dichloro-4-nitrophenol [7, 9]; EDTA [7]; Triton X-100 [7]; $FeCl_3$ [7]; $CuCl_2$ [7]; $ZnCl_2$ [7]; $MgCl_2$ [7]; $CaCl_2$ [7]; V_2O_5 [7]; $CoCl_2$ [7]; Tyrosine O^4-sulfate (uncompetitive to L-tyrosine) [8]; Thiol reagents [9]; Aryl sulfates [9]; Pentachlorophenol [9]

Cofactor(s)/prosthetic group(s)/activating agents

Dithiothreitol (activates, degree of activation is more marked with preparations previously stored at 0°C or –10°C) [3]; 2-Mercaptoethanol (activates, degree of activation is more marked with preparations previously stored at 0°C or–10°C) [3]; Glutathione (activates, degree of activation is more marked with preparations previously stored at 0°C or–10°C) [3]

Metal compounds/salts

$MnCl_2$ (activates) [7]

Turnover number (min^{-1})

1.2 (4-methylphenol) [1]; 1.4 (2-cyanoethyl-N-hydroxythioacetamide) [1]; 1.8 (phenol) [1, 9]; 2.8 (tyramine) [1, 9]; 4.2 (dopamine) [1]; 7.3 (epinephrine) [1, 9]; 8.4 (5-hydroxytryptophol) [1]; 11 (3-methylphenol, 3-methoxy-4-hydroxyphenolglycol) [1]; 14 (tyrosine methyl ester [1, 9], 2-cyanoethyl-N-hydroxythioacetimidate [9]) [1, 9]; 18 (N-acetylserotonin [1], 4-chlorophenol [9]) [1, 9]; 19 (4-chlorophenol) [1]; 27 (4-methoxyphenol [1, 9], 3-chlorophenol [1]) [1, 9]; 31 (4-nitrophenol) [1]; 42 (2-naphthol) [1, 9]; 48 (3-nitrophenol) [1]

Specific activity (U/mg)

0.20533 [8]; 0.55 [1]; More [3]

K_m-value (mM)

0.023 (3'-phosphoadenylylsulfate (+ tyrosine methyl ester)) 0.024 (3'-phosphoadenylylsulfate) [9]; 0.033 (L-tyrosine) [8]; 0.08 (L-tyrosine tert-butyl ester) [8]; 0.1 (2-naphthol [1, 9], L-tyrosine ethyl ester [8]) [1, 8, 9]; 0.133 (L-tyrosine benzyl ester) [8]; 0.15 (L-tyrosine methyl ester) [8]; 0.16 (dopamine) [1]; 0.17 (4-nitrophenol) [1]; 0.20 (3-methoxy-4-hydroxyphenylglycol [1], L-tyrosinamide [8]) [1, 8]; 0.21 (5-hydroxytryptophol) [1]; 0.34 (4-chlorophenol) [1, 9]; 0.35 (alanyltyrosine) [8]; 0.37 (L-tyrosine ethyl ester) [8]; 0.39–0.4 (m-chlorophenol) [1, 8]; 0.416 (3-nitrophenol) [8]; 0.43 (epinephrine) [1, 9]; 0.44 (3-nitrophenol) [1]; 0.46 (tyramine) [1, 9]; 0.5 (2-chloro-phenol) [8]; 0.55 (4-chlorophenol) [8]; 0.714 (L-4-hydroxyphenylglycine) [8]; 0.833 (tyrosylglycine) [8]; 0.91 (tyrosine methyl ester) [1, 9]; 0.92 (phenol) [1, 9]; 1.2 (4-methylphenol) [1]; 1.25 (4-nitrophenol) [8]; 1.4 (4-methoxyphe-nol) [1, 9]; 1.5 (N-acetylserotonin) [1]; 2.5 (glycyltyrosine) [8]; 6.9 (2-cyano-ethyl-N-hydroxythioacetamide) [1, 9]; More [3, 4, 7]

pH-optimum
5.5 (2-naphthol, no formation of sulfate esters from tyrosine methyl ester, epinephrine, octopamine, tyramine, serotonin, N-acetylserotonin [9], 2-naph-thol, sodium acetate buffer [1]) [1, 9]; 5.8 [7]; 7.0 [8]

pH-range
More [7, 8]

Temperature optimum (°C)
37 (assay at) [1]

Temperature range (°C)

---

3 ENZYME STRUCTURE

Molecular weight
26000 (Euglena gracilis, gel filtration) [8]
61000 (rat, gel filtration) [1, 9]

Subunits
Monomer (1 × 26000, Euglena gracilis, SDS-PAGE) [8]
Dimer (2 × 33500, rat, SDS-PAGE) [1, 9]

Glycoprotein/Lipoprotein
–

---

4 ISOLATION/PREPARATION

Source organism
Rat (male [2], cloned and expressed in E. coli [5]) [1–7, 9]; Euglena gracilis var. bacillaris (W10BSmL, aplastidic mutant) [8]

Source tissue
Liver [1–6, 9]; Brain (cerebral cortex) [7]

Localization in source
Microsomes [7]

Purification
Rat (male [2], partial [3], cloned and expressed in E. coli [5], aryl sulfo-transferase IV [9]) [1–3, 5, 6, 9]; Euglena gracilis var. bacillaris [8]

Crystallization
–

Cloned
(rat enzyme expressed in E. coli) [5]

Renatured
–

5 STABILITY

pH

Temperature (°C)

Oxidation

Organic solvent

General stability information
Instable to freezing [8]

Storage
4°C, 10 mM potassium phosphate, pH 6.8, 250 mM sucrose and 0.02% w/v NaN$_3$, no more than 10% loss of activity after 5 months [8]; 4°C, sodium phosphate buffer, pH 7.0, 0.25 M sucrose, 5 mM mercaptoethanol, 3 mM NaN$_3$, 5% loss of activity per week at protein concentration between 0.5 and 1 mg/ml [9]

6 CROSSREFERENCES TO STRUCTURE DATABANKS

PIR/MIPS code

Brookhaven code

7 LITERATURE REFERENCES

[1] Sekura, R.D., Jakoby, W.B.: Arch. Biochem. Biophys.,211,352–359 (1981)
[2] Duffel, M.W., Jakoby, W.B.: J. Biol. Chem.,256,11123–11127 (1981)
[3] Mattock, P., Jones, J.G.: Biochem. J.,116,797–803 (1970)
[4] Mattock, P., Barford, D.J., Basford, J.M., Jones, J.G.: Biochem. J.,116,805–810 (1970)
[5] Zheng, Y., Bergold, A., Duffel, M.W.: J. Biol. Chem.,269,30313–30319 (1994)
[6] Barford, D.J., Jones, J.G.: Biochem. J.,125,76P-77P (1971)
[7] Vargas, F., Frerot, O., Tuong, M.D.T., Schwartz, J.C.: Biochemistry,24,5938–5943 (1985)
[8] Saidha, T., Schiff, J.A.: Biochem. J.,298,45–50 (1994)
[9] Sekura, R.D., Duffel, M.W., Jakoby, W.B.: Methods Enzymol.,77,197–206 (1981) (Review)

1 NOMENCLATURE

EC number
2.8.2.10

Systematic name
3'-Phosphoadenylylsulfate:Renilla luciferin sulfotransferase

Recommended name
Renilla-luciferin sulfotransferase

Synonyms
Luciferin sulfotransferase
Luciferin sulfokinase
Sulfotransferase, luciferin
Luciferin sulfokinase (3'-phosphoadenylyl sulfate:luciferin sulfotransferase)
[1]

CAS Reg. No.
37278-33-4

2 REACTION AND SPECIFICITY

Catalyzed reaction
3'-Phosphoadenylylsulfate + Renilla luciferin →
→ adenosine 3',5'-bisphosphate + luciferyl sulfate

Reaction type
Sulfate group transfer

Natural substrates

Substrate spectrum
1 Luciferyl sulfate + adenosine 3',5'-bisphosphate [1, 2]

Product spectrum
1 Luciferin + 3'-phosphoadenylylsulfate [1]

Inhibitor(s)

Cofactor(s)/prosthetic group(s)/activating agents

Metal compounds/salts

Turnover number (min^{-1})

Specific activity (U/mg)

K_m-value (mM)

pH-optimum
    7.5 (assay at) [1]; 7.6 (assay at) [2]

pH-range

Temperature optimum (°C)
    25 (assay at) [2]; 30 (assay at) [1]

Temperature range (°C)

---

3 ENZYME STRUCTURE

Molecular weight

Subunits

Glycoprotein/Lipoprotein
    –

---

4 ISOLATION/PREPARATION

Source organism
    Renilla reniformis [1, 2]

Source tissue
    Cell [1, 2]

Localization in source

Purification
    Renilla reniformis (partial) [1, 2]

Crystallization
    –

Cloned
    –

Renatured
    –

5 STABILITY

pH

Temperature (°C)

Oxidation

Organic solvent

General stability information
Enzyme is labile, stabilization by presence of 0.05 ml/l beta-mercaptoethanol [2]

Storage
Requirement of beta-mercaptoethanol in storage buffer [2]; –80°C, stable for
at least several months [2]

6 CROSSREFERENCES TO STRUCTURE DATABANKS

PIR/MIPS code

Brookhaven code

7 LITERATURE REFERENCES

[1] Cormier, M.J., Hori, K., Karkhanis, Y.D.: Biochemistry,9,1184–1189 (1970)
[2] Anderson, J.M., Hori, K., Cormier, M.J.: Methods Enzymol.,57,244–257 (1978)

1 NOMENCLATURE

EC number
2.8.2.11

Systematic name
3'-Phosphoadenylylsulfate:galactosylceramide 3'-sulfotransferase

Recommended name
Galactosylceramide sulfotransferase

Synonyms
GSase [9]
Sulfotransferase, galactocerebroside
3'-Phosphoadenosine-5'-phosphosulfate-cerebroside sulfotransferase
Galactocerebroside sulfotransferase
Galactolipid sulfotransferase
Glycolipid sulfotransferase
Glycosphingolipid sulfotransferase

CAS Reg. No.
9081-06-5

---

2 REACTION AND SPECIFICITY

Catalyzed reaction
3'-Phosphoadenylylsulfate + a galactosylceramide →
→ adenosine 3',5'-bisphosphate + a galactosylceramidesulfate

Reaction type
Sulfate group transfer

Natural substrates
Galactocerebroside + 3'-phosphoadenylylsulfate (the product, sulfatides, are important components of myelin [5]) [1, 5]
More (enzyme regulates the sulfation of glycolipids [10], enzyme is responsible for the synthesis of the major mammalian testicular glycolipid, sulfogalactosylglycerol and is an early marker of differentiation during spermatogenesis [6, 8], enzyme may be regulated by a phosphorylation mechanism [8]) [6, 8, 10]

Substrate spectrum

1 Galactoglycerolipid + 3'-phosphoadenylylsulfate [8]
2 Galactocerebroside + 3'-phosphoadenylylsulfate [1, 3, 5]
3 Galactosylceramide + 3'-phosphoadenylylsulfate (biotinylated and light-sensitive azido derivatives of lysogalactosylceramide are synthesized by the crude enzyme preparation, these derivatives remain effective substrates for the testicular enzyme) [2, 10, 11]

Product spectrum

1 Sulfogalactoglycerolipid + adenosine 3',5'-bisphosphate [8]
2 Sulfatide + adenosine 3',5'-bisphosphate [1]
3 ?

Inhibitor(s)

Procion Red [7]; 1-Amino-4-bromoanthroquinone 2-sulfonic acid [7]; 1-Anilinonaphthalene 8-sulphonic acid [7]; Cibacron Blue F3GA [7]; Congo Red [7]; 2,4,5,7-Tetraiodofluorescein [7]; Orange A [7]; Green A [7]; Triazine aromatic dyes [7]; Mg^{2+} (1–10 mM: enhances activity, 20–35 mM: inhibition) [10]; Mn^{2+} (1–10 mM: enhances activity, 20–35 mM: inhibition) [10]; 3'-Phosphoadenosine 5'-phosphate [10]; 2-Deoxy-3'-phosphoadenosine 5'-phosphate [10]; Guanosine 3',5'-diphosphate [10]; ADP [11]; More (not: 2'-phosphoadenosine 5'-phosphate, fluorosulfonylbenzoyladenosine) [10]

Cofactor(s)/prosthetic group(s)/activating agents

More (a testicular protein kinase activity is capable of stimulating the activity in vitro [6], purification of an activator protein (MW 22000, glycoprotein, trimer: 3 x 8000, from human liver) [9]) [6, 9]; ATP (up to 4 mM stimulates [11], in vitro dependent on [8]) [8, 11]; Vitamin K1 (+ phosphate, activates purified preparation) [5]

Metal compounds/salts

Mg^{2+} (1–10 mM: enhances activity, 20–35 mM: inhibition [10], $MgCl_2$ stimulates [11]) [10, 11]; Ca^{2+} (1–10 mM and 20–35 mM: enhancement) [10]; Mn^{2+} (1–10 mM: enhances activity, 20–35 mM: inhibition [10], $MnCl_2$ stimulates [11]) [10, 11]; NaCl (stimulates) [11]; KCl (stimulates) [11]

Turnover number (min^{-1})

Specific activity (U/mg)

More (in vitro assay [11]) [5, 10, 11]

K_m-value (mM)

0.0009 (3'-phosphoadenylylsulfate) [3]; 0.0012 (3'-phosphoadenylylsulfate) [5]; 0.002 (galactocerebroside) [3]; 0.0052 (3'-phosphoadenylylsulfate) [10]; 0.026 (cerebroside) [5]

pH-optimum
5.8–6.0 [10]; 6.5 [11]; 7.0 [5]; 8.0 [8]

pH-range

Temperature optimum (°C)
30 (assay at) [11]

Temperature range (°C)

3 ENZYME STRUCTURE

Molecular weight
28000 (mouse, gel filtration) [5]
64000 (rat, nondenaturing PAGE) [10]

Subunits
Monomer (1 × 31000, mouse, SDS-PAGE) [5]

Glycoprotein/Lipoprotein
More (purified enzyme contains bound lipids, consisting primarily of cholesterol and phosphatidylcholine, removal of associated lipids results in loss of
activity) [10]

4 ISOLATION/PREPARATION

Source organism
Rat [1–4, 6–8, 10]; Mouse [5, 11]; Human [9]

Source tissue
Testis [2, 6, 8]; Brain [1–5, 7, 11]; Kidney [2, 10]; Liver [9]

Localization in source
Microsomes [1]

Purification
Rat [3, 4, 10]; Mouse [5]

Crystallization
–

Cloned
–

Renatured
–

5 STABILITY

pH

Temperature (°C)
60 (10 min, complete loss of activity) [10]

Oxidation

Organic solvent

General stability information
Ethylene glycol, 2-mercaptoethanol, 10 mM ATP and a mixture of phosphatidylcholine and phosphatidylethanolamine ratio 1:1 does not improve stability at –80°C [5]; ATP and Triton X-100 stabilize in vitro [8]

Storage
–80°C, vitamin K + phosphate activate purified preparation, stable for 14 days [5]; –20°C, 50% glycerol, 0.1% Triton X-100, pH 7.0–7.5, stable for more than 4 weeks [10]

6 CROSSREFERENCES TO STRUCTURE DATABANKS

PIR/MIPS code

Brookhaven code

7 LITERATURE REFERENCES

[1] McKhann, G.M., Levy, R., Ho, W.: Biochem. Biophys. Res. Commun.,20,109–113 (1965)
[2] Lingwood, C., Taylor, T.: Biochem. Cell Biol.,64,631–637 (1986)
[3] Tennekoon, G., McKhann, G.M.: J. Neurochem.,31,329–339 (1978)
[4] Sarlieve, L.L., Neskovic, N.M., Rebel, G., Mandel, P.: J. Neurochem.,26,211–215 (1976)
[5] Sundaram, K.S., Lev, M.: J. Biol. Chem.,267,24041–24044 (1992)
[6] Sakac, D., Lingwood, C.A.: Biochem. J.,261,423–429 (1989)
[7] Zaruba, M., Hilt, D., Tennekoon, G.: Biochem. Biophys. Res. Commun.,129,522–529 (1985)
[8] Taylor, T., Oda, K., Lingwood, C.: Biochim. Biophys. Acta,913,131–137 (1987)
[9] Mitsuyama, T., Gasa, S., Nojima, T., Taniguchi, N., Makita, A.: J. Biochem., 98,605–613 (1985)
[10] Tennekoon, G., Aitchinson, S., Zabura, M.: Arch. Biochem. Biophys.,240,932–944 (1985)
[11] Burkart, T., Siegrist, H.P., Herschkowitz, N.N., Wiesmann, U.N.: Biochim. Biophys. Acta,483,303–311 (1977)

1 NOMENCLATURE

EC number
2.8.2.12

Systematic name
3'-Phosphoadenylylsulfate:heparitin N-sulfotransferase

Recommended name
Heparitin sulfotransferase

Synonyms
N-HSST [4]
N-Heparan sulfate sulfotransferase [5]
Heparan sulfate N-deacetylase/N-sulfotransferase [4, 6]
Sulfotransferase, heparitin
Heparan sulfate 2-N-sulfotransferase
Heparan sulfate N-sulfotransferase
Heparan sulfate sulfotransferase
More (may be identical with EC 2.8.2.8)

CAS Reg. No.
37378-34-5

---

2 REACTION AND SPECIFICITY

Catalyzed reaction
3'-Phosphoadenylylsulfate + heparitin →
→ adenosine 3',5'-bisphosphate + N-sulfoheparitin

Reaction type
Sulfate group transfer

Natural substrates
More (N-sulfated residues of heparan sulfate participate in the binding of this polymer to proteins as basic fibroblast growth factor [4], the sulfation of the nitrogen of glucosamine in heparan sulfate is an obligatory step for subsequent epimerization of D-glucuronic to L-iduronic acid and of O-sulfation of the sugar chains [7], may be involved in biosynthesis of heparin and not of heparan sulfate [8]) [4, 7, 8]

Substrate spectrum

1 3'-Phosphoadenylylsulfate + heparitin [1–3, 7]
2 3'-Phosphoadenylylsulfate + heparan sulfate [2, 4, 6–10]
3 3'-Phosphoadenylylsulfate + N-desulfated heparan sulfate (chemically de-N-sulfated heparan sulfate is a better substrate than heparan sulfate [12]) [9, 12]
4 3'-Phosphoadenylylsulfate + N-,O-desulfated heparan sulfate [2, 3]
5 3'-Phosphoadenylylsulfate + oligosaccharides derived from N-desulfoheparan sulfate [3]
6 3'-Phosphoadenylylsulfate + N,O-desulfoheparan sulfate tetrasaccharides with the nonreducing terminus occupied by glucuronic acid (not iduronic acid) [3]
7 3'-Phosphoadenylylsulfate + N-desulfated heparin (best substrate [7], much poorer substrate than N-desulfated heparan sulfate [9]) [7, 9–11]
8 3'-Phosphoadenylylsulfate + N-deacetylated K5-polysaccharide (derived from E. coli K5-derived capsular polysaccharide) [8]
9 3'-Phosphoadenylylsulfate + dermatan sulfate (weak activity [10]) [9, 10]
10 3'-Phosphoadenylylsulfate + chondroitin 4-sulfate (weak activity) [10]
11 3'-Phosphoadenylylsulfate + N-acetylated heparan sulfate [11]
12 More (poor acceptors: N-desulfo-N-acetylheparan [3], heparin [3], N-desulfoheparin [3], no acceptors: N-acetylated heparan sulfate [7], N-acetylated heparin [7], chondroitin [7], chondroitin sulfate [7, 9], tyrosine-containing tripeptides [7], heparin [9, 10], hyaluronic acid [9], p-nitrophenol [10]) [3, 7, 9, 10]

Product spectrum

1 Adenosine 3',5'-bisphosphate + N-sulfoheparitin [1]
2 ?
3 Adenosine 3',5'-bisphosphate + heparan sulfate [9, 12]
4 ?
5 ?
6 More (the enzyme transfers sulfate to the 2-amino groups and to the 6-hydroxy groups of glucosamine units of the acceptor substrate, the ratio of the N/O-sulfation ranges between 3:1 and 2:1) [3]
7 ?
8 ?
9 ?
10 ?
11 ?
12 ?

Inhibitor(s)

More (NEM: no effect) [4]; NaCl (above 200 mM [4], 0.125 M [11]) [4, 11]; 3',5'-ADP [4]; EDTA [7]; PCMB [10]; Phenylmercuric acetate [10]; Cu^{2+} [10]; Zn^{2+} [10]

Cofactor(s)/prosthetic group(s)/activating agents

Estrogen (enhances activity with N,O-desulfated heparan sulfate as acceptor, progesterone suppresses the effect of estrogen) [2]

Metal compounds/salts

Mn^{2+} (divalent cation required [3, 11], Mn^{2+} most effective [3, 11], maximal
activation at: 5 mM [3], 10 mM (4- to 5-fold activation) [11], 62% of the activity with Mg^{2+} [7]) [3, 7, 11]; Mg^{2+} (activates at 10 mM [10], 52% of the activity with Mn^{2+} [3], metal ion required, maximal activity with 5 mM [7], little effect [11]) [3, 7, 10, 11]; Ca^{2+} (41% [3], 22% [7] of the activity with Mn^{2+} [3,
7], can partially replace Mn^{2+} in activation [11]) [3, 7, 11]; More (greatest
activity in presence of sodium phosphate buffer of ionic strength of 0.075, in
imidazole-HCl reaction rate is lower, maximum activity at ionic strength of
0.125) [10]

Turnover number (min^{-1})

Specific activity (U/mg)

1.95 (heparan sulfate) [7]; More [3]

K_m-value (mM)

0.0009 (N-deacetylase K5-polysaccharide, rat liver) [8]; 0.005
(3'-phosphoadenylylsulfate) [7]; 0.01 (3'-phosphoadenylylsulfate); 0.02
(3'-phosphoadenylylsulfate) [11]; 0.0224 (N-deacetylase K5-polysaccharide,
MST cells) [8]; 0.0407 (3'-phosphoadenylylsulfate (+ N-deacetylase
K5-polysaccharide), MST cells) [8]; 0.108 (3'-phosphoadenylylsulfate (+
N-deacetylase K5-polysaccharide), rat liver) [8]; 1.89 (N-desulfoheparan sulfate, calculated from disaccharide units) [3]; 2.5 (N,O-desulfoheparan sulfate tetrasaccharide, calculated from disaccharide units) [3]

pH-optimum

6.2 [3]; 6.7–7.2 [10]; 7.2 [7]; 7.5 [11]

pH-range

5.7–8.1 (5.7: 33% of activity maximum, 8.1: 54% of activity maximum) [7]

Temperature optimum (°C)

30 (assay at) [3]; 37 (assay at) [10]

Temperature range (°C)

3 ENZYME STRUCTURE

Molecular weight

92000 (rat, radiation inactivation analysis) [6]
97000 (rat, gel filtration) [7]
More (bovine, peaks with enzyme activity: 200000 and 110000 MW) [3]

Subunits
   Monomer (1 × 94000, rat, SDS-PAGE) [7]

Glycoprotein/Lipoprotein
   Glycoprotein [3, 7]

4 ISOLATION/PREPARATION

Source organism
   Mouse [8, 10, 11]; Chicken (hen) [1, 9]; Rabbit [2]; Bovine (calf [3], ox [12])
   [3, 12]; Rat (overexpressed in Chinese hamster ovary cells [4, 5], a single
   protein possesses both N-deacetylase and N-sulfotransferase activity [4, 6])
   [4–7]

Source tissue
   Oviduct [1]; Uterus (endometrium [2]) [2, 9]; Mastocytoma cells (MST cells
   [8]) [8, 11]; Arterial tissue [3]; Liver [4–7]; Lung [12]; Furth mouse mast cell
   tumor [10]

Localization in source
   Golgi vesicles (lumen [4], membrane [7]) [4, 6, 7]; Membranes [7]; Microsomes [11]; More (associated with postmicrosomal fraction) [10]

Purification
   Rat [7]; Bovine (N-desulfo-N-acetylheparan sulfate deacetylase activity co-
   purifies) [3]; Mouse [10]

Crystallization
   –

Cloned
   [4–6, 8]

Renatured
   –

5 STABILITY

pH

Temperature (°C)
   40 (2 min, no effect) [10]; 50 (1 min, 15% loss of activity) [11]; 55 (2 min,
   85% loss of activity) [10]; 70 (2 min, 87% loss of activity) [10]; 85 (2 min,
   complete loss of activity) [10]

Oxidation

Organic solvent

General stability information

Storage
-18°C, 20 h, stable [1]; -18°C, stable for at least 4 months [10]; -70°C, stable for at least 6 months in presence of 20% glycerol [7]

6 CROSSREFERENCES TO STRUCTURE DATABANKS

PIR/MIPS code

Brookhaven code

7 LITERATURE REFERENCES

[1] Suzuki, S., Trenn, R.H., Strominger, J.L.: Biochim. Biophys. Acta,50,169–170 (1961)
[2] Hiroshi, M., Isemura, M., Yosizawa, Z.: Int. J. Biochem.,17,1077–1083 (1985)
[3] Göhler, D., Niemann, R., Buddecke, E.: Eur. J. Biochem.,138,301–308 (1984)
[4] Wei, Z., Swiedler, S.J., Ishihara, M., Orellana, A., Hirschberg, C.B.: Proc. Natl. Acad. Sci. USA,90,3885–3888 (1993)
[5] Hashimoto, Y., Orellana, A., Gil, G., Hirschberg, C.B.: J. Biol. Chem.,267, 15744–15750 (1992)
[6] Mandon, E., Kempner, E.S., Ishihara, M., Hirschberg, C.B.: J. Biol. Chem., 269,11729–11733 (1994)
[7] Brandan, E., Hirschberg, C.B.: J. Biol. Chem.,263,2417–2422 (1988)
[8] Orellana, A., Hirschberg, C.B., Wei, Z., Swiedler, S.J., Ishihara, M.: J. Biol. Chem., 269,2270–2276 (1994)
[9] Johnson, A.H., Baker, J.R.: Biochim. Biophys. Acta,320,341–351 (1973)
[10] Eisenman, R.A., Balasubramanian, A.S., Marx, W.: Arch. Biochem. Biophys., 119,387–397 (1967)
[11] Jansson, L., Höök, M., Wasteson, A., Lindahl, U.: Biochem. J.,149,49–55 (1975)
[12] Foley, T., Baker, J.R.: Biochem. J.,124,25P-26P (1971)

1 NOMENCLATURE

EC number
2.8.2.13

Systematic name
3'-Phosphoadenylylsulfate:galactosylsphingosine sulfotransferase

Recommended name
Psychosine sulfotransferase

Synonyms
Sulfotransferase, psychosine
PAPS:psychosine sulphotransferase [1]
3'-Phosphoadenosine 5'-phosphosulfate-psychosine sulphotransferase [1]

CAS Reg. No.
37259-76-0

2 REACTION AND SPECIFICITY

Catalyzed reaction
3'-Phosphoadenylylsulfate + galactosylsphingosine →
→ adenosine 3',5'-bisphosphate + psychosine sulfate

Reaction type
Sulfate group transfer

Natural substrates
3'-Phosphoadenylylsulfate + galactosylsphingosine [1]

Substrate spectrum
1 3'-Phosphoadenylylsulfate + galactosylsphingosine (i.e. psychosine) [1]

Product spectrum
1 Adenosine 3',5'-bisphosphate + psychosine sulfate

Inhibitor(s)

Cofactor(s)/prosthetic group(s)/activating agents
Glutathione (stimulates) [1]; EDTA (stimulates) [1]; Tween 20 (required for activity) [1]

Metal compounds/salts

Enzyme Handbook © Springer-Verlag Berlin Heidelberg 1997
Duplication, reproduction and storage in data banks are only
allowed with the prior permission of the publishers

Turnover number (min^{-1})

Specific activity (U/mg)

K$_m$-value (mM)
  0.5 (psychosine) [1]

pH-optimum
  6.9 [1]

pH-range
  6.4–7.4 (6.4: about 50% of activity maximum, 7.4: about 60% of activity
  maximum) [1]

Temperature optimum (°C)
  37 (assay at) [1]

Temperature range (°C)

---

3 ENZYME STRUCTURE

Molecular weight

Subunits

Glycoprotein/Lipoprotein
  –

---

4 ISOLATION/PREPARATION

Source organism
  Mouse (maximal activity at 17–19 days after birth) [1]

Source tissue
  Brain [1]

Localization in source
  Microsomes [1]

Purification

Crystallization
  –

Cloned
  –

Renatured
  –

5 STABILITY

pH

Temperature (°C)
45 (5 min, 85% loss of activity, microsomal preparation) [1]; 55 (5 min, complete loss of activity, microsomal preparation) [1]

Oxidation

Organic solvent

General stability information

Storage
–20°C, enzyme in microsomes or a whole homogenate stable for at least 1 month [1]

---

6 CROSSREFERENCES TO STRUCTURE DATABANKS

PIR/MIPS code

Brookhaven code

---

7 LITERATURE REFERENCES

[1] Nussbaum, J.-L., Mandel, P.: J. Neurochem.,19,1789–1802 (1972)

1 NOMENCLATURE

EC number
    2.8.2.14

Systematic name
    3'-Phosphoadenylylsulfate:taurolithocholate sulfotransferase

Recommended name
    Bile-salt sulfotransferase

Synonyms
    Sulfotransferase, bile salt
    Bile salt sulfotransferase

CAS Reg. No.
    65802-92-8

2 REACTION AND SPECIFICITY

Catalyzed reaction
    3'-Phosphoadenylylsulfate + taurolithocholate →
    → adenosine 3',5'-bisphosphate + taurolithocholate sulfate (random mecha-
    nism [9], sequentially ordered Bi Bi reaction mechanism in which the bile
    salt is the 1st substrate [2])

Reaction type
    Sulfate group transfer

Natural substrates
    More (sulfation of bile salts) [1]

Substrate spectrum
    1 3'-Phosphoadenylylsulfate + taurolithocholate [1, 6–8]
    2 3'-Phosphoadenylylsulfate + 5alpha-cholanoic acid-3beta-ol [8]
    3 3'-Phosphoadenylylsulfate + 5-cholenic acid-3beta-ol [8, 9]
    4 3'-Phosphoadenylylsulfate + glycolithocholate [1–9]
    5 3'-Phosphoadenylylsulfate + lithocholate [1, 3, 4, 6–9]
    6 3'-Phosphoadenosine-5'-phosphosulfate + estrone (not [7–9]) [3]
    7 3'-Phosphoadenylylsulfate + dehydroepiandrosterone [3]
    8 3'-Phosphoadenylylsulfate + phenol [3]
    9 3'-Phosphoadenylylsulfate + chenodeoxycholate (sulfation at the 7-OH
        position [3, 7]) [3, 7, 9]
    10 3'-Phosphoadenylylsulfate + deoxycholate (sulfation at the 12-OH posi-
        tion) [3]

11 3'-Phosphoadenylylsulfate + glycochenodeoxycholate [7, 9]
12 3'-Phosphoadenylylsulfate + taurochenodeoxycholate [7, 8]
13 3'-Phosphoadenylylsulfate + taurocholate [8]
14 3'-Phosphoadenylylsulfate + isolithocholic acid (i.e. 5beta-cholanoic acid-3beta-ol) [8]
15 3'-Phosphoadenylylsulfate + cholic acid [9]
16 3'-Phosphoadenylylsulfate + glycocholate [7, 8]
17 More (no bile salt disulfate formation detected [3], the rates of reaction in decreasing order are monohydroxylated > dihydroxylated > trihydroxylated and glycoconjugates > tauroconjugates > unconjugates [7], no substrate: estradiol [8], testosterone [8, 9], dehydroepiandrosterone [8], cholesterol [8], phenol [8, 9], tyramine [8], serotonin [8], dihydroepiandrosterone [7], reacts with hydroxy groups of bile salts at both 3alpha and 3beta position [8]) [3, 7–9]

Product spectrum
1 Adenosine 3',5'-bisphosphate + taurolithocholate sulfate [1]
2 Adenosine 3',5'-bisphosphate + cholanic acid sulfate
3 Adenosine 3',5'-bisphosphate + cholenic acid sulfate
4 Adenosine 3',5'-bisphosphate + glycolithocholate 3-sulfate [2]
5 Adenosine 3',5'-bisphosphate + lithocholate sulfate [4]
6 Adenosine 3',5'-bisphosphate + estrone sulfate
7 Adenosine 3',5'-bisphosphate + dehydroepiandrosterone sulfate
8 Adenosine 3',5'-bisphosphate + phenyl sulfate
9 Adenosine 3',5'-bisphosphate + chenodeoxycholate 7-sulfate [9]
10 Adenosine 3',5'-bisphosphate + deoxycholate 12-sulfate [10]
11 Adenosine 3',5'-bisphosphate + glycochenodeoxycholate monosulfate [5]
12 Adenosine 3',5'-bisphosphate + taurochenodeoxycholate monosulfate + taurochenodeoxycholate disulfate [1]
13 Adenosine 3',5'-bisphosphate + taurocholate monosulfate + taurocholate disulfate + taurocholate trisulfate [1]
14 Adenosine 3',5'-bisphosphate + isolithocholate sulfate
15 Adenosine 3',5'-bisphosphate + cholic acid sulfate
16 Adenosine 3',5'-bisphosphate + glycocholate sulfate
17 ?

Inhibitor(s)
Taurodehydrocholate [8]; Chenodeoxycholate-7,12-one [8]; Adenosine 3',5'-diphosphate (PAP) [1, 6, 9]; Glycolithocholate [9]; 3-Oxolithocholate [9]; Isoadenosine 3',5'-diphosphate [9]; 3-Oxo-5-beta-cholanoate [2]; 3'-AMP [2]; NaN_3 [6, 8]; PCMB [1, 6,-9]; EDTA (slight [1]) [1, 6, 8]; NaF (slight) [1, 8]; ATP [1, 6, 8]; Iodoacetate [6–8]; Cu^{2+} [3, 6]; Fe^{2+} [3, 6]; Zn^{2+} [3]

Cofactor(s)/prosthetic group(s)/activating agents
More (sulfhydryl group required for activity) [1, 8]

Metal compounds/salts

Mg^{2+} (no absolute requirement, stimulates at low concentrations, < 2 mM [2], required [8], stimulates [3, 6]) [2, 3, 6, 8]; Mn^{2+} (stimulates) [3, 6]; Co^{2+} (stimulates) [3, 6]; Li^+ (0.5 mM, marginally enhances activity) [8]; K^+ (0.5 mM, marginally enhances activity) [8]; Fe^{2+} (enhances activity) [8]; Zn^{2+} (enhances activity) [8]; Cu^{2+} (enhances activity) [8]; Ca^{2+} (enhances activity) [8]

Turnover number (min^{-1})

Specific activity (U/mg)

More [1, 3, 6, 7]; 0.018 [9]

K_m-value (mM)

0.0007 (3'-phosphoadenylylsulfate) [9]; 0.0014 (3'-phosphoadenylylsulfate) [8]; 0.0018 (glycolithocholate) [3]; 0.002 (lithocholate [4], 3'-phosphoadenylylsulfate [7], glycolithocholate [9]) [4, 7, 9]; 0.0024 (3beta-hydroxy-5-cholenic acid) [9]; 0.0025 (lithocholic acid) [9]; 0.0033 (3'-phosphoadenylylsulfate [3], glycolithocholate [4]) [3, 4]; 0.008 (3'-phosphoadenylylsulfate) [1, 6]; 0.025 (chenodeoxycholic acid) [9]; 0.04 (taurolithocholate) [7]; 0.05 (taurolithocholate) [1, 6]; 0.071 (cholic acid) [9]; 0.077 (taurolithocholate) [8]

pH-optimum

6.5 [1]; 6.8 [8]; 7.0 [2]; 7–9.5 [4]

pH-range

5.5–10.5 (55% of activity maximum at pH 5.5 and 10.5) [4]

Temperature optimum (°C)

Temperature range (°C)

---

3 ENZYME STRUCTURE

Molecular weight

65000 (human, gel filtration) [9]
76000 (guinea pig, gel filtration) [8]
80000 (rat kidney, gel filtration) [7]
130000 (rat liver, gel filtration) [1, 6]
More (hamster liver, gel filtration, 3 peaks with sulfotransferase activity: 45000 (substrate: glycochenodeoxycholate), 60000 (substrate: glycolithocholate), 200000 (substrates: glycolithocholate and glycochenodeoxycholate), rat liver, gel filtration, 3 peaks with sulfotransferase activity: MW 32000, 64000 and 130000, 2-mercaptoethanol enhances the proportion of the highest MW 130000 form, the effect is duplicated by alkylation of sulfhydryl groups with iodoacetate and is interpreted as being due to intermolecular disulfide bonds) [5]

Subunits
? (x × 30000, Rhesus monkey, SDS-PAGE) [2]

Glycoprotein/Lipoprotein
  –

---

4 ISOLATION/PREPARATION

Source organism
  Guinea pig [8]; Rat [1, 5–7]; Rhesus monkey (female) [2]; Human [3, 4, 9];
  Hamster (female) [5]

Source tissue
  Liver [1–3, 5, 6, 8, 9]; Kidney [1, 4, 7]; More (not: brain, lung, heart, spleen,
  intestinal mucosa) [1]

Localization in source
  Cytosol [1–4, 6, 9]

Purification
  Rat [1, 6, 7]; Rhesus monkey [2]; Human (partial [3]) [3, 9]; Guinea pig [8]

Crystallization
  –

Cloned
  –

Renatured
  –

---

5 STABILITY

pH

Temperature (°C)

Oxidation

Organic solvent

General stability information
  Repeated freezing and thawing reduces activity considerably [3]

Storage
  4°C, unstable [3]

6 CROSSREFERENCES TO STRUCTURE DATABANKS

PIR/MIPS code

Brookhaven code

7 LITERATURE REFERENCES

[1] Chen, L.-J., Bolt, R.J., Admirand, W.H.: Biochim. Biophys. Acta,480,219–227 (1977)
[2] Barnes, S., Waldrop, R., Crenshaw, J., King, R.J., Taylor, K.B.: J. Lipid Res.,27, 1111–1123 (1986)
[3] Lööf, L., Hjerten, S.: Biochim. Biophys. Acta,617,192–204 (1980)
[4] Lööf, L., Wengle, B.: Biochim. Biophys. Acta,530,451–460 (1978)
[5] Barnes, S., Spenney, J.G.: Biochim. Biophys. Acta,704,353–360 (1982)
[6] Chen, L.J.: Methods Enzymol.,77,213–218 (1981) (Review)
[7] Chen, L.-J., Imperato, T.J., Bolt, R.J.: Biochim. Biophys. Acta,522,443–451 (1978)
[8] Chen, L.J.: Biochim. Biophys. Acta,717,316–321 (1982)
[9] Chen, L.J., Segel, I.H.: Arch. Biochem. Biophys.,241,371–379 (1985)

1 NOMENCLATURE

EC number
2.8.2.15

Systematic name
3'-Phosphoadenylylsulfate:phenolic-steroid sulfotransferase

Recommended name
Steroid sulfotransferase

Synonyms
Steroid alcohol sulfotransferase [2]

CAS Reg. No.

2 REACTION AND SPECIFICITY

Catalyzed reaction
3'-Phosphoadenylylsulfate + a phenolic steroid →
→ adenosine 3',5'-bisphosphate + steroid O-sulfate

Reaction type
Sulfate group transfer

Natural substrates

Substrate spectrum
1 3'-Phosphoadenylylsulfate + dehydroepiandrosterone (i.e. 3beta-hydroxy-androst-5-en-17-one) [1–3]
2 3'-Phosphoadenylylsulfate + androst-5-ene-3beta,17alpha-diol [1, 2]
3 3'-Phosphoadenylylsulfate + androst-5-ene-3beta,17beta-diol [1, 3]
4 3'-Phosphoadenylylsulfate + androsterone (i.e. 3alpha-hydroxy-5alpha-androstan-17-one) [1–3]
5 3'-Phosphoadenylylsulfate + epiandrosterone (i.e. 3beta-hydroxy-5alpha-androstan-17-one) [1–3]
6 3'-Phosphoadenylylsulfate + etiocholanone (i.e. 3alpha-hydroxy-5beta-androstan-17-one) [1, 3]
7 3'-Phosphoadenylylsulfate + pregnenolone (i.e. 3beta-hydroxypregn-5-en-20-one) [1–3]
8 3'-Phosphoadenylylsulfate + testosterone (i.e. 17beta-hydroxyandrost-4-en-3-one) [1–3]
9 3'-Phosphoadenylylsulfate + epitestosterone (i.e. 17alpha-hydroxyan-drost-4-en-3-one) [1]
10 3'-Phosphoadenylylsulfate + 11-deoxycorticosterone (i.e. 21-hydroxy-pregn-4-en-3,20-dione) [1]

11 3'-Phosphoadenylylsulfate + 17beta-estradiol (i.e. estra-1,3,5(10)-triene-3,17beta-diol) [1–5]

12 3'-Phosphoadenylylsulfate + 17alpha-estradiol (i.e. estra-1,3,5(10)-triene-3,17alpha-diol) [1, 4]

13 3'-Phosphoadenylylsulfate + estrone [1, 3]

14 3'-Phosphoadenylylsulfate + 2-naphthol [6]

15 3'-Phosphoadenylylsulfate + glycolithocholate [6]

Product spectrum

1 Adenosine 3',5'-bisphosphate + dehydroepiandrosterone 3-sulfate [2]

2 Adenosine 3',5'-bisphosphate + ?

3 Adenosine 3',5'-bisphosphate + ?

4 Adenosine 3',5'-bisphosphate + androsterone sulfate (i.e. 5alpha-androstan-17-one 3-sulfate)

5 Adenosine 3',5'-bisphosphate + epiandrosterone sulfate (i.e 5alpha-androstan-17-one 3-sulfate)

6 Adenosine 3',5'-bisphosphate + etiocholanone sulfate (i.e. 5beta-androstan-17-one 3-sulfate)

7 Adenosine 3',5'-bisphosphate + pregnenolone sulfate (i.e. pregn-5-en-20-one 3-sulfate)

8 Adenosine 3',5'-bisphosphate + testosterone sulfate (i.e. androst-4-en-3-one 17-sulfate)

9 Adenosine 3',5'-bisphosphate + epitestosterone sulfate (i.e. androst-4-en-3-one 17-sulfate)

10 Adenosine 3',5'-bisphosphate + 11-deoxycorticosterone sulfate (i.e. pregn-4-en-3-one 21-sulfate)

11 Adenosine 3',5'-bisphosphate + ?

12 Adenosine 3',5'-bisphosphate + ?

13 Adenosine 3',5'-bisphosphate + estrone sulfate (i.e. estra-1,3,5(10)-trien-17-one 3-sulfate)

14 Adenosine 3',5'-bisphosphate + 2-naphthyl sulfate

15 Adenosine 3',5'-bisphosphate + ?

Inhibitor(s)

Cofactor(s)/prosthetic group(s)/activating agents

Metal compounds/salts

Turnover number (min^{-1})

Specific activity (U/mg)
0.020 [1]; 0.133 [3]; 0.00071 [4]

K_m-value (mM)
0.0016 (dehydroepidandrosterone, 3'-phosphoadenylylsulfate) [3]

pH-optimum
5.5–6.0 (estradiol) [4]; 6.0 (estrone) [4]; 6.8 (dehydroepiandrosterone) [1]

pH-range

Temperature optimum (°C)

Temperature range (°C)

---

3 ENZYME STRUCTURE

Molecular weight
68000–70000 (human, sucrose density gradient centrifugation [2], gel filtration [3]) [2, 3]

Subunits
Dimer (2 × 34000–34500, human, SDS-PAGE) [1, 2]
? (x × 32500, rat, SDS-PAGE) [5]

Glycoprotein/Lipoprotein
–

---

4 ISOLATION/PREPARATION

Source organism
Human (foetus [1], adult [2]) [1–3]; Rat [4–6]

Source tissue
Adrenal gland [1, 2]; Liver [3–6]

Localization in source
Cytosol [1, 3, 4]

Purification
Human [1–3]; Rat [4–6]

Crystallization
–

Cloned
–

Renatured
–

5 STABILITY

pH

Temperature (°C)

Oxidation

Organic solvent

General stability information

Storage
   −70°C, 0.05 M Tris-HCl buffer, pH 7.5, 0.1 mM DTT, 2% v/v propylene glycol,
   1 mg/ml protein, 3.5 months stable [1]; 0°C, N_2–atmosphere, Tris-thiol buffer,
   4 weeks [2]

6 CROSSREFERENCES TO STRUCTURE DATABANKS

PIR/MIPS code

Brookhaven code

7 LITERATURE REFERENCES

[1] Adams, J.B., McDonald, D.: Biochim. Biophys. Acta,615,275–278 (1980)
[2] Adams, J.B., McDonald, D.: Biochim. Biophys. Acta,567,144–153 (1979)
[3] Falany, C.N., Vasquez, M.E., Kalb, J.M.: Biochem. J.,260,641–646 (1989)
[4] Sugiyama, Y., Stolz, A., Sugimoto, M., Kuhlenkamp, J., Yamada, T., Kaplowitz, N.:
    Biochem. J.,224,947–953 (1984)
[5] Takikawa, H., Stolz, A., Kaplowitz, N.: FEBS Lett.,207,193–197 (1986)
[6] Singer, S.S., Federspiel, M.J., Green, J., Lewis, W.G., Martin, V., Witt, K.R., Tappel, J.:
    Biochim. Biophys. Acta,700,110–117 (1982)

1 NOMENCLATURE

EC number
2.8.2.16

Systematic name
3'-Phosphoadenylylsulfate:thiol S-sulfotransferase

Recommended name
Thiol sulfotransferase

Synonyms
Sulfotransferase, phosphoadenylylsulfate-thiol
PAPS sulfotransferase [1]
Adenosine 3'-phosphate 5'-sulphatophosphate sulfotransferase [3]

CAS Reg. No.
70356-45-5

---

2 REACTION AND SPECIFICITY

Catalyzed reaction
3'-Phosphoadenylylsulfate + a thiol →
→ adenosine 3',5'-bisphosphate + an S-alkyl thiosulfate

Reaction type
Sulfate group transfer

Natural substrates
3'-Phosphoadenylylsulfate + a thiol (involved in assimilatory sulfate reduction) [2]

Substrate spectrum
1 3'-Phosphoadenylylsulfate + a thiol (i.e. 3'-phosphoadenosine 5'-phosphosulfate or PAPS, specific for PAPS, adenosine 5'-phosphosulfate (APS) is no substrate [2]) [1–3]
2 3'-Phosphoadenylylsulfate + 2',3'-dimercaptopropanol (not [2]) [1]
3 3'-Phosphoadenylylsulfate + dithioerythritol [1, 2]
4 3'-Phosphoadenylylsulfate + mercaptoethanol (at about 25% the rate with dithiols [1], not [2]) [1]
5 3'-Phosphoadenylylsulfate + glutathione (at about 25% the rate with dithiols [1], not [2]) [1]
6 3'-Phosphoadenylylsulfate + cysteine (at about 25% the rate with dithiols [1], not [2]) [1]

Product spectrum
    1 Adenosine 3',5'-bisphosphate + an S-alkyl thiosulfate (i.e. 3',5'-ADP) [2, 3]
    2 ?
    3 ?
    4 ?
    5 ?
    6 ?

Inhibitor(s)
    5'-AMP (kinetics [1], not [2]) [1]; 5'-ADP (kinetics [1], not [2]) [1]; Adenosine
    5'-phosphosulfate (i.e. APS, strong, reversible by higher PAPS-concentra-
    tions) [1]; 3',5'-ADP (product inhibition [2], not [1]) [2]; 2',5'-ADP (not [1]) [2];
    Mg^{2+} (above 50 mM, activation at 5–10 mM) [2]; More (no inhibition by
    2'-AMP, 3'-AMP, cAMP [1, 2], cysteic acid, cysteine or methionine [2]) [1, 2]

Cofactor(s)/prosthetic group(s)/activating agents
    Thioredoxin (activation [1], requirement [2]) [1, 2]

Metal compounds/salts
    Mg^{2+} (slight activation, 5–10 mM, severe inhibition above 50 mM) [2];
    Na_2SO_4 (activation) [1]

Turnover number (min^{-1})

Specific activity (U/mg)

K_m-value (mM)
    0.0083 (3'-phosphoadenylylsulfate) [1]; 0.02 (3'-phosphoadenylylsulfate) [2]

pH-optimum
    8 [2]; 8.5 [1]

pH-range
    7–10 (about 60% of maximal activity at pH 7 and about half-maximal activity
    at pH 10) [2]

Temperature optimum (°C)

Temperature range (°C)

3 ENZYME STRUCTURE

Molecular weight
    58000 (Synechococcus sp. 6301, gel filtration) [2]

Subunits

Glycoprotein/Lipoprotein
    –

4 ISOLATION/PREPARATION

Source organism
Cyanophora paradoxa [1]; Synechococcus sp. (6301) (cyanobacterium) [2];
Nicotiana tabacum (tobacco, var. Samsun) [3]

Source tissue
Cell (cell cultures [3]) [1–3]

Localization in source

Purification
Cyanophora paradoxa (partial) [1]; Synechococcus sp. (partial) [2]

Crystallization
–

Cloned
–

Renatured
–

5 STABILITY

pH

Temperature (°C)

Oxidation

Organic solvent

General stability information

Storage

6 CROSSREFERENCES TO STRUCTURE DATABANKS

PIR/MIPS code

Brookhaven code

7 LITERATURE REFERENCES

[1] Schmidt, A., Christen, U.: Z. Naturforsch.,34c,222–228 (1979)
[2] Schmidt, A., Christen, U.: Planta,140,239–244 (1978)
[3] Schwenn, J.D., Jender, H.G.: J. Chromatogr.,193,285–290 (1980)

1 NOMENCLATURE

EC number
2.8.2.17

Systematic name
3'-Phosphoadenylylsulfate:chondroitin 6'-sulfotransferase

Recommended name
Chondroitin 6-sulfotransferase

Synonyms
Chondroitin 6-O-sulfotransferase
Sulfotransferase, chondroitin 6-
3'-Phosphoadenosine 5'-phosphosulfate (PAPS):chondroitin sulfate sulfo-transferase [10]
Terminal 6-sulfotransferase [1]
More (not identical with EC 2.8.2.5)

CAS Reg. No.
37292-93-6; 83589-04-2

2 REACTION AND SPECIFICITY

Catalyzed reaction
3'-Phosphoadenylylsulfate + chondroitin →
→ adenosine 3',5'-bisphosphate + chondroitin 6'-sulfate

Reaction type
Sulfate group transfer

Natural substrates
3'-Phosphoadenylylsulfate + chondroitin 4-sulfate (may play a role in synthe-sis of proteoglycans) [1]
3'-Phosphoadenylylsulfate + chondroitin (involved in biosynthesis of chondroitin sulfate [2], is believed to act in the course of chondroitin sulfate synthesis in cooperation with, but shortly after the enzymes involved in the chain elongation reaction [7]) [2, 7]

Substrate spectrum
1 3'-Phosphoadenylylsulfate + chondroitin (native chondroitin is the best acceptor, chemically desulfated chondroitin sulfate has the lowest rate of sulfation [11], specificity towards desulfated chondroitin sulfate and oligosaccharides derived therefrom [7]) [2–11]

Enzyme Handbook © Springer-Verlag Berlin Heidelberg 1997
Duplication, reproduction and storage in data banks are only
allowed with the prior permission of the publishers

2 3'-Phosphoadenylylsulfate + chondroitin 4-sulfate (no activity with unsulfated chondroitin) [1]
3 3'-Phosphoadenylylsulfate + N-acetylgalactosamine 4-sulfate [1]
4 3'-Phosphoadenylylsulfate + 1-phospho-N-acetylgalactosamine 4-sulfate [1]
5 3'-Phosphoadenylylsulfate + UDP-N-acetylgalactosamine 4-sulfate [1]
6 3'-Phosphoadenylylsulfate + chondroitin-derived oligosaccharides (even-numbered oligosaccharides with a glucuronic acid at the nonreducing terminus are active as acceptors, their capacity decreasing with decreasing chain length, dodecasaccharide (64%), decasaccharide (55%), octasaccharide (17%), hexasaccharide (4%) of the chondroitin sulfation [7]) [4, 6, 7]
7 3'-Phosphoadenylylsulfate + corneal keratan sulfate (not [11]) [9]
8 More (does not sulfate a non-reducing terminal GalNAc residue when this residue is added concurrently with the sulfation [4], dermatan sulfate and keratan sulfate preparations which contain covalently bound chondroitin sulfate act as sulfate acceptors [11], not: UDP-N-acetylglucosamine [1], UDP-N-acetylgalactosamine [1], UDP-N-acetylgalactosamine 6-sulfate [1], UDPgalactose [1], GDPmannose [1], UDP-N-acetylglucosamine 6-sulfate [1], heparan sulfate [1], heparin [1, 11], hyaluronate [11]) [1, 4, 11]

Product spectrum
1 Adenosine 3',5'-bisphosphate + chondroitin 6'-sulfate [2–11]
2 Adenosine 3',5'-bisphosphate + chondroitin 4,6-bissulfate [1]
3 Adenosine 3',5'-bisphosphate + N-acetylgalactosamine 4,6-bissulfate [1]
4 Adenosine 3',5'-bisphosphate + 1-phospho-N-acetylgalactosamine 4,6-bissulfate [1]
5 Adenosine 3',5'-bisphosphate + UDP-N-acetylgalactosamine 4,6-bissulfate [1]
6 ?
7 More (sulfated product is degraded by keratanase but not by chondroitinase ABC) [9]
8 ?

Inhibitor(s)
More (no inhibition by detergents of enzyme in chicken cartilage system [3], not: iodoacetamide [11]) [3, 11]; $MgCl_2$ [1]; $CoCl_2$ [1]; $MnCl_2$ [1]; $CaCl_2$ [1]; Heparin [11]; ADP [11]; AMP [11]; ATP [11]; EDTA (in absence of added metal ion) [6]; PCMB (weak) [11]

Cofactor(s)/prosthetic group(s)/activating agents
Basic proteins (stimulate) [2]; Polyamines (stimulate) [2]; Protamine (stimulates, optimum concentration: 0.025 mg/ml) [2]; Histone (stimulates, optimum concentration: 0.5 mg/ml) [2]; Lysozyme (stimulates, optimum concentration: 3.0 mg/ml) [2]; Spermine (stimulates, optimum concentration: 0.6

mM) [2]; Spermidine (stimulates, optimum concentration: 2.5 mM) [2]; More (enzyme is present in supernatant and in particulate fraction, detergent activates indirectly by releasing enzyme into the medium) [8]

Metal compounds/salts

Mn^{2+} (divalent cation required, $Mn^{2+} > Mg^{2+}$, Ca^{2+} [7], stimulates [2], best activator [6]) [2, 6, 7]; Mg^{2+} (divalent cation required, $Mn^{2+} > Mg^{2+}$, Ca^{2+} [7], can partially replace Mn^{2+} in activation [6]) [6, 7]; Zn^{2+} (can partially replace Mn^{2+} in activation) [6]; Cu^{2+} (can partially replace Mn^{2+} in activation) [6]; Ca^{2+} (divalent cation required, $Mn^{2+} > Mg^{2+}$, Ca^{2+} [7], no effect [6]) [7]

Turnover number (min^{-1})

Specific activity (U/mg)

More [9, 11]

K_m-value (mM)

0.00011 (3'-phosphoadenylylsulfate (+ UDP-N-acetylgalactosamine 4-sulfate)) [1]; 0.00049 (chondroitin 4-sulfate) [1]; 0.018 (3'-phosphoadenylylsulfate) [11]; 0.027 (UDP-N-acetylgalactosamine 4-sulfate) [1]; 0.06 (native chondroitin) [11]; 0.15 (chondroitin [7], 1-phospho-N-acetylgalactosamine 4-sulfate [1]) [1, 7]; 0.36 (N-acetylgalactosamine 4-sulfate) [1]; 0.48 (dodecasaccharide derived from chondroitin) [7]; 0.90 (decasaccharide derived from chondroitin) [7]; 1.8 (desulfated chondroitin sulfate) [11]; 2.5 (octasaccharide derived from chondroitin) [7]; 5.0 (hexasaccharide derived from chondroitin) [7]; More [2, 4, 6, 7, 10]

pH-optimum

5.4 [5]; 5.5–8.5 [6]; 6.0 (double pH-optimum: 6.0 (100%) and 7.3 (65%)) [7]; 6.8 (imidazole buffer) [2]; 7.3 (double pH-optimum: 6.0 (100%) and 7.3 (65%)) [7]; 8.2 [11]

pH-range

4.0–9.0 (4.0: about 30% of activity maximum, 9.0: about 55% of activity maximum) [6]; More [2, 5]

Temperature optimum (°C)

37 (assay at) [1, 2, 6, 9]

Temperature range (°C)

3 ENZYME STRUCTURE

Molecular weight

150000 (chicken, gel filtration) [10]
160000 (chicken, gel filtration) [9]

Subunits
? (x × 38000, bovine, SDS-PAGE) [7]
Dimer (2 × 75000, chicken, SDS-PAGE) [9]

Glycoprotein/Lipoprotein
–

---

4 ISOLATION/PREPARATION

Source organism
Coturnix coturnix Japonica [1]; Chicken [2–6, 8–10]; Bovine (calf) [7];
Mouse [11]

Source tissue
Arterial tissue [7]; Liver [11]; Oviduct [1]; Epiphyseal cartilage (embryo)
[2–4, 8, 10]; Culture medium (embryo chondrocytes release enzyme into the
medium under culture conditions) [5, 9]

Localization in source
Golgi apparatus [5]; Microsomes [6]; Cytosol [7]; More (enzyme is present
in supernatant and in particulate fraction [8], embryo chondrocytes release
enzyme into the medium under culture conditions [5, 9]) [5, 8, 9]

Purification
Mouse (partial) [11]; Coturnix coturnix Japonica [1]; Bovine (calf) [7]; Chick-
en (partial [10]) [9, 10]

Crystallization
–

Cloned
–

Renatured
–

---

5 STABILITY

pH

Temperature (°C)

Oxidation

Organic solvent

General stability information

Storage
–20°C [9]

6 CROSSREFERENCES TO STRUCTURE DATABANKS

PIR/MIPS code
  PIR2:A57397 (precursor chicken)

Brookhaven code

---

7 LITERATURE REFERENCES

[1] Nakanishi, Y., Shimizu, M., Otsu, K., Kato, S., Tsuji, M., Suzuki, S.: J. Biol. Chem., 256,5443–5449 (1981)

[2] Habuchi, O., Miyashita, N.: Biochim. Biophys. Acta,717,414–421 (1982)

[3] Sugumaran, G., Silbert, J.E.: J. Biol. Chem.,263,4673–4678 (1988)

[4] Sugumaran, G., Cogburn, J.N., Silbert, J. E.: J. Biol. Chem.,261,12659–12664 (1986)

[5] Habuchi, O., Tsuzuki, M., Takeuchi, I., Hara, M., Matsui, Y., Ashikari, S.: Biochim. Biophys. Acta,1133,9–16 (1991)

[6] Delfert, D.M., Conrad, H.E.: J. Biol. Chem.,260,14446–14451 (1985)

[7] Hollmann, J., Niemann, R., Buddecke, E.: Biol. Chem. Hoppe-Seyler,367,5–13 (1986)

[8] Salac, M.L.B., Mourao, P.A.S.: Biochim. Biophys. Acta,1074,130–135 (1991)

[9] Habuchi, O., Matsui, Y., Kotoya, Y., Aoyama, Y., Yasuda, Y., Noda, M.: J. Biol. Chem., 268,21968–21974 (1993)

[10] Salac, M.L.B., Santos, J.A., Mourao, P.A.S.: Biochim. Biophys. Acta,883,605–609 (1986)

[11] Momburg, M., Stuhlsatz, H.W., Greiling, H.: Hoppe-Seyler's Z. Physiol. Chem.,353, 1351–1361 (1972)

1 NOMENCLATURE

EC number

2.8.2.18

Systematic name

3'-Phosphoadenylylsulfate:cortisol 21-sulfotransferase

Recommended name

Cortisol sulfotransferase

Synonyms

Sulfotransferase, glucocorticoid

Glucocorticoid sulfotransferase

Glucocorticosteroid sulfotransferase

CAS Reg. No.

71427-08-2

---

2 REACTION AND SPECIFICITY

Catalyzed reaction

3'-Phosphoadenylylsulfate + cortisol →

→ adenosine 3',5'-bisphosphate + cortisol 21-sulfate (mechanism [4])

Reaction type

Sulfate group transfer

Natural substrates

Substrate spectrum

1 3'-Phosphoadenylylsulfate + cortisol (i.e. PAPS, glucocorticoids are preferred substrates [2, 4]. No substrates are AMP, ADP, ATP or GTP [1]) [1–6]

2 3'-Phosphoadenylylsulfate + deoxycorticosterone [3, 4]

3 3'-Phosphoadenylylsulfate + dehydroepiandrosterone [2, 3]

4 3'-Phosphoadenylylsulfate + estradiol-17beta [3, 4]

5 3'-Phosphoadenylylsulfate + testosterone [3, 4]

6 More (sulfotransferase I is not as specific as sulfotransferase III) [3]

Product spectrum

1 Adenosine 3',5'-bisphosphate + cortisol 21-sulfate [1, 4, 5]

2 Adenosine 3',5'-bisphosphate + 4-pregnen-3,20-dione 21-sulfate

3 Adenosine 3',5'-bisphosphate + 5-androsten-17-one 3-sulfate

4 Adenosine 3',5'-bisphosphate + ?

5 Adenosine 3',5'-bisphosphate + 4-androsten-3-one 17-sulfate

6 ?

Enzyme Handbook © Springer-Verlag Berlin Heidelberg 1997
Duplication, reproduction and storage in data banks are only
allowed with the prior permission of the publishers

Inhibitor(s)

Zn^{2+} (5 mM, strong) [2–4]; Cd^{2+} (5 mM, strong) [2–4]; Dehydroepiandrosterone (kinetics [2], cortisol as substrate [2–4], strong [3, 4]) [2–4]; Dexamethasone (kinetics [2], weak [3, 4], cortisol as substrate [2–4]) [2–4]; Progesterone (kinetics [2], strong [3, 4], cortisol as substrate [2–4]) [2–4]; Diethylstilbestrol (strong [3, 4], cortisol as substrate [2–4]) [2–4]; Deoxycorticosterone (cortisol as substrate, strong) [3, 4]; Testosterone (cortisol as substrate, strong) [3, 4]; Corticosterone (cortisol as substrate, strong) [3, 4]; Estradiol-17beta (cortisol as substrate [3, 4], strong [3], weak [4]) [3, 4]; Tetrahydrocortisol (cortisol as substrate [3], not [4]) [3]; Cortisone (cortisol as substrate, weak) [4]; Cortisol 21-sulfate (product inhibition) [4]; Adenosine 3',5'-bisphosphate (product inhibition) [4]; Antibodies to sulfotransferase III (sulfotransferase I) [6]; More (no inhibition by adenine nucleotides) [2]

Cofactor(s)/prosthetic group(s)/activating agents

More (no activation by adenine nucleotides) [2]

Metal compounds/salts

Ba^{2+} (activation [2–4], slight [3], 5 mM [2–4], in decreasing orders of efficiency: Mn^{2+}, Cr^{2+}, Co^{2+}, Mg^{2+}, Ba^{2+}, Ca^{2+}, Ni^{2+} [2], Cr^{2+}, Mn^{2+}, Co^{2+}, Mg^{2+}, Ni^{2+}, Ca^{2+}, Fe^{3+}, Ba^{2+} [3], Cr^{2+}, Mg^{2+}, Ba^{2+}, Mn^{2+}, Ni^{2+}, Ca^{2+}, Co^{2+} [4]) [2–4]; Ca^{2+} (activation [2–4], slight [3], 5 mM [2–4], in decreasing orders of efficiency: Mn^{2+}, Cr^{2+}, Co^{2+}, Mg^{2+}, Ba^{2+}, Ca^{2+}, Ni^{2+} [2], Cr^{2+}, Mn^{2+}, Co^{2+}, Mg^{2+}, Ni^{2+}, Ca^{2+}, Fe^{3+}, Ba^{2+} [3], Cr^{2+}, Mg^{2+}, Ba^{2+}, Mn^{2+}, Ni^{2+}, Ca^{2+}, Co^{2+} [4]) [2–4]; Co^{2+} (activation [2–4], 5 mM [2–4], in decreasing orders of efficiency: Mn^{2+}, Cr^{2+}, Co^{2+}, Mg^{2+}, Ba^{2+}, Ca^{2+}, Ni^{2+} [2], Cr^{2+}, Mn^{2+}, Co^{2+}, Mg^{2+}, Ni^{2+}, Ca^{2+}, Fe^{3+}, Ba^{2+} [3], Cr^{2+}, Mg^{2+}, Ba^{2+}, Mn^{2+}, Ni^{2+}, Ca^{2+}, Co^{2+} [4]) [2–4]; Cr^{2+} (activation [2–4], 5 mM [2–4], in decreasing orders of efficiency: Mn^{2+}, Cr^{2+}, Co^{2+}, Mg^{2+}, Ba^{2+}, Ca^{2+}, Ni^{2+} [2], Cr^{2+}, Mn^{2+}, Co^{2+}, Mg^{2+}, Ni^{2+}, Ca^{2+}, Fe^{3+}, Ba^{2+} [3], Cr^{2+}, Mg^{2+}, Ba^{2+}, Mn^{2+}, Ni^{2+}, Ca^{2+}, Co^{2+} [4]) [2–4]; Mg^{2+} (activation [2–4], 5 mM [2–4], in decreasing orders of efficiency: Mn^{2+}, Cr^{2+}, Co^{2+}, Mg^{2+}, Ba^{2+}, Ca^{2+}, Ni^{2+} [2], Cr^{2+}, Mn^{2+}, Co^{2+}, Mg^{2+}, Ni^{2+}, Ca^{2+}, Fe^{3+}, Ba^{2+} [3], Cr^{2+}, Mg^{2+}, Ba^{2+}, Mn^{2+}, Ni^{2+}, Ca^{2+}, Co^{2+} [4]) [2–4]; Mn^{2+} (activation [2–4], 5 mM [2–4], in decreasing orders of efficiency: Mn^{2+}, Cr^{2+}, Co^{2+}, Mg^{2+}, Ba^{2+}, Ca^{2+}, Ni^{2+} [2], Cr^{2+}, Mn^{2+}, Co^{2+}, Mg^{2+}, Ni^{2+}, Ca^{2+}, Fe^{3+}, Ba^{2+} [3], Cr^{2+}, Mg^{2+}, Ba^{2+}, Mn^{2+}, Ni^{2+}, Ca^{2+}, Co^{2+} [4]) [2–4]; Ni^{2+} (activation [2–4], slight [3], 5 mM [2–4], in decreasing orders of efficiency: Mn^{2+}, Cr^{2+}, Co^{2+}, Mg^{2+}, Ba^{2+}, Ca^{2+}, Ni^{2+} [2], Cr^{2+}, Mn^{2+}, Co^{2+}, Mg^{2+}, Ni^{2+}, Ca^{2+}, Fe^{3+}, Ba^{2+} [3], Cr^{2+}, Mg^{2+}, Ba^{2+}, Mn^{2+}, Ni^{2+}, Ca^{2+}, Co^{2+} [4]) [2–4]; Fe^{3+} (slight activation, 5 mM [3], in decreasing order of efficiency: Cr^{2+}, Mn^{2+}, Co^{2+}, Mg^{2+}, Ni^{2+}, Ca^{2+}, Fe^{3+}, Ba^{2+} [3], not [4]) [3]

Turnover number (min⁻¹)

Specific activity (U/mg)
More [5]; 0.00098 [2]; 0.012 (sulfotransferase I) [3]; 0.016 [4]

K_m-value (mM)
0.00628 (3'-phosphoadenylylsulfate) [2]; 0.00648 (cortisol) [4]; 0.00678
(3'-phosphoadenylylsulfate) [4]; 0.00682 (cortisol) [2]; 0.007 (cortisol [3, 6],
3'-phosphoadenylylsulfate, sulfotransferase III [6], pH 6.8) [3, 6]; 0.011
(3'-phosphoadenylylsulfate [3, 6], pH 6.8, sulfotransferase I) [3, 6];
0.06–0.087 (3'-phosphoadenylylsulfate, guinea pig) [1]

pH-optimum
More (pI: 6.5) [6]; 6 (cortisol as substrate) [2–4, 6]; 6.5 (sulfotransferase I)
[5]; 6.7–6.9 (guinea pig) [1]; 7.4 (sulfotransferase II) [5]

pH-range
5.3–6.8 (about half-maximal activity at pH 5.3 and 6.8) [3]; 6–8.4 (about
half-maximal activity at pH 6 and 8.4, guinea pig) [3]

Temperature optimum (°C)
37.5 (assay at) [1, 2]

Temperature range (°C)

---

3 ENZYME STRUCTURE

Molecular weight
45700 (bovine, gel filtration, 2 enzyme species: MW 62100 and 45700) [5]
61500 (rat, sulfotransferase III, gel filtration) [2]
62100 (bovine, gel filtration, 2 enzyme species: MW 62100 and 45700) [5]
65000 (rat, sedimentation velocity experiments) [2]
68000 (rat, gel filtration) [3]
156000 (rat, sulfotransferase I, sedimentation velocity experiments) [3]
160000 (rat, sulfotransferase I) [6]

Subunits
Dimer (2 × 30000, rat, sulfotransferase III, SDS-PAGE) [6]
Hexamer (6 × 28000, rat, sulfotransferase I, SDS-PAGE) [6]

Glycoprotein/Lipoprotein
–

4 ISOLATION/PREPARATION

Source organism
Guinea pig (male and female albinos, Hartley strain) [1]; Bovine (Bos taurus [5]) [5, 6]; Rat (female Sprague-Dawley CD rats [1, 3–5], male CDR Fisher rats [2]) [1–6]; Chicken [6]; Hamster [6]; Gerbil [6]

Source tissue
Liver [1–6]; Blood [2]; More (tissue distribution) [2]

Localization in source
Cytosol [1–6]

Purification
Bovine (partial, 2 sulfotransferases: I and II) [5]; Guinea pig (partial) [1]; Rat (partial [2, 3], sulfotransferase III [2, 4], sulfotransferase I (restricted to female rats) [3]) [2–4]; More (the use of frozen liver or cytosol stored on ice for more than 1–2 h reduces the yield of sulfotransferase I greatly) [6]

Crystallization
–

Cloned
–

Renatured
–

5 STABILITY

pH

Temperature (°C)

Oxidation

Organic solvent

General stability information

Storage
–20°C, at least 3 months [2]; –20°C, partially purified preparation, 1 week [3]; –20°C, at least 3 weeks [4]; 0–4°C, in crude extracts, 3 days [3, 4]

6 CROSSREFERENCES TO STRUCTURE DATABANKS

PIR/MIPS code

Brookhaven code

7 LITERATURE REFERENCES

[1] Singer, S.S., Brill, B.: Biochim. Biophys. Acta,712,590–596 (1982)
[2] Singer, S.S., Gebhart, J., Hess, E.: Can. J. Biochem.,56,1028–1035 (1978)
[3] Singer, S.S.: Arch. Biochem. Biophys.,196,340–349 (1979)
[4] Singer, S.S., Bruns, L.: Can. J. Biochem.,58,660–666 (1980)
[5] Federspeil, M.J., Singer, S.S.: Comp. Biochem. Physiol. B,69B,511–516 (1981)
[6] Singer, S.S.: Biochem. Soc. Trans.,12,35–39 (1984) (Review)

1 NOMENCLATURE

EC number
2.8.2.19

Systematic name
3'-Phosphoadenylylsulfate:triglucosyl-1-O-alkyl-2-O-acylglycerol 6-sulfotrans-
ferase

Recommended name
Triglucosylalkylacylglycerol sulfotransferase

Synonyms
Sulfotransferase, triglucosylmonoalkylmonoacyl

CAS Reg. No.
83589-05-3

---

2 REACTION AND SPECIFICITY

Catalyzed reaction
3'-Phosphoadenylylsulfate + alpha-D-glucosyl-1,6-alpha-D-glucosyl-1,6-
alpha-D-glucosyl-1,3–1-O-alkyl-2-O-acylglycerol →
→ adenosine 3',5'-bisphosphate + 6-sulfo-alpha-D-glucosyl-1,6-alpha-D-glu-
cosyl-1,6-alpha-D-glucosyl-1,3–1-O-alkyl-2-O-acylglycerol

Reaction type
Sulfate group transfer

Natural substrates

Substrate spectrum
1 3'-Phosphoadenylylsulfate + alpha-D-glucosyl-1,6-alpha-D-glucosyl-
   1,6-alpha-D-glucosyl-1,3–1-O-alkyl-2-O-acylglycerol (i.e. PAPS + triglucosyl
   monoalkylmonoacylglycerol, transfers sulfate ester group from PAPS to
   C-6 of terminal glucosyl residue of triglucosyl monoalkylmonoacylglycerol
   [2]. No substrates are galactosylceramide [1, 2], glucosylceramide, lac-
   tosylceramide or glycosphingolipids [2]) [1, 2]
2 3'-Phosphoadenylylsulfate + triglucosyl monoalkylglycerol (reaction at
   75% the rate with triglucosyl monoalkylmonoacylglycerol) [1]

Product spectrum
1 Adenosine 3',5'-bisphosphate + 6-sulfo-alpha-D-glucosyl-1,6-alpha-D-glu-
   cosyl-1,6-alpha-D-glucosyl-1,3–1-O-alkyl-2-O-acylglycerol [1, 2]
2 ?

Inhibitor(s)
ADP [1, 2]; ATP [1, 2]; DTT [1, 2]

Cofactor(s)/prosthetic group(s)/activating agents
Triton X-100 (activation) [1, 2]

Metal compounds/salts
Mg^{2+} (activation) [1, 2]; F^- (activation) [1, 2]; Mn^{2+} (activation, can replace Mg^{2+} to some extent) [1, 2]

Turnover number (min^{-1})

Specific activity (U/mg)
More [1, 2]

K_m-value (mM)
0.00085 (3'-phosphoadenylylsulfate) [1]; 0.001 (3'-phosphoadenylylsulfate, submandibular gland) [2]; 0.0012 (3'-phosphoadenylylsulfate, parotid gland) [2]; 0.0588 (triglucosyl monoalkylmonoacylglycerol, parotid gland) [2]; 0.069 (triglucosyl monoalkylmonoacylglycerol) [1]; 0.0784 (triglucosyl monoalkylmonoacylglycerol, submandibular gland) [2]

pH-optimum
7.8 (imidazole-HCl buffer) [1, 2]

pH-range
6–9 (about half-maximal activity at pH 6 and 9) [1]

Temperature optimum (°C)
37 (assay at) [1, 2]

Temperature range (°C)

---

3 ENZYME STRUCTURE

Molecular weight

Subunits

Glycoprotein/Lipoprotein
–

---

4 ISOLATION/PREPARATION

Source organism
Rat (male Sprague-Dawley, 12 weeks old [1]) [1, 2]

Source tissue

Gastric mucosa [1]; Submandibular salivary gland [2]; Parotid salivary gland [2]

Localization in source

Cytosol (predominantly) [1, 2]; More (subcellular distribution) [1, 2]

Purification

Crystallization

–

Cloned

–

Renatured

–

---

5 STABILITY

pH

Temperature (°C)

Oxidation

Organic solvent

General stability information

Storage

---

6 CROSSREFERENCES TO STRUCTURE DATABANKS

PIR/MIPS code

Brookhaven code

---

7 LITERATURE REFERENCES

[1] Liau, Y.H., Zdebska, E., Slomiany, A., Slomiany, B.L.: J. Biol. Chem.,257,
12019–12023 (1982)
[2] Slomiany, B.L., Liau, Y.H., Zdebska, E., Murty, V.L.N., Slomiany, A.: Biochem.
Biophys. Res. Commun.,113,817–824 (1983)

1 NOMENCLATURE

EC number
2.8.2.20

Systematic name
3'-Phosphoadenylylsulfate:protein-tyrosine O-sulfotransferase

Recommended name
Protein-tyrosine sulfotransferase

Synonyms
Sulfotransferase, protein (tyrosine)
Tyrosylprotein sulfotransferase

CAS Reg. No.
87588-33-8

2 REACTION AND SPECIFICITY

Catalyzed reaction
3'-Phosphoadenylylsulfate + protein tyrosine →
→ adenosine 3',5'-bisphosphate + protein tyrosine-O-sulfate (mechanism
[4])

Reaction type
Sulfate group transfer

Natural substrates
3'-Phosphoadenylylsulfate + protein tyrosine (involved in post-translational
processing of specific PC12 cell proteins [1], enzyme of trans-Golgi network
catalyzing post-translational sulfation of a variety of secretory and membrane proteins [11], post-translational modification of biologically active
peptides and proteins [3]) [1, 3, 11]

Substrate spectrum
1 3'-Phosphoadenylylsulfate + acidic polypeptide tyrosine (polypeptides
from PC12 cells, referred to as p113, p105, p86 and p84, according to
their average MW [1], specifically sulfates Tyr-residues adjacent to acidic
amino acids [1, 6, 8]) [1, 6, 8]
2 3'-Phosphoadenylylsulfate + acidic amino acid polymer EAY (i.e. poly-EAY
or Glu_6,Ala_3,Tyr_1, model substrate) [2, 3, 5, 10, 12–14]

3 3'-Phosphoadenylylsulfate + synthetic peptides (structure based on sequences surrounding the sulfated Tyr of naturally sulfated proteins [4, 6, 7], e.g. CCK-(107–115) corresponding to C-terminal sulfation sites of preprocholecystokinin [7], Hir-(57–65) corresponding to the nine C-terminal amino acids of hirudin [7], C-terminal fragment Hir-(54–65) (not N-terminal fragment Hir-(1–11)) [9], peptides modelled after known or putative Tyr-sulfation sites of chromogranin B (i.e. cholecystokinin precursor, peptides CCK-1 and variants), secretogranin I (peptides Sgl–1 to Sgl–4), vitronectin, modeled after phosphorylation site of alpha-tubulin (tub-1) or autophosphorylation site of pp60src (pp60v-src(Tyr-416)) [8], highly site-specific [4], higher activity for those peptides with aspartyl residues on N-terminal side of Tyr-residue compared with glutamyl residues [9]) [4, 6–9]

4 3'-Phosphoadenylylsulfate + recombinant hirudin variants (HV2 (previously referred to as HV2(Lys-47)) [7] or rHV-1 [9], sulfated at physiological sulfation site: Tyr-63 [7]) [7, 9]

5 3'-Phosphoadenylylsulfate + tert-butoxycarbonylcholecystokinin [14]

6 3'-Phosphoadenylylsulfate + C-terminal peptide fragments of complement component C_4 [14]

7 More (structural determinants for substrate specificity) [8]

Product spectrum

1 Adenosine 3',5'-bisphosphate + acidic polypeptide tyrosine-O-sulfate [1]
2 ?
3 ?
4 ?
5 ?
6 ?
7 ?

Inhibitor(s)

EDTA [1, 10, 12]; Acidic amino acid polymer (in excess) [2]; NaCl (above 0.2 M, solubilized enzyme [2]) [2, 10, 12]; 2,6-Dichloro-4-nitrophenol (weak [10, 12], in excess, solubilized enzyme [2]) [2, 10, 12]; DTT [12]; NEM [10, 12]; Poly-EAY (substrate inhibition, above 0.01 mM [2] or 0.002 mM [13]) [2, 13]; Zn^{2+} [10]; Mg^{2+} [10]; Ca^{2+} [10]; Mn^{2+} (above 30 mM, activation below) [3]; Lubrol Px (weak, above 0.1% w/v, activation below) [3]; Sphingosine (strong, kinetics, enhanced by phosphatidylcholine or sphingomyelin. Phosphatidylinositol, phosphatidylserine or oleic acid reverses) [5]; Sphingomyelin (weak) [5]; Threosphingosine [5]; Erythrosphingosine (less effective than sphingosine) [5]; Psychosine (strong) [5]; Phosphatidylcholine (weak) [5]; Stearylamine [5]; More (no inhibition by Glu, Asp, Gln, Asn, Ser) [4]

Cofactor(s)/prosthetic group(s)/activating agents

Triton X-100 (requirement [2], activation [3, 13], microsomal membrane-bound enzyme, not solubilized enzyme [2], sulfation of poly-EAY [2],
not of tert-butoxycarbonylcholecystokinin [2]) [2, 3, 13]; Lubrol Px (activation
[3, 13], weak inhibition above 0.1% w/v [3]) [3, 13]; Tween 20 (activation,
0.5% w/v) [3]; Tween 80 (activation, 0.5% w/v) [3]; Nonidet P-40 (activation,
0.5% w/v) [3]; CHAPS (activation, 0.5% w/v) [3]; Oleic acid (activation, 0.2
mM) [5]; Phosphatidylserine (slight activation, 0.2 mM) [5]; Phosphatidylinositol (slight activation, 0.2 mM) [5]; Lysophosphatidylcholine (slight activation, 0.2 mM) [5]; More (no activation by octyl glucoside) [3]

Metal compounds/salts

Divalent cations (requirement) [1]; Mn^{2+} (requirement (rat liver [3]) [3, 5, 13,
14], 25 mM [3], 30 mM [13], activation [8, 10, 12], 5 mM [8], inhibits above
30 mM [3]) [3, 5, 8, 10, 12–14]; Mg^{2+} (requirement (sulfation of tert-butoxycarbonylcholecystokinin [14]) [6, 14], not [3, 13]) [6, 14]; Co^{2+} (requirement, 20 mM, rat liver [3], not [13]) [3]; NaCl (activation, sulfation of tert-butoxycarbonylcholecystokinin) [14]; NaF (activation) [3, 10, 12]; More (no activation by Ca^{2+}, Cd^{2+}, Cu^{2+} or Zn^{2+}) [3]

Turnover number (min^{-1})

Specific activity (U/mg)

0.0057 [6]

K_m-value (mM)

More (kinetic properties with synthetic peptides as substrates (with modified
lengths on NH_2- or COOH-termini, changes in K_m-values resulting from sub-
stitution of negatively charged amino acids [4]) [4, 6, 8], the K_m-values of
peptides with multiple Tyr-sulfation sites decrease exponentially with the
number of sites [8]) [4, 6, 8]; 0.00004 (peptide EAY) [5]; 0.000044 (peptide
SGI-3) [8]; 0.00025 (3'-phosphoadenylylsulfate) [5]; 0.00034 (peptide SGI-4)
[8]; 0.0014 (3'-phosphoadenylylsulfate) [6]; 0.0015–0.0016 (peptide EAY)
[10, 12, 13]; 0.0017 (peptide SGI-2) [8]; 0.0019–0.002 (3'-phosphoadenylyl-
sulfate (+ peptide CCKI [8] or peptide EAY [10])) [8, 10]; 0.0083
(3'-phosphoadenylylsulfate) [1]; 0.017 (peptide CCK-(107–115), leech) [7];
0.019 (peptide PKV) [6]; 0.021 (peptide CCKI) [8]; 0.035 (peptide CCKI) [8];
0.043 (peptide Sgl-1) [8]; 0.11 (peptide CCK-(107–115), bovine) [7]; 0.12
(peptide CCK-3) [8]; 0.14 (peptide tub-1) [8]; 0.15 (peptide CCK-2) [8]; 0.24
(peptide Hir-(57–65), bovine) [7]; 0.27 (recombinant hirudin HV2, leech) [7];
0.41 (peptide Hir-(57–65), leech) [7]; 5.6 (recombinant hirudin HV2, bovine)
[7]

pH-optimum
6 (peptide PKG) [6]; 6–6.5 (Golgi-enzyme, sulfation of tert-butoxy-carbonylcholecystokinin) [14]; 6.2 [12]; 6.3 [9]; 6.4 [13]; 6.4–6.6 (solubilized enzyme) [2]; 6.7 (sulfation of peptide EAY [3, 14], liver [3]) [3, 14]

pH-range
5.7–6.8 (about half-maximal activity at pH 5.7 and 6.8) [6]; 6.2–7.2 (sulfation of peptide EAY, liver, about half-maximal activity at pH 6.2 and 7.2) [3]

Temperature optimum (°C)
30 (assay at) [3, 6–8]; 37 (assay at) [1, 2, 5]

Temperature range (°C)

---

3 ENZYME STRUCTURE

Molecular weight
100000 (bovine, enzyme-detergent micelle, glycerol gradient centrifugation) [6]

Subunits
? (x × 50000–54000, bovine, SDS-PAGE) [6]

Glycoprotein/Lipoprotein
Yes (glycoprotein) [6]

---

4 ISOLATION/PREPARATION

Source organism
Bovine [6–9, 11]; Hirudo medicinalis (leech) [7]; Rat (male Sprague-Dawley [5]) [1–5, 10–12, 14]; Hamster [3]; Human [3, 13]; Mouse [3]; Rabbit [3]

Source tissue
Pheochromocytoma (fast-responding clone, cell line PC12) [1, 11]; Brain (cerebellum [3, 5]) [2, 3, 5]; Liver [3–5, 9, 13, 14]; Lung [3, 5]; Pituitary [3]; Salivary glands (submandibular [5, 12], leech [7]) [5, 7, 12]; Heart [5]; Adrenal medulla (bovine [7]) [6–8, 11]; Stomach (antrum and body mucosa) [10]; More (tissue distribution) [3]

Localization in source
Microsomes [2, 13]; Golgi apparatus (trans-most subcompartment [6]) [3–6, 8–12, 14]; Membrane-bound (integral membrane-protein [6, 11]) [1–9, 11]; More (subcellular distribution) [1, 2]

Purification
Bovine (treatment of membrane-bound enzyme with carbonate, followed by solubilization and affinity chromatography on a substrate peptide [6], partial [11]) [6, 11]; Rat (liver [3, 4], partial [3–5, 14]) [3–5, 14]; Human (partial) [13]

Crystallization

–

Cloned

–

Renatured

–

5 STABILITY

pH

Temperature (°C)

Oxidation

Organic solvent

General stability information
    Stable in cell lysates prepared by freeze-thawing in 100 mM HEPES-NaOH
    buffer, pH 7.4, 20 mM $MgCl_2$, 10 mM mercaptoethanol or in 5 mM EDTA or
    0.32 M sucrose or by hypoosmotic lysis using 5 mM EDTA, no activity in cell
    lysates prepared in the presence of Triton X-100 or by sonication [1]

Storage
    –80°C, in 10 mM HEPES-buffer, pH 7, 50 mM NaCl, 1% Triton X-100, 1 mM
    DTT, 25% glycerol, several months [4]; –80°C, in 25% glycerol, 30% loss of
    activity within 1 month [3]; –40°C, in 10 mM Tris-HCl, pH 7.4, 25% glycerol, 5
    mM mercaptoethanol, 1% Triton X-100, 3 days [2]; 4°C, detergent-solubili-
    zed enzyme preparation, $t_{1/2}$: 48 h [6]

6 CROSSREFERENCES TO STRUCTURE DATABANKS

PIR/MIPS code

Brookhaven code

7 LITERATURE REFERENCES

[1] Lee, R.W.H., Huttner, W.B.: J. Biol. Chem.,258,11326–11334 (1983)
[2] Vargas, F., Schwartz, J.-C.: FEBS Lett.,211,234–238 (1987)
[3] Rens-Domiano, S., Roth, J.A.: J. Biol. Chem.,264,899–905 (1989)
[4] Lin, W., Larsen, K., Hortin, G.L., Roth, J.A.: J. Biol. Chem.,267,2876–2879 (1992)
[5] Kasinathan, C., Sundaram, P., Slomiany, B.L., Slomiany, A.: Biochemistry,32,
    1194–1198 (1993)
[6] Niehrs, C., Huttner, W.B.: EMBO J.,9,35–42 (1990)

[7]   Niehrs, C., Huttner, W.B., Carvallo, D., Degryse, E.: J. Biol. Chem.,265,9314–9318
      (1990)
[8]   Niehrs, C., Kraft, M., Lee, R.W.H., Huttner, W.B.: J. Biol. Chem.,265,8525–8532
      (1990)
[9]   Suiko, M., Fernando, P.H.P., Sakakibara, Y., Nakajima, H., Liu, M.C., Abe, S.,
      Nakatsu, S.: Nucleic Acids Symp. Ser.,27,183–184 (1992)
[10] Kasinathan, C., Sundaram, P., Slomiany, B.L., Slomiany, A.: Enzyme,46,179–187
      (1993)
[11] Niehrs, C., Stinchcombe, J.C., Huttner, W.B.: Eur. J. Cell Biol.,58,35–43 (1992)
[12] Sundaram, P., Slomiany, A., Slomiany, B.L., Kasinathan, C.: Int. J. Biochem.,
      24,663–667 (1992)
[13] Lin, W.H., Roth, J.A.: Biochem. Pharmacol.,40,629–635 (1990)
[14] Rens-Domiano, S., Hortin, G.L., Roth, J.A.: Mol. Pharmacol.,36,647–653 (1989)

1 NOMENCLATURE

EC number
2.8.2.21

Systematic name
3'-Phosphoadenylylsulfate:keratan 6'-sulfotransferase

Recommended name
Keratan sulfotransferase

Synonyms
Sulfotransferase, keratan
3'-Phosphoadenylyl keratan sulfotransferase
Keratan sulfate sulfotransferase
3'-Phosphoadenylylsulfate:keratan sulfotransferase [1]
More (not identical with EC 2.8.2.5, EC 2.8.2.6 or EC 2.8.2.17)

CAS Reg. No.
62168-79-0

2 REACTION AND SPECIFICITY

Catalyzed reaction
3'-Phosphoadenylylsulfate + keratan $\rightarrow$
$\rightarrow$ adenosine 3',5'-bisphosphate + keratan 6'-sulfate

Reaction type
Sulfate group transfer

Natural substrates

Substrate spectrum
1 3'-Phosphoadenylylsulfate + keratan sulfate (partially desulfated [1], both
enzyme species react best with keratan sulfate segments exhibiting a relatively high degree of sulfation [1], activity decreases with increasing molecular mass and sulfation degree of keratan sulfate [2], very weak activity towards desulfated keratan sulfate [1], no activity with keratansul-
fate-derived oligosaccharides [1]) [1, 2]
2 3'-Phosphoadenylylsulfate + chitin dodecylsaccharide (sulfotransferase I
and II) [1]
3 3'-Phosphoadenylylsulfate + agarose (sulfotransferase I active, sulfotransferase II not) [1]

Product spectrum

1 Adenosine 3',5'-bisphosphate + keratan 6'-sulfate (specificity, sulfotransferase I: 60% of the sulfate ester groups formed are linked to the C-6
atom of galactosyl residues, the rest to the C-6 atom of N-acetylglucosamine, sulfotransferase II: 23% of the newly formed sulfate ester groups
are on galactosyl and 77% on N-acetylglucosaminyl residues [1]) [1, 2]

2 ?

3 ?

Inhibitor(s)

cAMP (weak) [2]; Cu^{2+} [2]; ATP [2]; ADP [2]; 5'-Adenylylsulfate [2]; 2'-AMP
(weak) [2]; 3'-AMP (weak) [2]; 5'-AMP (weak) [2]

Cofactor(s)/prosthetic group(s)/activating agents

Metal compounds/salts

Mn^{2+} (activates) [2]; Mg^{2+} (activates) [2]; Zn^{2+} (activates) [2]; Co^{2+} (activates) [2]

Turnover number (min^{-1})

Specific activity (U/mg)

More [1, 2]

K_m-value (mM)

0.025 (keratan sulfate) [2]

pH-optimum

6.0 (2 optima: pH 6.0 and 8.6 [2], sulfotransferase I [1]) [1, 2]; 8.5 (sulfotransferase II) [1]; 8.6 (2 optima: pH 6.0 and 8.6) [2]

pH-range

Temperature optimum (°C)

12 [2]; 15 [1]

Temperature range (°C)

15–37 (15°C: optimum, 37°C: 10% (sulfotransferase I), 14% (sulfotransferase II) of activity maximum) [1]

3 ENZYME STRUCTURE

Molecular weight

140000 (bovine, sulfotransferase II, gel filtration) [1]
220000 (bovine, sulfotransferase I, gel filtration) [1]
240000 (bovine, gel filtration) [2]

Subunits

Glycoprotein/Lipoprotein
  –

---

4 ISOLATION/PREPARATION

Source organism
  Bovine [1–3]

Source tissue
  Cornea [1–3]

Localization in source
  Microsomes [2]; Cytosol [2]; Membrane-bound [3]

Purification
  Bovine (sulfotransferase I and II [1]) [1–3]

Crystallization
  –

Cloned
  –

Renatured
  –

---

5 STABILITY

pH

Temperature (°C)

Oxidation

Organic solvent

General stability information

Storage

---

6 CROSSREFERENCES TO STRUCTURE DATABANKS

PIR/MIPS code

Brookhaven code

---

7 LITERATURE REFERENCES

[1] Rüter, E.-R., Kresse, H.: J. Biol. Chem.,259,11771–11776 (1984)
[2] Keller, R., Driesch, R., Stein, T., Momburg, M., Stuhlsatz, H.W., Greiling, H.:
    Hoppe-Seyler's Z. Physiol. Chem.,364,239–252 (1983)
[3] Keller, R., Stein, T., Weber, W., Kehrer, T., Stuhlsatz, H.W., Greiling, H.:
    Hoppe-Seyler's Z. Physiol. Chem.,364,253–260 (1983)

1 NOMENCLATURE

EC number
2.8.2.22

Systematic name
Arylsulfate:phenol sulfotransferase

Recommended name
Arylsulfate sulfotransferase

Synonyms
Sulfotransferase, arylsulfate
Arylsulfate-phenol sulfotransferase
Arylsulfotransferase
ASST [3]

CAS Reg. No.
158254-86-5

---

2 REACTION AND SPECIFICITY

Catalyzed reaction
An aryl sulfate + a phenol →
→ a phenol + an aryl sulfate (mechanism [1, 2])

Reaction type
Sulfate group transfer

Natural substrates
More (enzyme may play an important role in sulfo-conjugation of drugs and endogenous compounds [3], may play a role in metabolism and detoxification of phenolic compounds, through the enzymatic sulfation [4]) [3, 4]

Substrate spectrum
1 Phenol sulfate ester + phenol [1, 4]
2 Tyramine + 4-acetylphenyl sulfate (best donor with tyramine as acceptor) [1]
3 Tyramine + 4-methylumbelliferyl sulfate [1]
4 Tyramine + 4-nitrophenyl sulfate (ir [2]) [1–3]
5 Salicylamide + 4-nitrophenyl sulfate [2]
6 Phenolphthalein + 4-nitrophenyl sulfate [2]
7 4-Nitrophenyl sulfate + naphthol (best acceptor with p-nitrophenol as donor [1]) [1, 4]
8 4-Nitrophenyl sulfate + estradiol [1]

9 4-Nitrophenyl sulfate + phenol [1, 2, 4]
10 4-Nitrophenyl sulfate + tyrosine methyl ester [1]
11 4-Nitrophenyl sulfate + tyramine [1]
12 4-Nitrophenyl sulfate + epinephrine [1]
13 4-Methylumbelliferyl sulfate + phenol (55% of the activity with p-nitro-phenyl sulfate) [4]
14 4-Acetylphenyl sulfate + phenol (28% of the activity with 4-nitrophenyl sulfate) [4]
15 Estrone sulfate + phenol (11% of the activity with p-nitrophenyl sulfate) [4]
16 9-Phenanthrol + 4-nitrophenyl sulfate [4]
17 3-Chlorophenol + 4-nitrophenyl sulfate [4]
18 More (only the 4-position of catecholamines is specifically sulfated, naturally occuring phenolic compounds, e.g. flavine, chalcone, xanthone are sulfated, tyrosine-containing peptides, e.g. enkephalin, LH-RH, vasopressin, angiotensin, proctorin, cholecystokinin octopeptide, phyllocerulein are sulfated with high yield, hydroxyl groups of tyrosine residues in peptides such as angiotensin can act as acceptors, does not act on 3'-phosphoadenylylsulfate or adenosine 3',5'-bisphosphate) [1]

Product spectrum

1 Phenol + aryl sulfate
2 4-(2-Aminoethyl)phenyl sulfate + 4-acetylphenol
3 4-(2-Aminoethyl)phenyl sulfate + 4-methylumbelliferol
4 4-(2-Aminoethyl)phenyl sulfate (i.e. tyramine O-sulfate) + 4-nitrophenol [2]
5 Salicylamide O-sulfate + 4-nitrophenol
6 ? + 4-nitrophenol
7 Naphthyl sulfate + 4-nitrophenol
8 Estradiol sulfate + 4-nitrophenol
9 Phenyl sulfate + 4-nitrophenol
10 ? + 4-nitrophenol
11 4-(2-Aminoethyl)phenyl sulfate + 4-nitrophenol
12 ? + 4-nitrophenol
13 4-Methylumbelliferol + phenyl sulfate
14 4-Acetylphenol + phenyl sulfate
15 Estrone + phenyl sulfate
16 9-Phenanthrol sulfate + 4-nitrophenol
17 3-Chlorophenyl sulfate + 4-nitrophenol
18 ?

Inhibitor(s)

Hg^{2+} [2]; Cu^{2+} [2, 4]; Zn^{2+} [2, 4]; Fe^{2+} [2]; Ni^{2+} [2, 4]; EDTA (inactivation recovered by addition of metal ions) [2, 4]; Diethyl dicarbonate [2]; N-Tosyl-L-lysylchloromethane [2]; DTNB [2]; PCMB [2]; p-Chloromercuribenzenesulfonic acid [2]; Ca^{2+} [4]; Pb^{2+} [4]; Cd^{2+} [4]

Cofactor(s)/prosthetic group(s)/activating agents
Diethylpyrocarbamate [1]; Tosyllysine [1]; Chloromethyl ketone [1]

Metal compounds/salts
Mg^{2+} (increases activity) [2, 4]; Mn^{2+} (increases activity) [2]; Co^{2+} (increases activity) [4]; More (ineffective: Ca^{2+}, Ba^{2+}) [1]

Turnover number (min^{-1})

Specific activity (U/mg)
More [2]; 4.55 [3]; 8.82 [4]

K_m-value (mM)
0.104 (p-nitrophenyl sulfate (+ tyramine)) [2]; 0.11 (4-nitrophenyl sulfate (+ phenol)) [4]; 0.66 (phenol (+ 4-nitrophenyl sulfate)) [4]; 3.5 (tyramine (+ 4-nitrophenyl sulfate)) [2]

pH-optimum
8–9 [1]; 8.0 (tyramine, salicylamide) [2]; 8.5 (phenol) [2]; 9.0 (phenolphthalein) [2]; 10–10.5 [4]

pH-range

Temperature optimum (°C)
37 (assay at) [2, 4]

Temperature range (°C)

---

3 ENZYME STRUCTURE

Molecular weight
160000 (Klebsiella sp. K-36, gel filtration) [4]
315000 (Eubacterium sp. A-44 [1, 2], gel filtration [2]) [1, 2]

Subunits
Dimer (2 × 73000, Klebsiella sp. K-36, SDS-PAGE) [4]
Tetramer (4 × 80000, Eubacterium sp. A-44 [1, 2], SDS-PAGE [2]) [1, 2]

Glycoprotein/Lipoprotein
–

---

4 ISOLATION/PREPARATION

Source organism
Eubacterium sp. (A-44) [1–3]; Klebsiella sp. (K-36) [4]

Source tissue

Localization in source

Purification

Eubacterium sp. (purification using polyclonal antibodies [3]) [1–3]; Klebsiella sp. [4]

Crystallization

–

Cloned

–

Renatured

–

---

5 STABILITY

pH

5.5–7 (best value for storage) [2]

Temperature (°C)

45 (10 min, pH 5.5–7, 50% loss of activity) [2]; 50 (10 min, stable below) [4];
55 (10 min, pH 5.5–7, complete loss of activity) [2]; 60 (10 min, inactivation)
[4]

Oxidation

Organic solvent

General stability information

4-Nitrophenyl sulfate, stabilizes against thermal inactivation and against inactivation on storage under cold conditions [2]

Storage

4°C or –20°C, more than 50% loss of activity after 1 month [2]

---

6 CROSSREFERENCES TO STRUCTURE DATABANKS

PIR/MIPS code

Brookhaven code

---

7 LITERATURE REFERENCES

[1] Kobashi, K., Kim, D.H., Morikawa, T.: J. Protein Chem.,6,237–244 (1987)
[2] Kim, D.-H., Konishi, L., Kobashi, K.: Biochim. Biophys. Acta,872,33–41 (1986)
[3] Konishi-Imamura, L., Dohi, K., Sato, M., Kobashi, K.: J. Biochem.,115,1097–1100
    (1994)
[4] Kim, D.-H., Kim, H.-S., Kobashi, K.: J. Biochem.,112,456–460 (1992)

1 NOMENCLATURE

EC number
2.8.2.23

Systematic name
3'-Phosphoadenylylsulfate:heparin-glucosamine 3-O-sulfotransferase

Recommended name
Heparin-glucosamine 3-O-sulfotransferase

Synonyms
Glucosaminyl 3-O-sulfotransferase
Sulfotransferase, glucosaminyl 3-O-

CAS Reg. No.
118113-79-4

---

2 REACTION AND SPECIFICITY

Catalyzed reaction
3'-Phosphoadenylylsulfate + heparin-glucosamine →
→ adenosine 3',5'-bisphosphate + heparin-glucosamine 3-O-sulfate

Reaction type
Sulfate group transfer

Natural substrates
3-Phosphoadenylylsulfate + heparin-glucosamine (enzyme brings about the final stage in biosynthesis of heparin) [1]

Substrate spectrum
1 3-Phosphoadenylylsulfate + heparin-glucosamine (3-O-sulfation occurs only after the introduction of all other structural components required for the high affinity interaction with antithrombin, including the 6-O-sulfate groups on glucosamine unit 6) [1]

Product spectrum
1 Adenosine 3',5'-bisphosphate + heparin-glucosamine 3-O-sulfate [1]

Inhibitor(s)
More (3-O-sulfation may be restricted by other, as yet unidentified, inhibitory structural elements that are preferentially expressed in polysaccharide sequences selected for the generation of heparin with low affinity for antithrombin) [2]

Cofactor(s)/prosthetic group(s)/activating agents

Metal compounds/salts

Turnover number (min^{-1})

Specific activity (U/mg)

K_m-value (mM)

pH-optimum

pH-range

Temperature optimum (°C)

Temperature range (°C)

3 ENZYME STRUCTURE

Molecular weight

Subunits

Glycoprotein/Lipoprotein

–

4 ISOLATION/PREPARATION

Source organism
Mouse [1, 2]

Source tissue
Mastocytoma tissue [1, 2]

Localization in source
Microsomes [1, 2]

Purification

Crystallization
–

Cloned
–

Renatured
–

5 STABILITY

pH

Temperature (°C)

Oxidation

Organic solvent

General stability information

Storage

6 CROSSREFERENCES TO STRUCTURE DATABANKS

PIR/MIPS code

Brookhaven code

7 LITERATURE REFERENCES

[1] Kusche, M., Bäckström, G., Riesenfeld, J., Petitou, M., Choay, J., Lindahl, U.:
J. Biol. Chem.,263,15474–15484 (1988)
[2] Kusche, M., Torri, G., Casu, B., Lindahl, U.: J. Biol. Chem.,265,7292–7300 (1990)

1 NOMENCLATURE

EC number
2.8.2.24

Systematic name
3'-Phosphoadenylylsulfate:desulfoglucosinolate sulfotransferase

Recommended name
Desulfoglucosinolate sulfotransferase

Synonyms
Sulfotransferase, desulfoglucosinolate
PAPS-desulfoglucosinolate sulfotransferase
3'-Phosphoadenosine-5'-phosphosulfate:desulfoglucosinolate sulfotransferase

CAS Reg. No.
121479-85-4

---

2 REACTION AND SPECIFICITY

Catalyzed reaction
3'-Phosphoadenylylsulfate + desulfoglucotropeolin →
→ adenosine 3',5'-bisphosphate + glucotropeolin (mechanism [4])

Reaction type
Sulfate group transfer

Natural substrates
3'-Phosphoadenylylsulfate + desulfobenzylglucosinolate (involved with EC
2.4.1.195 in final steps of thioglycoside biosynthesis in cruciferous plants)
[3, 5]

Substrate spectrum
1 3'-Phosphoadenylylsulfate + desulfobenzylglucosinolate (absolute specificity for desulfoglucosinolate structure) [1–5]
2 3'-Phosphoadenylylsulfate + desulfo-p-hydroxy-benzylglucosinolate [4]
3 3'-Phosphoadenylylsulfate + desulfoallylglucosinolate [4]
4 More (no acceptors are quercetin, rutin, kaempferol, p-coumaric acid, ferulic acid, caffeic acid, phenylacetaldoxime [4], adenosine-5'-phospho-
   sulfate [5]) [4, 5]

Product spectrum
1 Adenosine 3',5'-bisphosphate + benzylglucosinolate [1–5]
2 ?
3 ?
4 ?

Inhibitor(s)
$PbNO_3$ [4]; $NiCl_2$ [4]; Co^{2+} (low concentration) [5]; Cu^{2+} (low concentration) [5]; Zn^{2+} (low concentration [5], $ZnCl_2$ [4]) [4, 5]; Mn^{2+} (10 mM, stimulating at 1 mM) [5]; Adenosine-3',5'-bisphosphate (product inhibition [4], kinetics [5]) [4, 5]; Iodoacetic acid (weak) [4, 5]; Iodoacetamide (weak [4]) [4, 5]; Iodosobenzoic acid [5]; N-Ethylmaleimide (2-mercaptoethanol protects) [4]; N-Pyrenylmaleimide [5]; N-Methylmaleimide [5]; PCMB [5]; Phenylmercuriacetate [5]; p-Chloromercuriphenylsulfonic acid (2-mercaptoethanol protects) [4]; DTNB (2-mercaptoethanol protects) [4]; More (fairly insensitive to EDTA, 2,2'-dipyridyl, 1,10-phenanthroline, DIECA, 10 mM each [4], no inhibition by 3'-AMP, 5'-AMP, 2',5'-AMP, 3',5'-cAMP, 5'-ADP, 5'-ATP, adenosine-5'-phosphosulfate, benzaldoxime, phenylacetohydroximate, phenylacetothiohydroximate, S-methylphenylacetothiohydroximate, benzylglucosinolate [5]) [4, 5]

Cofactor(s)/prosthetic group(s)/activating agents

Metal compounds/salts
Mg^{2+} (activation [4, 5], slight [5]) [4, 5]; Mn^{2+} (activation [4, 5], slight at 1 mM, 10 mM inhibits [5]) [4, 5]

Turnover number (min^{-1})

Specific activity (U/mg)
More [1]; 0.0000329 [5]; 0.0156 [4]

K_m-value (mM)
More (kinetic studies) [4]; 0.00078 (3'-phosphoadenylylsulfate) [5]; 0.0023 (desulfobenzylglucosinolate) [5]; 0.0065 (desulfoallylglucosinolate) [4]; 0.06 (3'-phosphoadenylylsulfate) [4]; 0.082 (desulfobenzylglucosinolate) [4]; 0.67 (desulfo-p-hydroxy-benzylglucosinolate) [4]

pH-optimum
More (pI: 4.84 [5], pI: 5.2 [4]) [4, 5]; 8.5–9.0 (maximal activity in Tris-HCl buffer) [5]; 9.0 [4]

pH-range
7.0–9.5 (about 65% of maximal activity at pH 7.0 and 9.5) [4]

Temperature optimum (°C)
30 [5]

Temperature range (°C)

25–40 (about 85% of maximal activity at 25°C and about half-maximal activity at 40°C) [5]

3 ENZYME STRUCTURE

Molecular weight

31000 (Lepidium sativum, FPLC gel filtration) [4]
44000 (Brassica juncea, gel filtration) [5]

Subunits

Monomer (1 × 31000–55000, Lepidium sativum, SDS-PAGE) [4]

Glycoprotein/Lipoprotein

–

4 ISOLATION/PREPARATION

Source organism

Brassica juncea (cv. Cutlass [1–3, 5], cv. Domo [1]) [1–3, 5]; Lepidium sativum (no.5089, curled cress) [4]; Brassica napus (rapeseed, cv. Westar) [1]; Brassica campestris (cv. R-500 [1]) [1, 4]; Brassica oleracea (savoy cabbage) [1]; Brassica nigra (cv. 526) [1]; Tropaeolum majus [4]; Sinapis alba [4]; Arabidopsis thaliana [4]

Source tissue

Seedlings (etiolated [4], tissue distribution [2]) [1, 2, 4]; Cell culture (derived from hypocotyls) [3, 5]

Localization in source

Cytoplasm (subcellular distribution [2]) [2, 3]

Purification

Lepidium sativum [4]; Brassica juncea (partial) [5]; Brassica napus (partial) [1]; More (persistently co-purified with EC 2.4.1.195) [3]

Crystallization

–

Cloned

–

Renatured

–

5 STABILITY

pH
6.0 ($t_{1/2}$: 1 h, 4°C) [3]; 8.0 (stable above) [3]; 10.5 ($t_{1/2}$: 1 h, 4°C) [3]

Temperature (°C)
40 (up to, stable [3], at least 1 h, inactivation above [5]) [3, 5]; 45 ($t_{1/2}$: 1 h, 20 mM Tris-HCl buffer, pH 7.5, 14 mM 2-mercaptoethanol) [3]

Oxidation

Organic solvent

General stability information
Bovine serum albumin, required for stabilizing [4]; $MgCl_2$, 2-mercaptoethanol, DTT and GSH do not stabilize [4]; Dilution inactivates [4]

Storage
–20°C, 0.25 M sucrose and bovine serum albumin, 6 months [4]; –20°C, 20 mM Tris buffer, pH 7.5, 14 mM 2-mercaptoethanol, 10% glycerol, at least 2 months [5]; 4°C, more than 50% loss of activity within 48 h [5]; 4°C, 14 mM 2-mercaptoethanol, $t_{1/2}$: about 2 weeks [5]

6 CROSSREFERENCES TO STRUCTURE DATABANKS

PIR/MIPS code

Brookhaven code

7 LITERATURE REFERENCES

[1] Jain, J.C., Reed, D.W., Groot Wassink, J.W.D., Underhill, E.W.: Anal. Biochem., 178,137–140 (1989)
[2] Jain, J.C., Michayluk, M.R., Groot Wassink, J.W.D., Underhill, E.W.: Plant Sci.,64, 25–29 (1989)
[3] Jain, J.C., Groot Wassink, J.W.D., Reed, D.W., Underhill, E.W.: J. Plant Physiol., 136,356–361 (1990)
[4] Glendening, T.M., Poulton, J.E.: Plant Physiol.,94,811–816 (1990)
[5] Jain, J.C., Groot Wassink, J.W.D., Kolenovsky, A.D., Underhill, E.W.: Phytochemistry,29,1425–1428 (1990)

1 NOMENCLATURE

EC number
2.8.2.25

Systematic name
3'-Phosphoadenylylsulfate:quercetin 3-sulfotransferase

Recommended name
Flavonol 3-sulfotransferase

Synonyms
Sulfotransferase, flavonol 3-

CAS Reg. No.
121855-10-5

2 REACTION AND SPECIFICITY

Catalyzed reaction
3'-Phosphoadenylylsulfate + quercetin →
→ adenosine 3',5'-bisphosphate + quercetin 3-sulfate

Reaction type
Sulfate group transfer

Natural substrates
More (involved in biosynthesis of polysulfated flavonols in Flaveria chlorae-folia [1, 4], in Flaveria bidentis [3]) [1, 3, 4]

Substrate spectrum
1. 3'-Phosphoadenylylsulfate + quercetin (Flaveria chloraefolia [1, 2, 4, 5], 58% of activity compared to rhamnetin [4, 5], Flaveria bidentis [1, 3]) [1–5]
2. 3'-Phosphoadenylylsulfate + isorhamnetin (Flaveria chloraefolia [1, 2, 4, 5], 94% of activity compared to rhamnetin [4, 5], 10% of activity compared to quercetin, Flaveria bidentis [3]) [1–5]
3. 3'-Phosphoadenylylsulfate + rhamnetin (Flaveria chloraefolia [1, 2, 4, 5], best substrate [4, 5], Flaveria bidentis, 75% of activity compared to quercetin [3]) [1–5]
4. 3'-Phosphoadenylylsulfate + patuletin (Flaveria chloraefolia [1, 2, 4, 5], 52% of activity compared to rhamnetin [4]) [1, 2, 4, 5]
5. 3'-Phosphoadenylylsulfate + kaempferol (Flaveria chloraefolia [1, 2, 4, 5], 48% of activity compared to rhamnetin [4, 5]) [1, 2, 4, 5]
6. 3'-Phosphoadenylylsulfate + eupatin (Flaveria chloraefolia) [1]

7 3'-Phosphoadenylylsulfate + ombuin (37% of activity compared to rhamnetin) [4]

8 3'-Phosphoadenylylsulfate + tamarixetin (31% of activity compared to rhamnetin) [4]

9 More (quercetagetin, gossypetin, myricetin or galangin are no substrates) [1, 2, 4]

Product spectrum

1 Adenosine 3',5'-bisphosphate + quercetin 3-sulfate (Flaveria chloraefolia [1, 2, 4, 5], Flaveria bidentis [1, 3]) [1-5]

2 Adenosine 3',5'-bisphosphate + isorhamnetin 3-sulfate (Flaveria chlorae-folia [1, 2, 4, 5], Flaveria bidentis [3]) [1-5]

3 Adenosine 3',5'-bisphosphate + rhamnetin 3-sulfate (Flaveria chloraefolia [1, 2, 4, 5], Flaveria bidentis [3]) [1-5]

4 Adenosine 3',5'-bisphosphate + patuletin 3-sulfate (Flaveria chloraefolia [1, 2, 4, 5]) [1, 2, 4, 5]

5 Adenosine 3',5'-bisphosphate + kaempferol 3-sulfate (Flaveria chloraefo-lia) [1, 2, 4, 5]

6 Adenosine 3',5'-bisphosphate + eupatin 3-sulfate [1]

7 ?

8 ?

9 ?

Inhibitor(s)

Quercetin 3-sulfate (noncompetitive with respect to quercetin or 3'-phosphoadenylylsulfate) [2]; 3',5'-Diphosphoadenosine (competitive with respect to 3'-phosphoadenylylsulfate, noncompetitive with respect to quercetin) [2]; More (EDTA, SH-reagents (1 and 10 mM), e.g. p-chloromercuribenzoate, iodoacetate or iodoacetamide are no inhibitors) [2, 4]

Cofactor(s)/prosthetic group(s)/activating agents

Metal compounds/salts

More (no divalent cation required) [2, 4]

Turnover number (min^{-1})

Specific activity (U/mg)

0.00027 (Flaveria bidentis [1]) [1, 4]

K_m-value (mM)

0.00018 (3'-phosphoadenylylsulfate (+ quercetin)) [2]; 0.0002 (3'-phosphoadenylylsulfate (+ rhamnetin) [1], (+ quercetin) [4], rhamnetin [1], quercetin [2, 4]) [1, 2, 4]; 0.0003 (quercetin, cloned enzyme) [3]; 0.0004 (3'-phosphoadenylylsulfate (+ quercetin), cloned enzyme) [3]

2

pH-optimum
6 (+ 8.5, two maxima) [2]; 6.5 (+ 8.5, two maxima) [1, 3, 4]; 8.5 (+ 6, two maxima [2], + 6.5, two maxima [1, 3, 4]) [1–4]

pH-range

Temperature optimum (°C)
30 (assay at) [2, 4]

Temperature range (°C)

---

3 ENZYME STRUCTURE

Molecular weight
35000 (Flaveria chloraefolia, gel filtration) [1, 2, 4]

Subunits
Monomer (1 × 34500, Flaveria chloraefolia, SDS-PAGE) [2]

Glycoprotein/Lipoprotein
–

---

4 ISOLATION/PREPARATION

Source organism
Flaveria chloraefolia [1, 2, 4, 5]; Flaveria bidentis [3]

Source tissue
Plant [1, 2]; Callus [3]; Shoot tips [4, 5]

Localization in source

Purification
Flaveria chloraefolia (partial) [1, 2, 4, 5]; Flaveria bidentis (partial) [3]

Crystallization
–

Cloned
[3, 5]

Renatured
–

5 STABILITY

pH

Temperature (°C)

Oxidation

Organic solvent

General stability information

Storage

6 CROSSREFERENCES TO STRUCTURE DATABANKS

PIR/MIPS code

PIR2:A42047 (Flaveria chloraefolia (fragments))

Brookhaven code

7 LITERATURE REFERENCES

[1] Varin, L.: Bull. Liaison-Groupe Polyphenols,14,248–257 (1988)
[2] Varin, L., Ibrahim, R.K.: J. Biol. Chem.,267,1858–1863 (1992)
[3] Varin, L., Gulick, P., Ibrahim, R.: Plant Physiol.,106,485–491 (1994)
[4] Varin, L., Ibrahim, R.K.: Plant Physiol.,90,977–981 (1989)
[5] Varin, L., DeLuca, V., Ibrahim, R.K., Brisson, N.: Proc. Natl. Acad. Sci. USA,89, 1286–1290 (1992)

1 NOMENCLATURE

EC number
    2.8.2.26

Systematic name
    3'-Phosphoadenylylsulfate:quercetin-3-sulfate 3'-sulfotransferase

Recommended name
    Quercetin-3-sulfate 3'-sulfotransferase

Synonyms
    Sulfotransferase, flavonol 3'-
    Flavonol 3'-sulfotransferase
    3'-Sulfotransferase [1]
    PAPS:flavonol 3-sulfate 3'-sulfotransferase [2]

CAS Reg. No.
    121855-11-6

2 REACTION AND SPECIFICITY

Catalyzed reaction
    3'-Phosphoadenylylsulfate + quercetin 3-sulfate →
    → adenosine 3',5'-bisphosphate + quercetin 3,3'-bissulfate

Reaction type
    Sulfate group transfer

Natural substrates
    More (sulfation at 3'-positions is second step in formation of polysulfated
    flavonoids in Flaveria chloraefolia) [1]

Substrate spectrum
    1 3'-Phosphoadenylylsulfate + quercetin 3-sulfate [1, 2]
    2 3'-Phosphoadenylylsulfate + tamarixetin 3-sulfate (33% of activity com-
       pared to quercetin 3-sulfate [2]) [1, 2]
    3 3'-Phosphoadenylylsulfate + patuletin 3-sulfate (33% of activity compared
       to quercetin 3-sulfate) [2]
    4 More (kaempferol 3-sulfate, isorhamnetin 3-sulfate or other flavonol agly-
       cones are no substrates) [1, 2]

Product spectrum
1 Adenosine 3',5'-bisphosphate + quercetin 3,3'-bissulfate [1, 2]
2 ?
3 ?
4 ?

Inhibitor(s)
More (EDTA up to 10 mM, or SH-group reagents e.g. p-chloromercuribenzoate, iodoacetate, iodoacetamide at 1 and 10 mM are no inhibitors) [2]

Cofactor(s)/prosthetic group(s)/activating agents

Metal compounds/salts

Turnover number (min^{-1})

Specific activity (U/mg)
0.0025 [2]

K_m-value (mM)
0.00029 (quercetin 3-sulfate) [1, 2]; 0.00035 (3'-phosphoadenylylsulfate) [1, 2]

pH-optimum
7.5 [1, 2]

pH-range

Temperature optimum (°C)

Temperature range (°C)

3 ENZYME STRUCTURE

Molecular weight
35000 (Flaveria chloraefolia, gel filtration) [1, 2]

Subunits

Glycoprotein/Lipoprotein
–

4 ISOLATION/PREPARATION

Source organism
Flaveria chloraefolia [1, 2]

Source tissue
Seed [1]; Shoot tips [2]

Localization in source

Purification
   Flaveria chloraefolia (partial) [1, 2]

Crystallization
   –

Cloned
   –

Renatured
   –

---

5 STABILITY

pH

Temperature (°C)

Oxidation

Organic solvent

General stability information

Storage
   4°C, half-life of 24 h [2]; –20°C, 1 mg/ml bovine serum albumin, half-life of 3 days [2]

---

6 CROSSREFERENCES TO STRUCTURE DATABANKS

PIR/MIPS code

Brookhaven code

---

7 LITERATURE REFERENCES

[1] Varin, L.: Bull. Liaison-Groupe Polyphenols,14,248–257 (1988)
[2] Varin, L., Ibrahim, R.K.: Plant Physiol.,90,977–981 (1989)

1 NOMENCLATURE

EC number
2.8.2.27

Systematic name
3'-Phosphoadenylylsulfate:quercetin-3-sulfate 4'-sulfotransferase

Recommended name
Quercetin-3-sulfate 4'-sulfotransferase

Synonyms
Sulfotransferase, flavonol 4'-
Flavonol 4'-sulfotransferase
PAPS:flavonol 3-sulfate 4'-sulfotransferase [3]

CAS Reg. No.
121855-12-7

2 REACTION AND SPECIFICITY

Catalyzed reaction
3'-Phosphoadenylylsulfate + quercetin 3-sulfate →
→ adenosine 3',5'-bisphosphate + quercetin 3,4'-bissulfate

Reaction type
Sulfate group transfer

Natural substrates
More (involved in biosynthesis of polysulfated flavonoids in Flaveria chloraefolia) [1]

Substrate spectrum
1 3'-Phosphoadenylylsulfate + quercetin 3-sulfate [1–3]
2 3'-Phosphoadenylylsulfate + isorhamnetin 3-sulfate (38% of activity compared to quercetin 3-sulfate [2, 3]) [1–3]
3 3'-Phosphoadenylylsulfate + kaempferol 3-sulfate (45% of activity compared to quercetin 3-sulfate [2, 3]) [1–3]
4 3'-Phosphoadenylylsulfate + patuletin 3-sulfate (12% of activity compared to quercetin 3-sulfate) [3]
5 More (tamarixetin 3-sulfate, flavonol aglycones, or adenosine 5'-phosphosulfate are no substrates) [1–3]

Product spectrum
1 Quercetin 3,4'-disulfate + adenosine 3',5'-bisphosphate [1–3]
2 Isorhamnetin 3,4'-disulfate + adenosine 3',5'-bisphosphate [1–3]
3 Kaempferol 3,4'-disulfate + adenosine 3',5'-bisphosphate [1–3]
4 Patuletin 3,4'-disulfate + adenosine 3',5'-bisphosphate [3]
5 ?

Inhibitor(s)
Phosphate buffer (complete inhibition) [3]; More (EDTA up to 10 mM,
SH-group reagents, e.g. p-chloromercuribenzoate, iodoacetate, iodoacet-
amide, all at 1 and 10 mM, are no inhibitors) [3]

Cofactor(s)/prosthetic group(s)/activating agents

Metal compounds/salts
More (no requirement for divalent cations) [3]

Turnover number (min^{-1})

Specific activity (U/mg)
0.002 [1, 3]

K_m-value (mM)
0.00036 (quercetin 3-sulfate) [1, 3]; 0.00038 (3'-phosphoadenylylsulfate) [1,
3]

pH-optimum
7.5 [1, 3]

pH-range

Temperature optimum (°C)
30 (assay at) [3]

Temperature range (°C)

3 ENZYME STRUCTURE

Molecular weight
35000 (Flaveria chloraefolia, gel filtration) [3]

Subunits

Glycoprotein/Lipoprotein
–

4 ISOLATION/PREPARATION

Source organism
Flaveria chloraefolia [1–3]

Source tissue
Seed [1]; Shoot tips [2]

Localization in source

Purification
Flaveria chloraefolia (partial) [1, 2]

Crystallization
–

Cloned
[2]

Renatured
–

5 STABILITY

pH

Temperature (°C)

Oxidation

Organic solvent

General stability information

Storage
4°C, half-life of 24 h [3]; –20°C, bovine serum albumin up to 1 mg/ml, half-life of 3 days [3]

6 CROSSREFERENCES TO STRUCTURE DATABANKS

PIR/MIPS code

Brookhaven code

7 LITERATURE REFERENCES

[1] Varin, L.: Bull. Liaison-Groupe Polyphenols,14,248–257 (1988)
[2] Varin, L., DeLuca, V., Ibrahim, R.K., Brisson, N.: Proc. Natl. Acad. Sci. USA,89, 1286–1290 (1992)
[3] Varin, L., Ibrahim, R.K.: Plant Physiol.,90,977–981 (1989)

1 NOMENCLATURE

EC number
2.8.2.28

Systematic name
3'-Phosphoadenylylsulfate:quercetin-3,3'-bissulfate 7-sulfotransferase

Recommended name
Quercetin-3,3'-bissulfate 7-sulfotransferase

Synonyms
Sulfotransferase, flavonol 7-
Flavonol 7-sulfotransferase
7-Sulfotransferase [1]
PAPS:flavonol 3,3'/3,4'-disulfate 7-sulfotransferase [1]

CAS Reg. No.
121855-13-8

2 REACTION AND SPECIFICITY

Catalyzed reaction
3'-Phosphoadenylylsulfate + quercetin 3,3'-bissulfate →
→ adenosine 3',5'-bisphosphate + quercetin 3,3',7-trissulfate

Reaction type
Sulfate group transfer

Natural substrates
More (involved in biosynthesis of polysulfated flavonols in Flaveria bidentis)
[1, 2]

Substrate spectrum
1 3'-Phosphoadenylylsulfate + quercetin 3,3'-disulfate (isoenzyme I and II
   [1]) [1, 2]
2 3'-Phosphoadenylylsulfate + quercetin 3,4'-disulfate (isoenzyme I and II
   [1], best substrate [2]) [1, 2]
3 3'-Phosphoadenylylsulfate + isorhamnetin 3-sulfate (isoenzyme I and II
   [1]) [1, 2]
4 More (quercetin, quercetin 3-sulfate, quercetin 3'-sulfate, flavones (apige-
   nin, luteolin), phenylpropanoids (p-coumaric, caffeic or ferulic acids) are
   no substrates) [1]

Enzyme Handbook © Springer-Verlag Berlin Heidelberg 1997

Product spectrum
1 Quercetin 3,3',7-trisulfate + adenosine 3',5'-bisphosphate [1, 2]
2 Quercetin 3,4',7-trisulfate + adenosine 3',5'-bisphosphate [1, 2]
3 Isorhamnetin 3,7-disulfate + adenosine 3',5'-bisphosphate [1, 2]
4 ?

Inhibitor(s)
Flavonol substrate (< 0.0015 mM, inhibition) [1]; More (EDTA, SH-group reagents at 1–10 mM, no inhibition) [1]

Cofactor(s)/prosthetic group(s)/activating agents

Metal compounds/salts
More (no requirement for a divalent cation at 1–10 mM) [1]

Turnover number (min^{-1})

Specific activity (U/mg)
0.00038 (isoenzyme II) [1]; 0.00058 (isoenzyme I) [1]

K_m-value (mM)
0.0002 (quercetin 3,3'-disulfate, isoenzyme II) [1]; 0.00024 (quercetin 3,3'-disulfate, isoenzyme I) [1]; 0.00026 (quercetin 3,4'-disulfate) [2]; 0.00033 (3'-phosphoadenylylsulfate (+ quercetin 3,3'-disulfate), isoenzyme I) [1]; 0.00038 (3'-phosphoadenylylsulfate (+ quercetin 3,4'-disulfate)) [2]; 0.00046 (3'-phosphoadenylylsulfate (+ quercetin 3,3'-disulfate), isoenzyme II) [1]

pH-optimum
7.5 [1, 2]

pH-range
6.5–8.5 (about 40–45% of maximal activity at pH 6.5 and 8.5) [1]

Temperature optimum (°C)

Temperature range (°C)

---

3 ENZYME STRUCTURE

Molecular weight
35000 (Flaveria bidentis, gel filtration) [1, 2]

Subunits

Glycoprotein/Lipoprotein
–

4 ISOLATION/PREPARATION

Source organism
  Flaveria bidentis [1, 2]

Source tissue
  Shoot tips [1]; Seed [2]

Localization in source

Purification
  Flaveria bidentis (partial) [1, 2]

Crystallization
  –

Cloned
  –

Renatured
  –

---

5 STABILITY

pH

Temperature (°C)

Oxidation

Organic solvent

General stability information

Storage

---

6 CROSSREFERENCES TO STRUCTURE DATABANKS

PIR/MIPS code

Brookhaven code

---

7 LITERATURE REFERENCES

[1] Varin, L., Ibrahim, R.K.: Plant Physiol.,95,1254–1258 (1991)
[2] Varin, L.: Bull. Liaison-Groupe Polyphenols,14,249–257 (1988)

---

1 NOMENCLATURE

EC number
2.8.3.1

Systematic name
Acetyl-CoA:propanoate CoA-transferase

Recommended name
Propionate CoA-transferase

Synonyms
Coenzyme A transferase, propionate
Propionate coenzyme A-transferase
Propionate-CoA:lactoyl-CoA transferase
Propionyl CoA:acetate CoA transferase
Propionyl-CoA transferase

CAS Reg. No.
9026-15-7

2 REACTION AND SPECIFICITY

Catalyzed reaction
Acetyl-CoA + propanoate →
→ acetate + propanoyl-CoA

Reaction type
Coenzyme A transfer

Natural substrates
Acetyl-CoA + propanoate (reaction of (S)-alanine fermentation pathway) [2]

Substrate spectrum
1 Acetyl-CoA + propanoate (best substrate [2], specific for monocarboxylic acids [2]) [1, 2]
2 Acetyl-CoA + butanoate [2]
3 Acetyl-CoA + lactate ((R)-lactate preferred over (S)-lactate) [2]
4 Acetyl-CoA + acrylate [2]

Product spectrum
1 Acetate + propanoyl-CoA [1, 2]
2 Acetate + butanoyl-CoA
3 ?
4 ?

Inhibitor(s)

Cofactor(s)/prosthetic group(s)/activating agents

Metal compounds/salts

Turnover number (min^{-1})

Specific activity (U/mg)

K_m-value (mM)

pH-optimum

pH-range

Temperature optimum (°C)

Temperature range (°C)

3 ENZYME STRUCTURE

Molecular weight
224000 (Clostridium propionicum, gel filtration) [2]

Subunits
Tetramer (4 × 67000, Clostridium propionicum, SDS-PAGE) [2]

Glycoprotein/Lipoprotein
–

4 ISOLATION/PREPARATION

Source organism
Clostridium kluyverii [1]; Clostridium propionicum [2]

Source tissue
Cell [1, 2]

Localization in source

Purification
Clostridium propionicum [2]

Crystallization
–

Cloned
–

Renatured
–

5 STABILITY

pH

Temperature (°C)

Oxidation

Organic solvent

General stability information

Storage
    4°C, in saturated ammonium sulfate, several months [2]

6 CROSSREFERENCES TO STRUCTURE DATABANKS

PIR/MIPS code

Brookhaven code

7 LITERATURE REFERENCES

[1] Stadtman, E.R.: Fed. Proc.,11,291 (1952)
[2] Schweiger, G., Buckel, W.: FEBS Lett.,171,79–84 (1984)

1 NOMENCLATURE

EC number
  2.8.3.2

Systematic name
  Succinyl-CoA:oxalate CoA-transferase

Recommended name
  Oxalate CoA-transferase

Synonyms
  Succinyl-beta-ketoacyl-CoA transferase
  Oxalate coenzyme A-transferase
  Coenzyme A-transferase, oxalate

CAS Reg. No.
  9026-17-9

---

2 REACTION AND SPECIFICITY

Catalyzed reaction
  Succinyl-CoA + oxalate →
  → oxalyl-CoA + succinate

Reaction type
  Coenzyme A transfer

Natural substrates

Substrate spectrum
  1 Succinyl-CoA + oxalate [1]

Product spectrum
  1 Oxalyl-CoA + succinate [1]

Inhibitor(s)

Cofactor(s)/prosthetic group(s)/activating agents

Metal compounds/salts

Turnover number (min^{-1})

Specific activity (U/mg)

K_m-value (mM)

---

pH-optimum

pH-range

Temperature optimum (°C)

Temperature range (°C)

3 ENZYME STRUCTURE

Molecular weight

Subunits

Glycoprotein/Lipoprotein

–

4 ISOLATION/PREPARATION

Source organism
  Pseudomonas oxalaticus [1]

Source tissue
  Cell [1]

Localization in source

Purification

Crystallization

–

Cloned

–

Renatured

–

5 STABILITY

pH

Temperature (°C)

Oxidation

Organic solvent

General stability information

Storage

6 CROSSREFERENCES TO STRUCTURE DATABANKS

PIR/MIPS code

Brookhaven code

7 LITERATURE REFERENCES

[1] Quayle, J.R., Keech., D.B., Taylor, G.A.: Biochem. J.,78,225–236 (1961)

1 NOMENCLATURE

EC number
2.8.3.3

Systematic name
Acetyl-CoA:malonate CoA-transferase

Recommended name
Malonate CoA-transferase

Synonyms
Coenzyme A-transferase, malonate
Malonate coenzyme A-transferase
More (the bifunctional enzyme from Pseudomonas ovalis also catalyzes the
reaction of malonate decarboxylase, EC 4.1.1.9 [1])

CAS Reg. No.
9026-18-0

---

2 REACTION AND SPECIFICITY

Catalyzed reaction
Acetyl-CoA + malonate →
→ acetate + malonyl-CoA

Reaction type
Coenzyme A transfer

Natural substrates
Acetyl-CoA + malonate (inducible enzyme) [1]

Substrate spectrum
1 Acetyl-CoA + malonate [1]
2 Malonyl-CoA + malonate [1]
3 Methylmalonyl-CoA + malonate [1]
4 Propionyl-CoA + malonate [1]
5 Butyryl-CoA + malonate [1]
6 More (succinyl-CoA or palmityl-CoA cannot replace acetyl-CoA) [1]

Product spectrum
1 Acetate + malonyl-CoA
2 Malonyl-CoA + malonate
3 Malonyl-CoA + methylmalonate
4 Malonyl-CoA + propanoate
5 Malonyl-CoA + butanoate
6 ?

Inhibitor(s)

Cofactor(s)/prosthetic group(s)/activating agents

Metal compounds/salts

Turnover number (min^{-1})

Specific activity (U/mg)

K_m-value (mM)

pH-optimum

pH-range

Temperature optimum (°C)
  30 (assay at) [1]

Temperature range (°C)

---

3 ENZYME STRUCTURE

Molecular weight
  170000 (Pseudomonas ovalis, gel filtration) [1]

Subunits
  Tetramer ($1 \times 70000 + 1 \times 40000 + 2 \times 30000$, alphabeta$_1$(beta$_2$)$_2$, Pseudomonas ovalis, SDS-PAGE) [1]

Glycoprotein/Lipoprotein
  –

---

4 ISOLATION/PREPARATION

Source organism
  Pseudomonas ovalis (IAM 1177) [1]; Pseudomonas fluorescens (strain 23) [2]

Source tissue
  Cell [1, 2]

Localization in source

Purification
  Pseudomonas ovalis (partial) [1]

Crystallization
  –

Cloned

–

Renatured

–

5 STABILITY

pH

Temperature (°C)

Oxidation

Organic solvent

General stability information

Storage

6 CROSSREFERENCES TO STRUCTURE DATABANKS

PIR/MIPS code

Brookhaven code

7 LITERATURE REFERENCES

[1] Takamura, Y., Kitayama, Y.: Biochem. Int.,3,483–491 (1981)
[2] Hayaishi, O.: J. Biol. Chem.,215,125–136 (1955)

1 NOMENCLATURE

EC number
2.8.3.5

Systematic name
Succinyl-CoA:3-oxo-acid CoA-transferase

Recommended name
3-Oxoacid CoA-transferase

Synonyms
Coenzyme A-transferase, 3-oxoacid
3-Ketoacid CoA-transferase
3-Ketoacid coenzyme A transferase
3-Ketoacid coenzyme A-transferase
3-Oxoacid coenzyme A-transferase
3-Oxo-CoA transferase
Acetoacetate succinyl-CoA transferase
Acetoacetyl coenzyme A-succinic thiophorase
Succinyl coenzyme A-acetoacetyl coenzyme A-transferase
Succinyl-CoA transferase

CAS Reg. No.
9027-43-4

2 REACTION AND SPECIFICITY

Catalyzed reaction
Succinyl-CoA + a 3-oxo acid →
→ succinate + a 3-oxoacyl-CoA (mechanism [1, 14])

Reaction type
Coenzyme A transfer

Natural substrates
Succinyl-CoA + acetoacetate (ketolytic enzyme, uniquely involved in complete oxidation of ketone bodies [6], liver enzyme: produces a substrate circle between acetoacetyl-CoA and acetoacetate [15]) [6, 15]

Substrate spectrum

1 Succinyl-CoA + acetoacetate (r [1, 3, 4, 6, 7, 12–14]) [1–15]
2 Succinyl-CoA + maleate [12]
3 More (catalyzes exchange reactions in the absence of cosubstrates: succinate/succinyl-CoA and acetoacetate/acetoacetyl-CoA [1], no substrates are diacids with connecting chain lengths of 3 or more methylene groups [12]) [1, 12]

Product spectrum

1 Succinate + acetoacetyl-CoA (via enzyme-coenzyme A covalent complex [2]) [1–6, 9–15]
2 Succinate + maleyl-CoA
3 ?

Inhibitor(s)

Acetoacetyl-CoA (in the absence of succinate, cysteine restores, succinate or 0.1 M NaCl protects [1], EDTA, trisodium citrate and diphosphate protect, too, addition of Cu^{2+}, Mn^{2+}, Ca^{2+} or Zn^{2+} (decreasing order) restores inactivating activity of acetoacetyl-CoA [1], product inhibition, kinetics [14]) [1, 14]; Succinyl-CoA (product inhibition) [14]; Acetoacetate (product inhibition [1], substrate inhibition (above 1 mM [7]) [7, 8], kinetics [1]) [1, 7, 8]; Succinate (product inhibition [1, 6, 12, 14], kinetics [1, 14]) [1, 6, 12, 14]; NaCl (kinetics) [1]; Oxalate (kinetics) [12]; Malonate (kinetics [12], 0.1 M, weak [1]) [1, 12]; 2,2-Difluorosuccinate (strong) [12]; Perfluorosuccinate (strong) [12]; 3-Sulfopropanoate [12]; Monomethylsuccinate [12]; Succinamate [12]; Maleamate (r) [12]; N-Ethylmaleamate (r) [12]; Maleimide (succinate or acetoacetate protects) [12]; N-Ethylmaleimide (succinate or acetoacetate protects) [12]; 2-Nitro-5-(thiocyanate)benzoate (kinetics, methyl methanethiosulfonate and 5,5'-dithiobis(2-nitro-benzoate) protect, DTT removes this protection) [4]; 2,4-Dinitrophenylacetate (at pH 7.9, less inactivating activity at pH 7, acetoacetyl-CoA protects) [1]; p-Nitrophenylacetate (at pH 7.9, less inactivating activity at pH 7) [1]; Acetylimidazole (equally efficient at pH 7 and 7.9) [1]; Acetic anhydride [1]; Acetylene dicarboxylate (weak) [12]; Monovalent anions (decreasing order of effectiveness: SCN^-, ClO_4^-, I^-, Br^-, Cl^-, not F^-) [1]; SO_4^{2-} (0.1 M, weak) [1]; Malate (0.1 M, weak) [1]; Glutarate (0.1 M, weak) [1]; HPO_4^{2-} (0.1 M, weak) [1]; Citrate (0.1 M, weak) [1]; More (no inhibition by glutarate, adipate, cis- or trans-cyclobutane-1,2-dicarboxylate, cis- or trans-cyclohexane-1,2-dicarboxylate, methylsuccinate, mercaptosuccinate, malate, aspartate, succinimide or iodoacetamide) [12]

Cofactor(s)/prosthetic group(s)/activating agents

Metal compounds/salts

Turnover number (min^{-1})
2240 (succinyl-CoA) [1]; 56000 (acetoacetyl-CoA) [1]

Specific activity (U/mg)
2.88 [12]; 3 [1]; 3.9–4.5 [3]; 10.9 (heart) [9]; 15.6 (brain) [9]; 19.5 (skeletal muscle) [9]; 24.1 (kidney) [9]; 145 [2]; 161 [7]; 200 [13]; 280 [4]

K_m-value (mM)
More (kinetic study [1]) [1, 14]; 0.006 (acetoacetyl-CoA (+ succinate)) [3]; 0.025 (succinate (+ acetoacetyl-CoA)) [7]; 0.04 (acetoacetyl-CoA) [12]; 0.059 (acetoacetyl-CoA (+ succinate)) [7]; 0.07 (acetoacetate (+ succinyl-CoA)) [7]; 0.156 (succinyl-CoA (+ acetoacetate)) [7]; 0.2 (acetoacetate (+ succinyl-CoA [1, 6]), heart [6]) [1, 6]; 0.21 (acetoacetate, kidney, skeletal muscle) [6]; 0.28 (succinyl-CoA) [12]; 0.31 (acetoacetate, brain) [6]; 0.44 (acetoacetate) [12]; 0.72 (acetoacetyl-CoA (+ succinate)) [1]; 4.2 (succinyl-CoA (+ acetoacetate)) [1]; 28 (succinate (+ acetoacetyl-CoA)) [12]; 35 (maleate (+ acetoacetyl-CoA)) [12]; 36 (succinate (+ acetoacetyl-CoA)) [1]

pH-optimum
More (4 isozymes with pI: 5.72, 5.93, 6.2 and 6.5 [3], pI: 4.8 (pig brain enzyme), pI: 5.5 (pig heart and kidney enzymes), pI: 6.3 (sheep brain enzyme) [13], pI: 6.8 (heart enzyme [6]) [6, 12], pI: 7.6 [6], pI: 8.2 (sheep heart) [13], pI: 9 [13]) [3, 6, 12, 13]; 7.4 (assay at) [10]; 8–8.7 [1]; 8.1 (assay at) [14]

pH-range
7.1–8.7 (about 50% of activity at pH 7.1, pH 8–8.7: optimum) [1]

Temperature optimum (°C)
10 (assay at, all fish enzymes except Salmo gairdneri) [15]; 25 (assay at) [1, 4, 12, 15]; 30 (assay at) [2, 13, 14]; 37 (assay at) [11]

Temperature range (°C)

3 ENZYME STRUCTURE

Molecular weight
More (amino acid composition) [3, 13]
78000 (pig, gel filtration [1], PAGE [3]) [1, 3]
80000 (pig, gel filtration) [3]
90000 (rat, gel filtration) [7]
92000 (pig, sedimentation equilibrium centrifugation) [3]
100000 (rat, gel filtration) [6]
102000 (sheep, analytical ultracentrifugation) [13]
105000 (pig, SDS-PAGE after cross-linking with dimethyl dodecanediimidate) [3]
110000 (sheep, gel filtration) [13]
113000 (pig, gel filtration) [2]

Subunits

Dimer (2 × 45600, pig, sedimentation equilibrium in 6 M guanidine chloride
[3], 2 × 52000–63000, pig, SDS-PAGE [3], 2 × 52197, pig, deduced from
amino acid sequence [5], 2 × 53000, rat, SDS-PAGE [7], 2 × 55000, pig,
SDS-PAGE [2], 2 × 55000–58000, rat, SDS-PAGE [9], 2 × 56000, sheep,
SDS-PAGE [13]) [2, 3, 5, 7, 9, 13]

Glycoprotein/Lipoprotein

Glycoprotein (total sugar content: 1.6%) [3]

---

4 ISOLATION/PREPARATION

Source organism

Pig (piglet [5]) [1–5, 13]; Rat (male adult Buffalo [6, 9, 12], male Wistar [10])
[6–12, 15]; Sheep [13, 14]; Mouse [11, 15]; Rabbit [15]; Columba livia (domestic pigeon) [15]; Gallus gallus (domestic fowl) [15]; Lacerta viridis
(green lizard) [15]; Salmo gairdneri (rainbow trout) [15]; Scombrus scombrus (mackerel) [15]; Clupea harengus (herring) [15]; Dicentrarcus labrax
(bass) [15]; Pleuronecthes platessa (plaice) [15]; Scylliorhinus canicula
(dogfish) [15]; Raja clavata (ray) [15]

Source tissue

Heart [1–5, 8, 9, 12, 13]; Brain [6–9, 13]; Kidney [6, 8, 9, 13, 14]; Skeletal
muscle [6, 8, 9]; Stomach (glandular mucosa) [10]; Intestines (duodenum,
jejunum, ileum, caecum, colon, muscle) [10]; Neural cells (Neuroblastoma
N2a cell line (mouse), Glioma C6 cell line (rat), cell suspension culture) [11];
Liver [6, 15]; More (distribution in gastro-intestinal tract) [10]

Localization in source

Mitochondria (predominant [6]) [5–7, 10–12]; More (subcellular distribution)
[6]

Purification

Pig (partial [1], 4 isozymes, separable by isoelectric focussing [3]) [1–3];
Rat (partial [6, 9, 12]) [6, 7, 9, 12]; Sheep [13]

Crystallization

–

Cloned

(pig, cDNA clone of mature mitochondrial and cytoplasmic precursor to mitochondrial enzyme) [5]

Renatured

–

5 STABILITY

pH
3.1 (below, 1 min at 25°C, inactivation) [1]; 5 (slow loss of activity at 25°C) [1]; 10.7 (slow loss of activity at 25°C) [1]

Temperature (°C)
25 (1 min, inactivation at pH-values below 3.1 or in 0.1 M NaOH, slow loss of activity at pH 5 and 10.7) [1]

Oxidation

Organic solvent

General stability information
Enzyme is susceptible to proteolytic cleavage to produce a nicked but active enzyme, PMSF and EDTA protect [5]; Deoxycholate does not stabilize [6]; Degradation of 3-oxo acid CoA-transferase in glioma and neuroblastoma cells [11]

Storage
–20°C, 1.4 mg protein/ml, 0.02 M potassium phosphate buffer, pH 7.4, $t_{1/2}$: 9 months [1]; –20°C, at least 1 month [13]; Frozen, less than 10% loss of activity within 2 months [9]; Frozen, partially purified preparation in phosphate solution, less than 10% loss of activity within 1–2 weeks [12]; Storage as ammonium sulfate suspension leads to rapid loss of activity [13]

---

6 CROSSREFERENCES TO STRUCTURE DATABANKS

PIR/MIPS code
PIR2:A41771 (pig)

Brookhaven code

---

7 LITERATURE REFERENCES

[1] Hersh, L.B., Jencks, W.P.: J. Biol. Chem.,242,3468–3480 (1967)
[2] Edwards, M.R., Singh, M., Tubbs, P.K.: FEBS Lett.,37,155–158 (1973)
[3] White, H., Jencks, W.P.: J. Biol. Chem.,251,1708–1711 (1976)
[4] Kindman, L.A., Jencks, W.P.: Biochemistry,20,5183–5187 (1981)
[5] Lin, T., Bridger, W.A.: J. Biol. Chem.,267,975–978 (1992)
[6] Fenselau, A., Wallis, K.: Biochem. J.,142,619–627 (1974)
[7] Russell, J.J., Patel, M.S.: J. Neurochem.,38,1446–1452 (1982)
[8] Fenselau, A., Wallis, K.: Life Sci.,15,811–818 (1974)
[9] Fenselau, A., Wallis, K.: Biochem. Biophys. Res. Commun.,62,350–356 (1975)
[10] Hanson, P.J., Carrington, J.M.: Biochem. J.,200,349–355 (1981)
[11] Haney, P.M., Bolinger, L., Raefsky, C., Patel, M.S.: Biochem. J.,224,67–74 (1984)
[12] Fenselau, A., Wallis, K.: Biochemistry,13,3884–3889 (1974)
[13] Sharp, J.A., Edwards, M.R.: Biochem. J.,173,759–765 (1978)
[14] Sharp, J.A., Edwards, M.R.: Biochem. J.,213,179–185 (1983)
[15] Zammit, V.A., Beis, A., Newsholme, E.A.: FEBS Lett.,103,212–215 (1979)

1 NOMENCLATURE

EC number
2.8.3.6

Systematic name
Succinyl-CoA:3-oxoadipate CoA-transferase

Recommended name
3-Oxoadipate CoA-transferase

Synonyms
Coenzyme A-transferase, 3-oxoadipate
3-Oxoadipate coenzyme A-transferase
3-Oxoadipate succinyl-CoA transferase

CAS Reg. No.
9026-16-8

2 REACTION AND SPECIFICITY

Catalyzed reaction
Succinyl-CoA + 3-oxoadipate →
→ succinate + 3-oxoadipyl-CoA

Reaction type
Coenzyme A transfer

Natural substrates
Succinyl-CoA + 3-oxoadipate (mediates penultimate step in conversion of
protocatechuate to succinate and acetyl-CoA via 3-oxoadipate pathway [2],
together with EC 4.1.1.44 and EC 3.1.1.24 a component of 3-oxoadipate
pathway [3]) [2, 3]

Substrate spectrum
1 Succinyl-CoA + 3-oxoadipate (r [1]) [1–5]
2 More (no substrates are acetoacetate, oxalacetate or acetyl-CoA, and in
the reverse reaction malonate, fumarate, oxalate or acetate) [1]

Product spectrum
1 Succinate + 3-oxoadipyl-CoA [1]
2 ?

Inhibitor(s)
p-Chloromercuribenzoate (one molecule per alphabeta-protomer inactivates, DTT restores) [4]

Cofactor(s)/prosthetic group(s)/activating agents

Metal compounds/salts

Turnover number (min^{-1})

Specific activity (U/mg)
   21.3 (Pseudomonas putida) [2]; 24.2 (Acinetobacter calcoaceticus, transferase I) [2]

K_m-value (mM)

pH-optimum

pH-range

Temperature optimum (°C)
   25 (assay at) [1]

Temperature range (°C)

3 ENZYME STRUCTURE

Molecular weight
   108000 (Acinetobacter calcoaceticus, gel filtration) [2]
   109000 (Pseudomonas putida, gel filtration) [2]

Subunits
   Tetramer (2×25600 (alpha$_2$) + 2×26500 (beta$_2$), Acinetobacter calcoaceticus, SDS-PAGE, 2×24200 (alpha$_2$) + 2×25300 (beta$_2$), Pseudomonas
   putida, SDS-PAGE) [2]
   More (amino acid composition) [2]

Glycoprotein/Lipoprotein
   –

4 ISOLATION/PREPARATION

Source organism
   Pseudomonas putida (strains PRS 2260 [2], PRS2000 (wild-type),
   PRS2241(pHRP100) or PRS3004(pHRP100) [5]) [2–5]; Pseudomonas fluorescens [1]; Acinetobacter calcoaceticus (mutant strain ADP152, 2 trans-
   ferases, I: induced by protocatechuate [4], II: induced by cis,cis-muconate)
   [2, 4]; More (enzymes from Pseudomonas putida and Acinetobacter are im-
   munologically related, not identical) [2]

Source tissue
   Cell [1–5]

Localization in source
Soluble [1]

Purification
Pseudomonas fluorescens (partial) [1]; Pseudomonas putida [2]; Acineto-bacter calcoaceticus (transferase I) [2]

Crystallization
−

Cloned
(Pseudomonas putida pcaI and pcaJ-genes encoding the two subunits of the enzyme, expressed in E. coli) [5]

Renatured
−

---

5 STABILITY

pH

Temperature (°C)
22 (1 h, stable in the presence of DTT, Acinetobacter calcoaceticus) [4]; 40 ($t_{1/2}$: 26 min, Pseudomonas putida) [2]; 40–45 (30 min, stable, Acinetobacter calcoaceticus) [2]; 45 ($t_{1/2}$: 4 min, Pseudomonas putida) [2]; 50 (30 min, 10% loss of activity (Acinetobacter calcoaceticus), $t_{1/2}$: 1 min (Pseudomonas putida)) [2]; 60 (10 min, stable at neutral pH) [1]; 80 (5 min, inactivation at neutral pH) [1]

Oxidation

Organic solvent

General stability information
DTT stabilizes [4]

Storage
−10°C, at least 3 months [1]; 0°C, in the presence of DTT, 20% loss of activi-ty within 5 h [4]

---

6 CROSSREFERENCES TO STRUCTURE DATABANKS

PIR/MIPS code
PIR2:A44570 (alpha chain Acinetobacter calcoaceticus); PIR2:A42985 (alpha chain Pseudomonas putida); PIR2:B44570 (beta chain Acinetobacter calcoaceticus); PIR2:B42985 (beta chain Pseudomonas putida)

Brookhaven code

---

7 LITERATURE REFERENCES

[1] Katagiri, M., Hayaishi, O.: J. Biol. Chem.,226,439–448 (1957)
[2] Yeh, W.-K., Ornston, L.N.: J. Biol. Chem.,256,1565–1569 (1981)
[3] Yeh, W.-K., Ornston, L.N.: J. Bacteriol.,149,374–377 (1982)
[4] Yeh, W.-K., Ornston, L.N.: Arch. Microbiol.,138,102–105 (1984)
[5] Parales, R.E., Harwood, C.S.: J. Bacteriol.,174,4657–4666 (1992)

1 NOMENCLATURE

EC number
2.8.3.7

Systematic name
Succinyl-CoA:citramalate CoA-transferase

Recommended name
Succinate-citramalate CoA-transferase

Synonyms
Coenzyme A-transferase, citramalate
Itaconate CoA-transferase
Citramalate CoA-transferase
Succinyl coenzyme A-citramalyl coenzyme A transferase
More (cf. EC 2.8.3.11)

CAS Reg. No.
9033-60-7 (indistinguishable in Chemical Abstracts from EC 2.8.3.11)

2 REACTION AND SPECIFICITY

Catalyzed reaction
Succinyl-CoA + citramalate →
→ succinate + citramalyl-CoA

Reaction type
Coenzyme A transfer

Natural substrates

Substrate spectrum
1 Succinyl-CoA + citramalate (no substrate: acetyl-CoA) [1]
2 Succinyl-CoA + itaconate [1]

Product spectrum
1 Succinate + citramalyl-CoA [1]
2 Succinate + itaconyl-CoA [1]

Inhibitor(s)

Cofactor(s)/prosthetic group(s)/activating agents

Metal compounds/salts

Turnover number (min^{-1})

Specific activity (U/mg)

K_m-value (mM)

pH-optimum

pH-range

Temperature optimum (°C)

Temperature range (°C)

3 ENZYME STRUCTURE

Molecular weight

Subunits

Glycoprotein/Lipoprotein

–

4 ISOLATION/PREPARATION

Source organism
  Pseudomonas sp. (strain B2aba) [1]

Source tissue
  Cell [1]

Localization in source

Purification
  Pseudomonas sp. (partial) [1]

Crystallization
  –

Cloned
  –

Renatured
  –

5 STABILITY

pH

Temperature (°C)

Oxidation

Organic solvent

General stability information

Storage

6 CROSSREFERENCES TO STRUCTURE DATABANKS

PIR/MIPS code

Brookhaven code

7 LITERATURE REFERENCES

[1] Cooper, R.A., Kornberg, H.L.: Biochem. J.,91,82–91 (1964)

1 NOMENCLATURE

EC number
2.8.3.8

Systematic name
Acyl-CoA:acetate CoA-transferase

Recommended name
Acetate CoA-transferase

Synonyms
Coenzyme A-transferase, acetate
Acetate coenzyme A-transferase
Butyryl CoA:acetate CoA transferase
Butyryl coenzyme A transferase
Succinyl-CoA:acetate CoA transferase

CAS Reg. No.
37278-35-6

2 REACTION AND SPECIFICITY

Catalyzed reaction
Acyl-CoA + acetate →
→ a fatty acid anion + acetyl-CoA

Reaction type
Coenzyme A transfer

Natural substrates
Butanoyl-CoA + acetate (provides ability to grow on various fatty acids) [1]

Substrate spectrum
1 Butanoyl-CoA + acetate (r, best substrate) [1]
2 Pentanoyl-CoA + acetate (r) [1]

Product spectrum
1 Butanoate + acetyl-CoA [1]
2 Pentanoate + acetyl-CoA

Inhibitor(s)

Cofactor(s)/prosthetic group(s)/activating agents

Metal compounds/salts

Turnover number (min^{-1})

Specific activity (U/mg)
   0.1–1 (in crude cell extract) [1]

K$_m$-value (mM)

pH-optimum

pH-range

Temperature optimum (°C)

Temperature range (°C)

---

3 ENZYME STRUCTURE

Molecular weight

Subunits

Glycoprotein/Lipoprotein
   –

---

4 ISOLATION/PREPARATION

Source organism
   E. coli (K12, constitutive mutant strain V10) [1]

Source tissue
   Cell [1]

Localization in source

Purification

Crystallization
   –

Cloned
   –

Renatured
   –

5 STABILITY

pH

Temperature (°C)

Oxidation

Organic solvent

General stability information

Storage

6 CROSSREFERENCES TO STRUCTURE DATABANKS

PIR/MIPS code

Brookhaven code

7 LITERATURE REFERENCES

[1] Vanderwinkel, Furmanski, P., Reeves, H.C., Ajl, S.J.: Biochem. Biophys. Res. Commun.,33,902–908 (1968)

1 NOMENCLATURE

EC number
2.8.3.9

Systematic name
Butanoyl-CoA:acetoacetate CoA-transferase

Recommended name
Butyrate-acetoacetate CoA-transferase

Synonyms
Coenzyme A-transferase, butyryl coenzyme A-acetoacetate
Butyryl-CoA-acetoacetate CoA-transferase

CAS Reg. No.
66231-37-6

2 REACTION AND SPECIFICITY

Catalyzed reaction
Butanoyl-CoA + acetoacetate →
→ butanoate + acetoacetyl-CoA (mechanism [4])

Reaction type
Coenzyme A transfer

Natural substrates
Butanoate + acetoacetyl-CoA (reaction of lysine degradation pathway in
Clostridium) [1]

Substrate spectrum
1 Butanoate + acetoacetyl-CoA (r [1–3], reaction at 68% the rate of acetate [3]. Reverse reaction: at 25% the rate of forward reaction [1], succinyl-
   CoA, epsilon-acetyl-CoA or decanoyl-CoA cannot replace butyryl-CoA
   [3]. No substrates are monofluoroacetate or hexanoate [1]) [1–3]
2 Vinylacetate + acetoacetyl-CoA (best substrate) [1]
3 Acetate + acetoacetyl-CoA (r [1–3], best substrates [3], reaction at 24%
   the rate of butyrate [1], decanoyl-CoA cannot replace acetoacetyl-CoA,
   and succinate, malonate or 2,3-butandienoate cannot replace acetate
   [3]) [1–3]
4 Butanoate + acetyl-CoA (r, reaction at 56% the rate of acetoacetate, reverse reaction at 86% the rate of acetoacetyl-CoA) [3]
5 DL-3-Hydroxybutanoate + acetyl-CoA (reaction at 9% the rate of acetoacetate) [3]

6 Propanoate + acetoacetyl-CoA (r [1, 2], reaction at 21% the rate of butyrate [1]) [1, 2]

7 3-Mercaptopropanoate + acetoacetyl-CoA (reaction at 71% the rate of butyrate) [1]

8 Crotonate + acetoacetyl-CoA (r, reaction at 16% the rate of butyrate) [1]

9 Crotonate + butyryl-CoA [1]

10 Formate + acetoacetyl-CoA (reaction at 7% the rate of acetate) [3]

11 Octanoate + acetoacetyl-CoA (reaction at 3% the rate of acetate) [3]

12 Pentanoate + acetoacetyl-CoA (reaction at 19% the rate of butyrate) [1]

13 Isobutanoate + acetoacetyl-CoA (reaction at 19% the rate of butyrate) [1]

14 4-Pentenoate + acetoacetyl-CoA (reaction at 21% the rate of acetate) [3]

15 3-Methylvinylacetate + acetoacetyl-CoA (reaction at 44% the rate of butyrate) [1]

16 DL-3-Hydroxybutanoate + acetoacetyl-CoA (reaction at 30% the rate of butyrate) [1]

17 DL-2-Hydroxypentanoate + acetoacetyl-CoA (reaction at 7.9% the rate of butanoate) [1]

18 Isopentanoate + acetoacetyl-CoA (reaction at 8.3% the rate of butanoate) [1]

19 Acrylate + acetoacetyl-CoA (reaction at 10% the rate of butanoate) [1]

20 Monochloroacetate + acetoacetyl-CoA (reaction at 5.4% the rate of butyrate) [1]

21 4-Hydroxybutanoate + acetoacetyl-CoA (reaction at 14% the rate of butanate) [1]

22 3-Hydroxypropanoate + acetoacetyl-CoA (reaction at 8.6% the rate of butanoate) [1]

23 DL-2-Hydroxybutanoate + acetoacetyl-CoA (reaction at 19% the rate of butanoate) [1]

24 Acetate + crotonyl-CoA (reaction at 4% the rate of acetoacetyl-CoA) [3]

Product spectrum

1 Butanoyl-CoA + acetoacetate [1–3]

2 Vinylacetyl-CoA + acetoacetate

3 Acetyl-CoA + acetoacetate [2, 3]

4 Butanoyl-CoA + acetate [3]

5 3-Hydroxybutanoyl-CoA + acetate

6 Propanoyl-CoA + acetoacetate [2]

7 3-Mercaptopropanoyl-CoA + acetoacetate

8 Crotonyl-CoA + acetoacetate [1]

9 Crotonyl-CoA + butanoate

10 Formyl-CoA + acetoacetate

11 Octanoyl-CoA + acetoacetate

12 Pentanoyl-CoA + acetoacetate

13 Isobutanoyl-CoA + acetoacetate

14 4-Pentenoyl-CoA + acetoacetate

15 3-Methylvinylacetyl-CoA + acetoacetate
16 3-Hydroxybutanoyl-CoA + acetoacetate
17 2-Hydroxypentanoyl-CoA + acetoacetate
18 Isopentanoyl-CoA + acetoacetate
19 Acrylyl-CoA + acetoacetate
20 Monochloroacetyl-CoA + acetoacetate
21 4-Hydroxybutanoyl-CoA + acetoacetate
22 3-Hydroxybutanoyl-CoA + acetoacetate
23 2-Hydroxybutanoyl-CoA + acetoacetate
24 Acetyl-CoA + crotonate

Inhibitor(s)

PCMB (not [1]) [3]; N-Ethylmaleimide (not [1]) [3]; Iodoacetamide (not [1]) [3]; Acyl-CoA substrates (partial inactivation, in the absence of carboxylic acid substrates, 2-mercaptoethanol slightly enhances inactivation, EDTA, cysteine or acetoacetate protects) [3]; epsilon-Acetyl-CoA (acetoacetyl-CoA and acetate as substrates) [3]; Borohydride (kinetics) [3]; $MnCl_2$ (activation, 1.2 mM, inhibits at higher concentrations, only with acetoacetate or acetoacetyl-CoA as substrate) [1]; $CaCl_2$ (activation, 6 mM, inhibits at higher concentrations, only with acetoacetate or acetoacetyl-CoA as substrate) [1]; $MgCl_2$ (activation, 12 mM, inhibits at higher concentrations, only with acetoacetate or acetoacetyl-CoA as substrate) [1]; KCl (activation, 0.08 M, inhibits at higher concentrations, only with acetoacetate or acetoacetyl-CoA as substrate) [1]; NH_4Cl (activation, 0.08 M, inhibits at higher concentrations, only with acetoacetate or acetoacetyl-CoA as substrate) [1]; NaCl (activation, 0.17 M, inhibits at higher concentrations, only with acetoacetate or acetoacetyl-CoA as substrate) [1]; LiCl (activation, 0.2 M, inhibits at higher concentrations, only with acetoacetate or acetoacetyl-CoA as substrate) [1]; Tris-HCl (activation, 0.25 M, inhibits at higher concentrations, only with acetoacetate or acetoacetyl-CoA as substrate) [1]; More (no inhibition by Na_2AsO_2 [1, 3], diamide [3], DTNB or DTT [1]) [1, 3]

Cofactor(s)/prosthetic group(s)/activating agents

Metal compounds/salts

$MnCl_2$ (activation, 1.2 mM, inhibits at higher concentrations, only with acetoacetate or acetoacetyl-CoA as substrate) [1]; $CaCl_2$ (activation, 6 mM, inhibits at higher concentrations, only with acetoacetate or acetoacetyl-CoA as substrate) [1]; $MgCl_2$ (activation, 12 mM, inhibits at higher concentrations, only with acetoacetate or acetoacetyl-CoA as substrate) [1]; KCl (activation, 0.08 M, inhibits at higher concentrations, only with acetoacetate or acetoacetyl-CoA as substrate) [1]; NH_4Cl (activation, 0.08 M, inhibits at higher concentrations, only with acetoacetate or acetoacetyl-CoA as substrate) [1]; NaCl (activation, 0.17 M, inhibits at higher concentrations, only with acetoacetate or acetoacetyl-CoA as substrate) [1]; LiCl (activation, 0.2 M, inhibits at higher concentrations, only with acetoacetate or acetoacetyl-CoA as substrate) [1]; Tris-HCl (activation, 0.25 M, inhibits at higher concentrations, only with acetoacetate or acetoacetyl-CoA as substrate) [1]

Turnover number (min^{-1})

Specific activity (U/mg)
More [2]; 117–160 [3]; 329–822 [1]

K$_m$-value (mM)
0.0034 (acetoacetyl-CoA (+ crotonate)) [1]; 0.023 (butanoyl-CoA (+ crotonate)) [1]; 0.034–0.083 (butanoyl-CoA (+ acetoacetate), pH 6.3–8.1) [1]; 0.035 (acetoacetyl-CoA) [4]; 0.135 (acetyl-CoA (+ acetoacetate)) [1]; 0.167 (crotonyl-CoA (+ acetoacetate)) [1]; 0.205 (propanoyl-CoA (+ acetoacetate)) [1]; 0.26 (acetyl-CoA) [4]; 0.33 (acetyl-CoA) [3]; 0.8 (butanoate) [1]; 1 (acetoacetate) [1]; 2 (crotonate) [1]; 3.2 (propanoate) [1]; 7 (pentanoate) [1]; 17 (vinylacetate) [1]

pH-optimum
6.5–8 [2]; 6.6–7.9 [1]

pH-range
More (active over a wide range) [2]; 4–9 [1]

Temperature optimum (°C)
24–25 (assay at) [1, 3, 4]

Temperature range (°C)

3 ENZYME STRUCTURE

Molecular weight
90000 (Clostridium sp. SB4, PAGE) [1]
97800 (E. coli, sucrose density gradient centrifugation) [3]
99000 (E. coli, gel filtration) [3]
108000 (Clostridium sp. SB4, gel filtration) [1]

Subunits
Tetramer (2 × 23000 + 2 × 25000, alpha$_2$beta$_2$, Clostridium SB4, SDS-PAGE [1], 2 × 26000 + 2 × 23000, alpha$_2$beta$_2$, E. coli C22, SDS-PAGE [3]) [1, 3]

Glycoprotein/Lipoprotein
More (contains no amino sugars or hexose) [3]

4 ISOLATION/PREPARATION

Source organism
Clostridium sp. (SB4) [1]; E. coli (C22) [3, 4]; Bovine [2]

Source tissue
Cell [1]; Rumen epithelium [2]

Localization in source
Membrane-associated [3]

Purification
Clostridium (partial) [1]; E. coli [3]

Crystallization
–

Cloned
–

Renatured
–

---

5 STABILITY

pH
6.5–7 (1 h, at 30°C, most stable, about 20% loss of activity at pH 6 and 8) [1]

Temperature (°C)
30 (1 h, in potassium phosphate buffer, pH 6.5–7, most stable, about 20% loss of activity at pH 6 and 8) [1]

Oxidation

Organic solvent

General stability information
Considerably less stable in Tris-HCl than in phosphate buffer [1]

Storage
–15°C, 3.1 mg protein/ml, 20 mM potassium phosphate, pH 7, 1 year, 0.0062 mg protein/ml, 10 mM phosphate buffer, pH 7, $t_{1/2}$: 12 days [1]; 4°C, 0.0062 protein mg/ml, 10 mM phosphate buffer, pH 7, $t_{1/2}$: 7 days [1]; Diluted enzyme solutions have reduced storage stability [1]

---

6 CROSSREFERENCES TO STRUCTURE DATABANKS

PIR/MIPS code
PIR2:JN0488 (alpha chain Clostridium acetobutylicum); PIR2:JN0489 (beta chain Clostridium acetobutylicum); PIR2:B49346 (small chain Clostridium acetobutylicum)

Brookhaven code

7 LITERATURE REFERENCES

[1] Barker, H.A., Jemg, I.-M., Neff, N., Robertson, J.M., Tam, F.K., Hosaka, S.:
    J. Biol. Chem.,253,1219–1225 (1978)
[2] Emmanuel, B., Milligan, L.P.: Can. J. Anim. Sci.,63,355–360 (1983)
[3] Sramek, S.J., Frerman, F.E.: Arch. Biochem. Biophys.,171,14–26 (1975)
[4] Sramek, S.J., Frerman, F.E.: Arch. Biochem. Biophys.,171,27–35 (1975)

1 NOMENCLATURE

EC number
2.8.3.10

Systematic name
Acetyl-CoA:citrate CoA-transferase

Recommended name
Citrate-CoA transferase

Synonyms
Coenzyme A-transferase, citrate
More (the enzyme is a component of EC 4.1.3.6, cf. EC 2.8.3.11)

CAS Reg. No.
65187-14-6

---

2 REACTION AND SPECIFICITY

Catalyzed reaction
Acetyl-CoA + citrate →
→ acetate + (3S)-citryl-CoA

Reaction type
Coenzyme A transfer

Natural substrates
Acetyl-thioacyl carrier protein + citrate (biologically significant reaction of transferase subunit as part of citrate lyase, the isolated transferase subunit represents not only an acetyl-thioacyl carrier protein:citrate acyl carrier protein transferase but also an acetyl-CoA:citrate CoA-transferase) [1]

Substrate spectrum
1 Acetyl-CoA + citrate (r, higher transferase activity in the isolated state than as part of the enzyme complex, propionyl-CoA, butyryl-CoA or acetyl-dephospho-CoA can replace acetyl-CoA to some extent. No substrate: acetyl-4'-phosphopantetheine) [1]
2 Acetyl-thioacyl carrier protein + citrate [1]
3 More (also catalyzes a citrate independent exchange reaction of acetyl residues between acetyl-thioacyl carrier protein or acetyl-CoA and acetate) [1]

Product spectrum
   1 Acetate + (3S)-citryl-CoA (no enzyme-CoA intermediate) [1]
   2 Acetate + (3S)-citryl-acetyl-thioacyl carrier protein [1]
   3 ?

Inhibitor(s)
   Iodoacetate [1]

Cofactor(s)/prosthetic group(s)/activating agents

Metal compounds/salts

Turnover number (min^{-1})
   1300 (acetyl-CoA) [1]; 1600 ((3S)-citryl-CoA) [1]

Specific activity (U/mg)
   9.1 (acetyl-CoA, integrated into enzyme complex) [1]; 25 (acetyl-CoA, isolated transferase) [1]

K_m-value (mM)
   0.2 ((3S)-citryl-CoA) [1]; 1.3 (acetyl-CoA, propionyl-CoA, butyryl-CoA, acetyl-dephospho-CoA, isolated transferase) [1]; 3.3 (propionyl-CoA, integrated
   into enzyme complex) [1]; 6.7 (acetyl-CoA, integrated into enzyme complex)
   [1]; 12.5 (acetyl-dephospho-CoA, integrated into enzyme complex) [1]

pH-optimum

pH-range

Temperature optimum (°C)
   25 (assay at) [1]

Temperature range (°C)

3 ENZYME STRUCTURE

Molecular weight

Subunits

Glycoprotein/Lipoprotein
   –

4 ISOLATION/PREPARATION

Source organism
   Klebsiella aerogenes [1]

Source tissue
  Cell [1]

Localization in source

Purification
  Klebsiella aerogenes [1]

Crystallization
  –

Cloned
  –

Renatured
  –

5 STABILITY

pH

Temperature (°C)

Oxidation

Organic solvent

General stability information

Storage
  –70°C, stable at [1]

6 CROSSREFERENCES TO STRUCTURE DATABANKS

PIR/MIPS code

Brookhaven code

7 LITERATURE REFERENCES

[1] Dimroth, P., Loyal, R., Eggerer, H.: Eur. J. Biochem.,80,479–488 (1977)

1 NOMENCLATURE

EC number
2.8.3.11

Systematic name
Acetyl-CoA:citramalate CoA-transferase

Recommended name
Citramalate CoA-transferase

Synonyms
Coenzyme A-transferase, citramalate
More (the enzyme is a component of EC 4.1.3.22, cf. EC 2.8.3.10 and
EC 2.8.3.7)

CAS Reg. No.
9033-60-7 (indistinguishable from EC 2.8.3.7 in Chemical Abstracts)

---

2 REACTION AND SPECIFICITY

Catalyzed reaction
Acetyl-CoA + citramalate →
→ acetate + (3S)-citramalyl-CoA

Reaction type
Coenzyme A transfer

Natural substrates
Acetyl-thioacyl carrier protein + citramalate (biological significant reaction,
alpha-subunit of citramalate lyase enzyme complex) [1]

Substrate spectrum
1 Acetyl-CoA + citramalate (r, no substrate: citrate) [1]
2 Acetyl-thioacyl carrier protein + citramalate (substrate is the acetylated
 acyl carrier protein of citrate lyase or citramalate lyase enzyme complex)
 [1]
3 More (also catalyzes citramalate dependent exchange reaction between
 acetate and acetyl-CoA) [1]

Product spectrum
1 Acetate + (3S)-citramalyl-CoA [1]
2 ?
3 ?

Inhibitor(s)

Cofactor(s)/prosthetic group(s)/activating agents

Metal compounds/salts

Turnover number (min^{-1})

Specific activity (U/mg)

K$_m$-value (mM)

pH-optimum

pH-range

Temperature optimum (°C)
    25 (assay at) [1]

Temperature range (°C)

3 ENZYME STRUCTURE

Molecular weight

Subunits

Glycoprotein/Lipoprotein
    –

4 ISOLATION/PREPARATION

Source organism
    Clostridium tetanomorphum [1]

Source tissue
    Cell [1]

Localization in source

Purification
    Clostridium tetanomorphum [1]

Crystallization
    –

Cloned
    –

Renatured
    –

5 STABILITY

pH

Temperature (°C)

Oxidation

Organic solvent

General stability information

Storage

6 CROSSREFERENCES TO STRUCTURE DATABANKS

PIR/MIPS code

Brookhaven code

7 LITERATURE REFERENCES

[1] Dimroth, P., Buckel, W., Loyal, R., Eggerer, H.: Eur. J. Biochem.,80,469–477 (1977)

1 NOMENCLATURE

EC number
2.8.3.12

Systematic name
Acetyl-CoA:(E)-glutaconate CoA-transferase

Recommended name
Glutaconate CoA-transferase

Synonyms
Coenzyme A-transferase, glutaconate
(E)-Glutaconate CoA-transferase

CAS Reg. No.
79078-99-2

---

2 REACTION AND SPECIFICITY

Catalyzed reaction
Acetyl-CoA + (E)-glutaconate →
→ acetate + glutaconyl-1-CoA (mechanism [1])

Reaction type
Coenzyme A transfer

Natural substrates
Acetate + glutaconyl-1-CoA (involved in glutamate fermentation via hydroxyglutarate pathway, reaction prior to glutaconate decarboxylation) [1]

Substrate spectrum
1 Acetyl-CoA + (E)-glutaconate (r [1], broad substrate specificity [1], removal of 3'-phospho group from acetyl-CoA leads to reduced activity [1]. No substrates: acetyl-4'-phosphopantetheine, (Z)-glutaconate, C_4-dicarboxylic acids, reverse reaction: glutamate, 2-oxoglutarate, succinate, malate or citrate [1]) [1–3]
2 Glutaconyl-CoA + glutarate [1]
3 Acetyl-CoA + glutarate (r) [1]
4 Acetyl-CoA + (R)-2-hydroxyglutarate [1, 2]
5 Glutaconyl-CoA + (R)-2-hydroxyglutarate [1, 2]
6 Glutaconyl-CoA + (S)-2-hydroxyglutarate (reaction at 60% the rate of (R)-isomer or glutarate) [1]
7 Glutaconyl-CoA + 3-hydroxyglutarate (reaction at 20% the rate of glutarate) [1]

    8 Glutaconyl-CoA + adipate (reaction at 60% the rate of glutarate) [1]
    9 Glutaconyl-CoA + propionate [1]
   10 Glutaconyl-CoA + butanoate [1]
   11 Acetyl-CoA + acrylate (poor substrate) [1]
   12 Acetyl-CoA + crotonate (poor substrate) [1]
   13 Acetyl-CoA + isocrotonate (poor substrate) [1]

Product spectrum

    1 Acetate + glutaconyl-1-CoA [1]
    2 Glutaconate + glutaryl-CoA [1]
    3 Acetate + glutaryl-CoA [1]
    4 Acetate + (R)-2-hydroxyglutaryl-1-CoA (in vivo [2], in vitro both possible
      isomers: (R)-2-hydroxyglutaryl-1-CoA and (R)-2-hydroxyglutaryl-5-CoA)
      [1, 2]
    5 Glutaconate + (R)-2-hydroxyglutaryl-1-CoA (both possible isomers:
      (R)-2-hydroxyglutaryl-1-CoA and (R)-2-hydroxyglutaryl-5-CoA) [1]
    6 ?
    7 ?
    8 ?
    9 Propanoyl-CoA + glutaconate
   10 Butanoyl-CoA + glutaconate
   11 ?
   12 ?
   13 ?

Inhibitor(s)

Sodium borohydride [1]; 4-Hydroxymercuribenzoate (2-mercaptoethanol partially restores) [1]; More (no inhibition by iodoacetate, IAA or DTNB) [1]

Cofactor(s)/prosthetic group(s)/activating agents

Metal compounds/salts

Turnover number (min^{-1})

Specific activity (U/mg)

More [1, 3]

K_m-value (mM)

0.015 (glutaryl-CoA (+ acetate)) [1]; 0.017 (glutaconyl-CoA (+ glutarate)) [1]; 0.17 (acetyl-CoA (+ glutaconate)) [1]; 0.2 (glutaconate (+ acetyl-CoA)) [1]; 0.7 (glutarate (+ glutaconyl-CoA)) [1]; 1.1 (acetyl-dephospho-CoA (+ glutaconate)) [1]; 1.5 ((R)-2-hydroxyglutarate (+ glutaconyl-CoA)) [1]; 8 (adipate (+ glutaconyl-CoA)) [1]; 10 (acrylate (+ acetyl-CoA)) [1]; 13 (3-hydroxyglutarate (+ glutaconyl-CoA)) [1]; 14 ((S)-2-hydroxyglutarate (+ glutaconyl-CoA)) [1]; 16 (propionate (+ glutaconyl-CoA)) [1]; 26 (acetate (+ glutaconyl-CoA)) [1]; 100 (isocrotonate (+ acetyl-CoA)) [1]; 150 (butyrate (+ glutaconyl-CoA)) [1]; 500 (crotonate (+ acetyl-CoA)) [1]

pH-optimum
7 (glutaconyl-CoA + acetate) [1]

pH-range
5.6–7.4 (about 70% of maximal activity at pH 5.6 and 7.4, about 10% of maximal activity at pH 8.5) [1]

Temperature optimum (°C)
25 (assay at) [1]; 37 (assay at) [2]

Temperature range (°C)

---

3 ENZYME STRUCTURE

Molecular weight
275000 (Acidaminococcus fermentans, gel filtration) [1]

Subunits
Octamer ($4 \times 32000 + 4 \times 34000$, $alpha_4beta_4$, Acidaminococcus fermentans, SDS-PAGE) [1]

Glycoprotein/Lipoprotein
–

---

4 ISOLATION/PREPARATION

Source organism
Acidaminococcus fermentans [1–3]; Clostridium sporosphaeroides [1]; Clostridium symbiosum (HB25) [1]; More (not in Clostridium tetanomorphum H1) [1]

Source tissue
Cell [1–3]

Localization in source

Purification
Acidaminococcus fermentans (and from recombinant E. coli [3]) [1, 3]

Crystallization
(Acidaminococcus fermentans [1, 3] and from recombinant E. coli strain [3]) [1, 3]

Cloned
(Acidaminococcus fermentans gctAB-genes, expressed in E. coli strain DH5alpha) [3]

Renatured
–

---

5 STABILITY

pH

Temperature (°C)

Oxidation

Organic solvent

General stability information

Storage
    4°C, crystalline suspension in ammonium sulfate, 2 years [1]

6 CROSSREFERENCES TO STRUCTURE DATABANKS

PIR/MIPS code
    PIR2:S51052 (Acidaminococcus fermentans)

Brookhaven code

7 LITERATURE REFERENCES

[1] Buckel, W., Dorn, U., Semmler, R.: Eur. J. Biochem.,118,315–321 (1981)
[2] Klees, A.-G., Buckel, W.: Biol. Chem. Hoppe-Seyler,372,319–324 (1991)
[3] Mack, M., Bendrat, K., Zelder, O., Eckel, E., Linder, D., Buckel, W.: Eur. J. Biochem., 226,41–51 (1994)

1 NOMENCLATURE

EC number ·
2.8.3.13

Systematic name
Succinate:(S)-3-hydroxy-3-methylglutarate CoA-transferase

Recommended name
Succinate-hydroxymethylglutarate CoA-transferase

Synonyms
Coenzyme A-transferase, hydroxymethylglutarate
Dicarboxyl-CoA:dicarboxylic acid coenzyme A transferase (this seems to be
a more appropriate name than the recommended or systematic name, due
to the substrate specificity as documented in [3])

CAS Reg. No.
80237-90-7

---

2 REACTION AND SPECIFICITY

Catalyzed reaction
Succinyl-CoA + (S)-3-hydroxy-3-methylglutarate →
→ succinate + 3-hydroxy-3-methylglutaryl-CoA

Reaction type
Coenzyme A transfer

Natural substrates

Substrate spectrum
1 Succinyl-CoA + 3-hydroxy-3-methylglutarate (r [2, 3], poor substrate:
butyryl-CoA [1]. No substrates are acetyl-CoA, acetoacetyl-CoA, acetoacetate [1, 3], ATP plus CoA [1]) [1–3]
2 Malonyl-CoA + 3-hydroxy-3-methylglutarate (r, forward reaction at 34%,
reverse reaction at 23% the rate of the CoA-transfer between adipyl-CoA
and succinate [3]) [1, 3]
3 Adipyl-CoA + succinate (r, best substrates, reverse reaction at 38% the
rate of forward reaction. No substrates are acetyl-CoA, acetoacetyl-CoA,
propionyl-CoA, hexanoyl-CoA, tiglyl-CoA, methylcrotonyl-CoA, palmitoyl-CoA, fumarate, malate, oxaloacetate, 2-oxoglutarate, 2-oxohexa-
nedioate, or 3-oxohexandioate, citrate, isocitrate, propionate, hexanoate
or 2-hydroxybutanoate) [3]

4 Adipyl-CoA + glutarate (r, forward reaction at 82%, reverse reaction at 68% the rate of the CoA-transfer between adipyl-CoA and succinate) [3]

5 Adipyl-CoA + 3-hydroxy-3-methylglutarate (r, forward reaction at 64%, reverse reaction at 32% the rate of the CoA-transfer between adipyl-CoA and succinate) [3]

6 Adipyl-CoA + malonate (r, forward reaction at 40%, reverse reaction at 43% the rate of the CoA-transfer between adipyl-CoA and succinate) [3]

7 Succinyl-CoA + glutarate (r, forward reaction at 58%, reverse reaction at 55% the rate of the CoA-transfer between adipyl-CoA and succinate) [3]

8 Succinyl-CoA + malonate (r, forward reaction at 29%, reverse reaction at 22% the rate of the CoA-transfer between adipyl-CoA and succinate) [3]

9 Glutaryl-CoA + 3-hydroxy-3-methylglutarate (r, forward and reverse reaction at 50% the rate of the CoA-transfer between adipyl-CoA and succinate) [3]

10 Glutaryl-CoA + malonate (r, forward reaction at 33%, reverse reaction at 41% the rate of the CoA-transfer between adipyl-CoA and succinate) [3]

11 Adipyl-CoA + methylmalonate (r, poor substrate, succinyl-CoA, glutaryl-CoA, malonyl-CoA or 3-hydroxy-3-methylglutaryl-CoA can replace adipyl-CoA) [3]

Product spectrum

1 Succinate + 3-hydroxy-3-methylglutaryl-CoA [1–3]
2 Malonate + 3-hydroxy-3-methylglutaryl-CoA [3]
3 Adipate + succinyl-CoA [3]
4 Adipate + glutaryl-CoA [3]
5 Adipate + 3-hydroxy-3-methylglutaryl-CoA [3]
6 Adipate + malonyl-CoA [3]
7 Succinate + glutaryl-CoA [3]
8 Succinate + malonyl-CoA [3]
9 Glutarate + 3-hydroxy-3-methylglutaryl-CoA [3]
10 Glutarate + malonyl-CoA [3]
11 Adipate + methylmalonyl-CoA [3]

Inhibitor(s)

Acetoacetate [1, 2]; Succinate (product inhibition [2]) [1, 2]; Malonate [1]; Acetate [2]; Acetyl-CoA [2]; Coenzyme A [2]; Carnitine [2]; $ZnCl_2$ [2]; ClO_4^- (at high concentrations) [2]; F^- (at high concentrations) [2]; I^- (at high concentrations) [2]; Cl^- (at high concentrations) [2]

Cofactor(s)/prosthetic group(s)/activating agents

More (no activation by mercaptoethanol) [2]

Metal compounds/salts

Turnover number (min^{-1})

Specific activity (U/mg)
0.019 [1]; 0.249 [2]

K$_m$-value (mM)
More (kinetic properties) [2]; 0.07 (glutaryl-CoA (+ adipate), adipyl-CoA (+ glutarate)) [3]; 0.11 (adipyl-CoA (+ succinate)) [3]; 0.13 (malonyl-CoA (+ glutarate)) [3]; 0.15 (succinyl-CoA (+ glutarate)) [3]; 0.16 (adipyl-CoA (+ 3-hydroxy-3-methylglutarate)) [3]; 0.18 (glutaryl-CoA (+ succinate), glutarate (+ adipyl-CoA), 3-hydroxy-3-methylglutarate-CoA (+ adipate)) [3]; 0.22 (succinyl-CoA (+ 3-hydroxy-3-methylglutarate)) [1]; 0.26 (succinyl-CoA (+ adipate), glutarate (+ malonyl-CoA)) [3]; 0.27 (glutaryl-CoA (+ malonate)) [3]; 0.28 (succinyl-CoA (+ 3-hydroxy-3-methylglutarate)) [3]; 0.32 (3-hydroxy-3-methylglutarate-CoA (+ malonate)) [3]; 0.33 (malonyl-CoA (+ adipate)) [3]; 0.37 (malonyl-CoA (+ 3-hydroxy-3-methylglutarate)) [1]; 0.38 (adipyl-CoA (+ malonate)) [3]; 0.39 (3-hydroxy-3-methylglutarate-CoA (+ succinate)) [3]; 0.4 (adipyl-CoA (+ methylmalonate)) [3]; 0.44 (3-hydroxy-3-methylglutarate-CoA (+ glutarate)) [3]; 0.45 (glutarate (+ succinyl-CoA), glutaryl-CoA (+ methylmalonate)) [3]; 0.5 (succinate (+ glutaryl-CoA), glutarate (+ 3-hydroxy-3-methylglutarate-CoA)) [3]; 0.52 (malonyl-CoA (+ 3-hydroxy-3-methylglutarate)) [3]; 0.55 (malonyl-CoA (+ succinate)) [3]; 0.62 (3-hydroxy-3-methylglutarate-CoA (+ methylmalonate)) [3]; 0.65 (methylmalonyl-CoA (+ malonate)) [3]; 0.66 (adipate (+ 3-hydroxy-3-methylglutarate-CoA)) [3]; 0.68 (succinate (+ malonyl-CoA)) [3]; 0.7 (malonyl-CoA (+ methylmalonate)) [3]; 0.72 (adipate (+ glutaryl-CoA), methylmalonyl-CoA (+ glutarate or adipate)) [3]; 0.73 (glutaryl-CoA (+ 3-hydroxy-3-methylglutarate)) [3]; 0.74 (methylmalonyl-CoA (+ 3-hydroxy-3-methylglutarate)) [3]; 0.75 (malonate (+ succinyl-CoA)) [3]; 0.76 (3-hydroxy-3-methylglutarate (+ adipyl-CoA)) [3]; 0.83 (succinyl-CoA (+ methylmalonate), adipate (+ malonyl-CoA)) [3]; 0.84 (glutarate (+ methylmalonyl-CoA)) [3]; 0.87 (methylmalonyl-CoA (+ succinate)) [3]; 0.88 (succinate (+ 3-hydroxy-3-methylglutarate)) [3]; 0.95 (succinate (+ adipyl-CoA)) [3]; 1.01 (malonate (+ 3-hydroxy-3-methylglutarate-CoA)) [3]; 1.2 (succinate (+ methylmalonyl-CoA)) [3]; 1.25 (3-hydroxy-3-methylglutarate (+ glutaryl-CoA)) [3]; 1.32 (adipate (+ succinyl-CoA)) [3]; 1.42 (3-hydroxy-3-methylglutarate (+ succinyl-CoA)) [3]; 1.51 (adipate (+ methylmalonyl-CoA)) [3]; 1.55 (malonate (+ glutaryl-CoA)) [3]; 1.63–1.64 (malonate (+ succinyl-CoA or methylmalonyl-CoA)) [3]; 1.7 (3-hydroxy-3-methylglutarate) [1]; 1.71 (malonate (+ adipyl-CoA)) [3]; 1.8 (3-hydroxy-3-methylglutarate (+ malonyl-CoA)) [3]; 1.95 (methylmalonate (+ 3-hydroxy-3-methylglutarate-CoA)) [3]; 2.2 (methylmalonate (+ malonyl-CoA)) [3]; 2.33 (3-hydroxy-3-methylglutarate (+ methylmalonyl-CoA)) [3]; 2.44 (methylmalonate (+ glutaryl-CoA)) [3]; 2.7 (methylmalonate (+ succinyl-CoA)) [3]; 3.05 (methylmalonate (+ adipyl-CoA)) [3]

pH-optimum
    7.8 [2]

pH-range

Temperature optimum (°C)
    30 (assay at) [1, 3]

Temperature range (°C)

---

3 ENZYME STRUCTURE

Molecular weight
    42000 (rat, gel filtration) [1]
    48000 (rat, sucrose density gradient centrifugation) [2]
    52000 (rat, gel filtration) [2]

Subunits
    ? (x × 12000–14000, rat, SDS-PAGE) [2]

Glycoprotein/Lipoprotein
    –

---

4 ISOLATION/PREPARATION

Source organism
    Rat (Wistar albino) [1–3]

Source tissue
    Liver [1–3]

Localization in source
    Mitochondria (predominantly matrix and some activity in inner mitochondrial
    membrane, submitochondrial distribution [1]) [1–3]

Purification
    Rat (partial) [1, 2]

Crystallization
    –

Cloned
    –

Renatured
    –

5 STABILITY

pH
  6 (below, irreversible loss of activity) [1]

Temperature (°C)

Oxidation

Organic solvent

General stability information
  Glycerol, 10%, is a better stabilizer than bovine serum albumin at 3 mg/ml
  [1]

Storage
  –20°C, 20% loss of activity within 2 weeks [1]; –20°C, 0.1 M KH_2PO_4 buffer,
  pH 7.8, 20% loss of activity within 1 month, mercaptoethanol or DTT does
  not improve stability [2]

6 CROSSREFERENCES TO STRUCTURE DATABANKS

PIR/MIPS code

Brookhaven code

7 LITERATURE REFERENCES

[1] Deana, R., Rigoni, F., Donella Deana, A., Galzigna, L.: Biochim. Biophys. Acta,
    662,119–124 (1981)
[2] Francesconi, M.A., Donella-Deana, A., Furlanetto, V., Cavallini, L., Palatini, P., Deana,
    R.: Biochim. Biophys. Acta,999,163–170 (1989)
[3] Deana, R.: Biochem. Int.,26,767–773 (1992)

1 NOMENCLATURE

EC number
2.8.3.14

Systematic name
Acetyl-CoA:5-hydroxypentanoate CoA-transferase

Recommended name
5-Hydroxypentanoate CoA-transferase

Synonyms
5-Hydroxyvalerate CoA-transferase
Coenzyme A-transferase, 5-hydroxyvalerate
5-Hydroxyvalerate coenzyme A transferase

CAS Reg. No.
111684-68-5

2 REACTION AND SPECIFICITY

Catalyzed reaction
Acetyl-CoA + 5-hydroxypentanoate →
→ acetate + 5-hydroxypentanoyl-CoA

Reaction type
Coenzyme A transfer

Natural substrates

Substrate spectrum
  1 5-Hydroxypentanoyl-CoA + acetate (best substrate) [1]
  2 Propanoyl-CoA + acetate (r) [1]
  3 Acetyl-CoA + acetate [1]
  4 Butanoyl-CoA + acetate [1]
  5 Pentanoyl-CoA + acetate [1]
  6 5-Hydroxypentanoyl-CoA + (Z)-5-hydroxy-2-pentenoate [1]
  7 Pentanoyl-CoA + (Z)-5-hydroxy-2-pentenoate [1]
  8 Butanoyl-CoA + (Z)-5-hydroxy-2-pentenoate [1]
  9 Propanoyl-CoA + (Z)-5-hydroxy-2-pentenoate [1]
  10 Acetyl-CoA + (Z)-5-hydroxy-2-pentenoate (r) [1]
  11 Acetyl-CoA + 3-pentenoate (no substrates: (E)-2-pentenoate, (E)-5-hy-
      droxy-2-pentenoate, 2,4-pentadienoate) [1]
  12 Acetyl-CoA + 4 pentenoate [1]

Enzyme Handbook © Springer-Verlag Berlin Heidelberg 1997
Duplication, reproduction and storage in data banks are only
allowed with the prior permission of the publishers

Product spectrum
1 Acetyl-CoA + 5-hydroxypentanoate (via enzyme-CoA thiolester) [1]
2 Acetyl-CoA + propanoate
3 Acetyl-CoA + acetate
4 Acetyl-CoA + butanoate
5 Acetyl-CoA + pentanoate
6 5-Hydroxypentanoate + (Z)-5-hydroxy-2-pentenoyl-CoA
7 Pentanoate + (Z)-5-hydroxy-2-pentenoyl-CoA
8 Butanoate + (Z)-5-hydroxy-2-pentenoyl-CoA
9 Propanoate + (Z)-5-hydroxy-2-pentenoyl-CoA
10 Acetate + (Z)-5-hydroxy-2-pentenoyl-CoA [1]
11 Acetate + 3-pentenoyl-CoA [1]
12 Acetate + 4-pentenoyl-CoA [1]

Inhibitor(s)
ATP (kinetics) [1]; CTP [1]; UTP (weak) [1]; GTP (weak) [1]; ADP (weak) [1]; Sodium boranate (only in the presence of propionyl-CoA) [1]; More (no inhibition by AMP) [1]

Cofactor(s)/prosthetic group(s)/activating agents

Metal compounds/salts

Turnover number (min^{-1})

Specific activity (U/mg)
38–46 [1]

K_m-value (mM)
More (kinetic study) [1]; 0.0038 (5-hydroxypentanoyl-CoA (+ (Z)-5-hydroxy-2-pentenoate)) [1]; 0.009 (propanoyl-CoA (+ (Z)-5-hydroxy-2-pentenoate)) [1]; 0.026 (acetyl-CoA (+ (Z)-5-hydroxy-2-pentenoate)) [1]; 0.029 (butanoyl-CoA (+ (Z)-5-hydroxy-2-pentenoate)) [1]; 0.03 ((Z)-5-hydroxy-2-pentanoyl-CoA (+ acetate)) [1]; 0.037 (pentanoyl-CoA (+ (Z)-5-hydroxy-2-pentenoate)) [1]; 0.07 (butanoyl-CoA (+ acetate)) [1]; 0.1 (5-hydroxypentanoyl-CoA (+ acetate)) [1]; 0.12 (propanoyl-CoA (+ acetate)) [1]; 0.45 (pentanoyl-CoA (+ acetate)) [1]; 3.6 ((Z)-5-hydroxy-2-pentenoate (+ acetyl-CoA)) [1]; 20 (acetate (+ propanoyl-CoA)) [1]

pH-optimum

pH-range

Temperature optimum (°C)
25 (assay at) [1]

Temperature range (°C)

3 ENZYME STRUCTURE

Molecular weight
155000–160000 (Clostridium aminovalericum T2–7, PAGE) [1]
180000–195000 (Clostridium aminovalericum T2–7, FPLC gel filtration) [1]

Subunits
Tetramer (4 × 47000, Clostridium aminovalericum T2–7, SDS-PAGE) [1]

Glycoprotein/Lipoprotein
–

4 ISOLATION/PREPARATION

Source organism
Clostridium aminovalericum (T2–7) [1]

Source tissue
Cell [1]

Localization in source

Purification
Clostridium aminovalericum [1]

Crystallization
–

Cloned
–

Renatured
–

5 STABILITY

pH

Temperature (°C)

Oxidation

Organic solvent

General stability information

Storage
–20°C, several months [1]; 4°C, 20 mM potassium phosphate buffer, pH 6.8,
$t_{1/2}$: 2 months [1]

6 CROSSREFERENCES TO STRUCTURE DATABANKS

PIR/MIPS code

Brookhaven code

7 LITERATURE REFERENCES

[1] Eikmanns, U., Buckel, W.: Biol. Chem. Hoppe-Seyler,371,1077–1082 (1990)